Where the Rivers Ran Red

The Indian Fights of George Armstrong Custer

by

Michael N. Donahue
Michael N. Donahue

Books by Michael N. Donahue

Drawing Battle Lines-The Map Testimony of Custer's Last Fight
Winner of the G. Joseph Sills Jr. Book Award for Best Book of the Year
by the Custer Battlefield Historical & Museum Association.
Winner of the John M. Carroll Award for Best Book of the Year by the Little Bighorn Associates.

Where the Rivers Ran Red
First published May, 2018 as a 250-copy Collector's Edition, casebound with dust jacket.
Now available to the public as a trade publication.

Praise for *Where the Rivers Ran Red*

"Michael Donahue's years of walking and riding Custer's battlefields, combined with his meticulous research, especially in the use primary eyewitness accounts, has led to a straight-forward narrative that will explode long-held assumptions and change the way battle students view the deadly encounters. This book is certain to take its rightful place among the classics on Custer and his frontier military exploits and be a source of lively debate among battle students now and in the future."

~ David Harrington, author, lecturer and former acting superintendent,
Little Bighorn Battlefield National Monument

"Michael Donahue knows the Little Bighorn story better than anyone I know. He has not only studied every detail of the battle and its larger cultural context, but serving as a seasonal ranger over these many years, he has walked the terrain as few historians have been able to do. Thus, this book will be a welcome challenge to preconceived notions about the battle, as well as an enduring source for research on the mysteries of June 25, 1876."

~ Vincent A. Heier Author, *Little Bighorn*

Fig. 1. "I am not impetuous or impulsive. I resent that. Everything that I have ever done has been the result of the study that I have made of imaginary military situations that might arise. When I became engaged in campaign or battle and a great emergency arose, everything that I had ever read or studied focused in my mind as if the situation were under a magnifying glass, and my decision was the instantaneous result. My mind worked instantaneously but always as the result of everything I had ever studied being brought to bear on the situation."

George Armstrong Custer

First Edition published by San Juan Publishing Group, 2018.
Fourth Trade Edition published by Loose Chippings Studio and Gallery, 2020.
Printed in the United States.
10 9 8 7 6 5 4 3 2 1

Library of Congress Control Number: 2018914142
ISBN: 978-0-578-41569-7

10-13-21-12/8/2021

Loose Chippings Studio and Gallery

11599 Highview Drive
Belton, Texas 76513

michaeldonahueart.com

Cover image: *Custer's Last Stand-Where in the Hell is Benteen?* By Michael N. Donahue.
Limited edition prints available at author's website: michaeldonahueart.com
From the collection of Dawn and Ted Hopkins.

Book and cover design, Kathryn R. Burke.

Dedicated to my mother and father—
Patsy and Frank Donahue.
They always supported my dreams and goals
even when they did not understand them.

Table of Contents

List of Illustrations

Foreword

This publication vividly takes you back in time to George Custer and the primary characters and events that shaped his last legendary battle confrontation. From the Washita to Last Stand Hill, the narrative screams of historic authenticity. Rich in detail from Frederick Benteen to Crazy Horse, Sitting Bull to Marcus Reno, Donahue has written a blockbuster account destined to ignite renewed interest in even the most ardent students of the battle.

Where the Rivers Ran Red is the culminating effort of 45 years of intense fascination, meticulous research, and confirming discoveries by Michael Donahue in his journey to uncover the mysteries of the Battle of the Little Bighorn.

Thousands of authors have examined this subject. Many have made substantial contributions in widening our understanding of what happened on that dusty Montana hillside 14 decades ago. However, I believe no one has demonstrated more determination and skill in examining primary sources of information, analyzing data, cross referencing sources, and solving cultural misconceptions about this battle than Michael Donahue. He is a historical detective. His intense curiosity, vast knowledge of the battle, and constant willingness to examine potential new material is second to none.

Combine these assets with his 30 years of unfettered battlefield access as a Ranger/Interpreter at Little Bighorn Battlefield National Monument, and he heads the lineup of those gifted authors who have written previously about this epic story.

I have read this book and it is a riveting narrative of Custer's Indian fights interwoven with expanding insights into his evolving understanding of effective tactical strategies in manipulating Indian warfare to his advantage. Once again, Donahue's work is the first of its kind to examine Custer's experience on the Plains with aggressive Indian confrontations and how these fights impacted his subsequent military action at Little Bighorn.

Steve Adelson
Ranger/Interpreter Little Bighorn Battlefield National Monument
Author of *Little Bighorn, Voices From A Distant Wind*

Acknowledgments

With humble gratitude, I would like to thank my Little Bighorn Battlefield colleagues for their help and support with this volume. Many kudos go to Ken Woody, Neil Mangum, John Doerner, Kitty Deer Nose, Sharon Small, Cindy Hagen, Marvin Dawes, Jerry Jasmer, Chris Ziegler, David Harrington and Denise Swanke. Others who helped with this effort were Dick Zahn, Kevin Bowles Mohr, Joel Shockley, and the staff of the Washita Battlefield. Sincere appreciation to Jim Court, Todd Hively, Tom Reynolds, Ike Eichler, Scott and Tanya Hardgrove, Mark Hilderbrand, Mark Jacobsen, Jim Hatzell, Bob Reese, Michael O'Keefe, Tom Heski, Scott Burgan, Dave Miller, Chip Watts, Sandy Barnard, James Brust, Margot Liberty, Jason Pitsch, Vance Haynes, Jim Donovan, and Robert Utley.

I appreciate the use of photographs by Chris Kortlander, Putt Thompson, Judith Justus, and John Turner. I would like to recognize Lisa Keys and Sara Keckeisen of the Kansas State Historical Society as well as Rachel Mosman at the Oklahoma Historical Society, for their valuable help.

I am indebted to friend, archeologist, and historian Doug Scott who has always been willing to help me better understand the archeology of the site and for allowing me to participate when possible to uncover a few more bits of the 1876 event. His sharing of maps and archeological data is another great avenue for better understanding the battle.

Thanks to my longtime friend Dr. Richard Fox, Jr. whose book, *Archaeology, History, and Custer's Last Battle*, set a new standard for battle research and was the framework that allowed me to reach higher in order to realize and grasp my own conclusions that sometimes challenge his own. He has always been a supportive mentor whose ideas are highly prized, and I always listened to them with tremendous value as they escaped his mouth between puffs of smoke coming from his pipe while we sat, sharing the clear night air, near his cabin in the Rosebud Valley.

A special testimonial is extended to my generous friend and fellow artist George Kush for his willingness to share his unique collection of unpublished material and photos with me. George has a wealth of knowledge and is a tremendous collector as well as a talented artist. We have spent many hours re-fighting the battle and bouncing ideas off of one another.

I want to thank Kathryn R. Burke of San Juan Publishing for her expertise and guidance in designing this book. I appreciate her eye for detail and suggestions for making this an aesthetically pleasing volume. She and her staff, including Carol Bucy, Candy Burton, Bryce Chismire, and Modene Gaulke, provided editorial suggestions, page layout, content and copy editing, indexing, and image preparation.

I would like to acknowledge Vince Heier and C. Lee Noyes for their enthusiasm in reviewing the rough draft. Also, I am thankful for my friend and fellow ranger Steve Adelson for his flattering foreword. Special thanks go to Lauren Saunders whose comments and edits on the massive first draft were so helpful. I give my greatest thanks to Carolyn Wilson Bennett for her keen eye in editing, constructive comments and constant moral support of this book. I wish to acknowledge my Temple College colleague Brenda Nicholas who skillfully helped me get this volume ready for publication. She was gracious enough to make suggestions even with her busy teaching schedule.

My appreciation goes to Bob Tiernan and Dale Kosman for their constructive comments. I would like to give a special acknowledgement to my friend Peter Russell in England for his help and suggestions for the volume.

This book would not exist if it were not for the generosity and support of all of the above individuals.

Prologue

Where the Rivers Ran Red

Washita River, Oklahoma Territory
Friday, December 11, 1868

She was beautiful even in the cold blue paleness of violent death. Her chin exhibited a tiny dimple and there was a sprinkling of freckles across her small nose. Her calico dress and new leather gaiters stuck, frozen to her petite figure, but the state of her clothing indicated she had not been an Indian captive long. The woman's prostrate body lay near a cottonwood log in the snow, her tiny hands clutching her last meal—a meager piece of bread. It was strangely quiet except for the gurgling water flowing under a thin sheet of ice on the nearby Washita River. The only other sound was the crunching of cavalry boots penetrating the crusty snow beneath the gray, skeletal cottonwood trees.

The young woman had died suddenly without warning. Her death was an example of overkill, since the back of her skull had been crushed from a tomahawk blow, and one or more bullets had penetrated her head. Much of her long, dark hair was missing—a coveted souvenir of an angry warrior. Nearby lay a curly-headed, two-year-old, light-skinned boy with a dark bloody bruise on his head. All indications suggested that someone had taken him by the feet and slammed his tiny head against a nearby tree. Speculation ran wild about the dead Indian woman lying near the pair, carefully wrapped in a buffalo hide. Had she tried to intervene on their behalf?

Ohioan Clara Harrington had married Richard Blinn in 1865, and the couple had a son who they named Willie. Settling for a short time in Kansas, the Blinns migrated westward into current day Colorado. Richard and his brother-in-law created a business shipping goods to western government outposts.

In an ironic twist, a wagon train consisting of eight supply wagons was traveling from Colorado to Fort Dodge, Kansas, heading eastward along the Sante Fe Trail.[1] On October 9, 1868, a Cheyenne war party numbering over 100 warriors attacked the train near Sand Creek, capturing Clara and Willie.[2] Richard Blinn along with ten men escaped and survived the Indian siege.[3]

The young lady, known for being a jokester and having a zest for life, wanted to survive. She had scribbled a frantic plea for help that was discovered under a bush only four miles from her burned supply wagon.

Dear Dick, Willie and I are prisoners. They are going to keep us. If you live, save us if you can. We are with them. Clara Blinn

On the other side of the small card was another frantic appeal: *Dick, If you love us, save us.*[4]

The distraught husband could do little, and army scouts had been unable to locate any trace of the duo. Before the end of October, Clara turned 21 as they moved eastward, somewhere in Cheyenne captivity. Yet, luck seemed to be on her side. Almost one month later, she found herself over 200 miles away in the Arapaho camp of Yellow Bear, where an employee of a Fort Cobb trader spotted her and handed her a pencil and paper. She jotted down an urgent plea that made its way back to the trader. Clara believed she was still being held by the Cheyenne who had obviously traded her.

Saturday, Nov. 7th 68

Kind Friend,

Whoever you may be, I thank you for your kindness to me and my child. You want me to let you know my wishes. If you could only buy us from the Indians with ponies or anything, and let me come and stay with you until I get word to my friends, they would pay you, and I would work and do all that I could for you. If it is not too far to their camp, and you are not afraid to come, I pray that you will try. They tell me as near as I can understand they expect traders to come and they will sell us to them. Can you find out by this man and let me know if it is white men? If it is Mexicans I am afraid they would sell us into slavery in Mexico. If you can do nothing for me write to W. T. Harrington, Ottawa, Franklin Co., Kansas—my father. Tell him we are with the Cheyennes and they say when the white men make peace we can go home. Tell him to write to the Governor of Kansas about it and for them to make peace. Send this to him.

We were taken on the 9th of Oct. on the Arkansas below Fort Lyon. I can't tell whether they killed my husband or not. My name is Mrs. Clara Blinn, my little boy Willie Blinn. He is two years old. Do all you can for me. Write to the Peace Commissioners to make peace this fall. For our sakes, do all you can and God will bless you. If you can, let me hear from you again. Let me know what you think about it. Write to my father and send him this.

Goodbye.
Mrs. R. F. Blinn
near the Antelope Hills

I am as well as can be expected, but my baby is very weak.[5]

The letter was Clara's last desperate appeal. Trader William Griffinstein, forwarded the communication to the Army commander of Fort Cobb, Colonel William Hazen, who requested permission from his superiors to authorize making such a trade, and negotiations were in progress. While Clara had pleaded for peace so she could be released, the military had decided on war.

A bearded man, who looked like a western scout dressed in fringed, cream-colored buckskin pants, peered down at the frozen corpses. It was a dramatically different image from his flamboyant military dress as a Civil War brigadier general. The cavalry officer could not help but compare the beautifully shaped lips and dark hair of the dead woman to those of his beloved wife Libbie. He had once ordered a trusted officer to shoot his wife rather than let her suffer "a fate worse than death" if she were ever about to become an Indian captive.

Just ten days earlier, the well-known officer's first victory over the southern tribes had ended with a spectacular parade in front of his boss and mentor—Major General Philip Sheridan. The diminutive Civil War hero Sheridan relished the fact that his subordinate had punished some of the raiding tribes in a dawn attack along the Washita River, yet it was a victory that rang hollow due to 18 soldiers missing in action and assumed dead. During the chaotic fray, Major Joel Elliott disappeared with 17 enlisted men as he chased scattering Cheyenne in the direction of downstream camps where Clara and Willie Blinn were held captive.

Sheridan wanted to locate the missing troopers and to see the Washita battlefield with his own eyes. As the army officers and troops searched the distant perimeter of a charred village site destroyed by the 7th Cavalry on November 27, a gruesome sight greeted them in the tall, dead grass. They found Elliott with his men, butchered and gutted like animals, in a disorganized circle on the snow-covered ground. Frozen, severed body parts were matched with the nude torsos as some kind of gruesome game.

Sheridan was irked as he looked at his subordinate officer. Just a year earlier, his protégé had endured an ill-conceived marathon race across Kansas to see his wife. He had nonchalantly outdistanced some soldiers during the ride, and Indians had killed those who lagged behind. The buckskin-clad officer believed that they should have put up more of a fight rather than running, and thought someone else could bury them. Sheridan had vouched for the officer after his courtmartial and brought him back into service before his punishment was complete. This seemed like a repeat performance.[6]

As the two officers continued down the Washita River, their attention was drawn to the pitiful bodies of Clara and Willie. Sheridan snipped off the hem of Clara's red-violet dress and then placed it along with a lock of Willie's hair into a captured Arapaho beaded bag that he later sent to their family.

Both officers suddenly realized that the attack on a nearby, isolated Cheyenne village ten days before had initiated the deaths of these innocents. The Victorian cavalier in buckskins had been unable to save the damsel in distress; it would haunt him and force him to reconsider the aggressive charges for which he was famous. He adopted a warrior tactic of taking captives as he chased the murderers of the Blinns. Two could play this game.

It was a bad day for George Armstrong Custer, and it would not be his last.

Chapter 1

Under the Magnifying Glass

His family called him "Autie," his early childhood attempt at a middle name he could not pronounce: Armstrong. His dedicated wife, "Libbie" sometimes shortened it to "Aut" or later, "The General." Fellow West Point cadets named him "Fannie" because of his girlish looks, while other classmates called him "Cinnamon" due to the scent of the oil he doused in his curly blonde hair. During the Indian wars, his 7th Cavalry soldiers resorted to the name of "Old Iron Ass" because of his tireless ability to stay in the saddle. Other troopers labeled him "Old Leather Britches," as he often donned a distinctive buckskin suit on the plains. The Lakota named him *Pahaska* or "Long Hair" due to the legendary length of his curls during earlier years, while to the Cheyenne he was the sneaky *Ouchess* or "Creeping Panther." He seemed to be blessed by the cosmos, and his Arikara, as well as Crow, scouts addressed him as "Son of the Morning Star." The mere mention of his name today is enough to incite accusatory hatred or passionate defense. He was George Armstrong Custer.

Fig. 2. George Custer modeling his buckskin attire as he appeared about the time of the Washita fight. His flamboyant dress brought him attention in battle during the Civil War; however, this Indian Wars' clothing became more functional and practical during the harsh conditions on the plains.

Courtesy of Christopher Kortlander & the Custer Battlefield Museum, Garryowen, Montana

Custer was a complex individual with a paradoxical personality described in extreme and contradictory terminology: prankster, serious, womanizer, dedicated husband, naive, visionary, empathetic, cruel, cowardly, and heroic. Such contrasting descriptions of one man might lead one to believe he was schizophrenic in personality, or that observers were extremely biased. The dichotomy of his persona could be attributed to changes throughout his 36 years, and most historians agree that he went through an early mid-life crisis after his Civil War service. So much material has been written, dramatized, painted, and filmed about the man that he has often become an exaggerated caricature of someone far removed from reality.

Officer and friend William Ludlow presented one of the most objective and revealing descriptions of Custer's personality ever written:

> He (Custer) was the personification of bravery and dash. . . . If he had only added discretion to his valor he would have been a perfect soldier. His impetuosity very often ran away with his judgment. He was impatient of control. He liked to act independently of others and take all the risk and the glory to himself. He frequently got himself into trouble by assuming more authority than really belonged to his rank.
>
> Custer was more dependent than most men on the kind approval of his fellows. He was even vain; he loved display in dress and in action. . . . He was a man of great energy and remarkable endurance. . . . He was never idle when he had a pretext for doing anything. Whatever he did he did thoroughly. He would overshoot the mark, but never fall short.
>
> He had a remarkable memory. . . . He was rather verbose in writing, and had no gifts as a speaker. . . . As he was apt to overdo in action, so he was apt to exaggerate in statement, not from any willful disregard of truth, but because he saw things bigger than they really were. He did not distort the truth; he magnified it. He was a natural optimist. He took rose-colored views of everything. . . . He had a historical memory, but not a historical mind. He was no philosopher. . . . He was not a student nor a deep thinker. He was fond of fun, genial, and pleasant in his manner; a loving and devoted husband. He loved to take part in events rather than brood over them.[1]

George Custer was born on December 5, 1839, in New Rumley, Ohio, son of widower Emmanuel, a local blacksmith, and his second wife Maria Ward Kirkpatrick. After a short stint as a schoolteacher, George was fortunate enough to secure an appointment to the Military Academy at West Point, where his record excelled in demerits and low academic rank.

During the Civil War, his ability to garner valuable intelligence and his fearless personal initiative gained him important positions under Union generals George McClellan, Alfred Pleasonton, and Philip Sheridan. His sixth sense of appraising favorable terrain, his ability to perceive the weakness of the Confederates, and his unrelenting saber charges led to meteoric fame and the rank of brigadier general two days before the Battle of Gettysburg. While General Winfield Scott Hancock repulsed the deadly charge of George Pickett on the third day of the battle, it was Custer, outnumbered two-to-one by J. E. B. Stuart's rebel cavalry, who stopped the assault from the backside of Cemetery Ridge, some distance away.

These two actions would be a turning point in favor of the Union, as well as the high water mark of the war. A fellow soldier proclaimed, "There was in him an indescribable something—call it caution, call it sagacity, call it the real military instinct—it may have been genius—by whatever name entitled, it nearly always impelled him to do the right thing." [2]

Custer stood out among the standard uniformed blue lines of Union officers as he always led from the front. His ensemble included black velvet trousers tucked into his knee-high polished boots, a coat of the same material, and a blue sailor's shirt, with a large collar embroidered with two white stars that hung over his jacket lapel. A large, floppy black hat with two stars on the crown protected his fair complexion and his strawberry blonde curls bounced as he trotted his horse. His men could readily see the blood red tie with a Tiffany pin that Custer had designed to hold it in place. His adoring Michigan cavalry units would adopt his red cravat and follow his screams of, "Come on you Wolverines!" in some of the 50 fights where he slashed through gray rebel lines that fled before him.[3] In the unflinching stare of his crystal blue eyes was something scary—it was the look of a natural born killer that was unmoved by the gore of men torn apart from the canister shot around him. Yet the same eyes could cry a flood of tears when leaving his mother, or watching a romantic theater production.

In 1867, the recently appointed Lieutenant Colonel was thrown into the arena of Indian warfare, along with his newly-formed 7th Cavalry, and in partnership with his colleague from Gettysburg, General Winfield Hancock.[4] Hancock was approved to use military force to control the resisting tribes of the Lakota, Kiowa, Arapaho, and Cheyenne residing and raiding in Kansas. His attempted parley with the chiefs ended when the frightened tribes escaped during the night as an embarrassed Custer surrounded empty tipis along the Pawnee Fork of the Arkansas River. The angry Hancock exacerbated the situation by burning the abandoned lodges left behind by Cheyenne Dog Soldiers and Oglala Lakota villagers. These warriors quickly embarked on a series of bloody raids across Kansas in retaliation, and Custer was sent on an unsuccessful pursuit that took a devastating toll on both the men and the regiment's horses.

Some enlisted soldiers, pushed beyond their mental and physical limits, wanted out of their five-year commitment and desertions skyrocketed, endangering the remaining command. One afternoon, a huge dust cloud rose as thirteen enlisted men boldly took the 'grand bounce' in broad daylight and an infuriated Custer decided to make an example of them. He yelled at a nearby officer,

"I want you to get on your horse and go after those deserters and shoot them down!" [5]

Six deserters on foot were quickly overtaken; three of them were shot, and one died. The whole episode had been for effect, and desertions stopped, but Custer did not foresee the possible legal repercussion. To compound his problems, he missed his wife Elizabeth, "Libbie," the daughter of Daniel Bacon in Monroe, Michigan. She was intelligent and beautiful. During the early stages of the Civil War, Judge Daniel Bacon had found Custer an unacceptable suitor for his cultured daughter due to his risky profession and social status. As Custer's prestige and rank climbed during the war, her father eventually allowed the two to marry. Whenever possible, Custer kept her near him, and the long campaign had deprived him of this connection. Hearing reports of deadly cholera spreading through stage stations and regional forts, the de facto commander of the 7th feared for his beloved wife, who was at Fort Riley, Kansas. Having been apart from her for over six weeks, Custer left his troops in the field and rode away with a small group of soldiers on horseback in order to see her. Warriors chased and killed one soldier and wounded another man who had been unable to keep up with Custer. The officer journeyed onward, traveling over 200 miles in a little over 55 hours, to find his wife safe, and according to Libbie, the surprise reunion produced "one long, perfect day." [6]

Custer's side trip to see his wife, as well as his actions against the deserters, would halt his military career, and he was court-martialed for leaving his command and shooting deserters. In September of 1867, he received a punishment of a one-year suspension from the army without pay. Sheridan generously allowed the Custers to stay in his own quarters at Fort Leavenworth for a time, but the couple eventually moved to Libbie's hometown of Monroe. Believing he was innocent of the charges, it was a bitter pill of inactivity that the embarrassed Custer had to swallow, and he was anxious for redemption.

Near the Washita River,
Oklahoma Territory
Friday, November 27, 1868, 1:00 a.m.

Lieutenant Colonel George Custer slowly rode a black stallion through a moonlit night highlighted with a smattering of wispy clouds. He held the reins of his steed with tan-colored gauntlets covering knuckles that were unusually large in proportion to his lean 5-foot, 10-inch figure. Weighing about 170 pounds at his heaviest, Custer was bigger than many of the men in his command. His wind-burned, ruddy cheekbones were covered with freckles; his long aquiline nose and face chafed red from exposure to the bitter coldness. He proudly rode with a light, dotted bandana around his neck and a small, brimless fur cap balanced on his head. His clothing consisted of a dusky white buckskin suit, and when not concealed with an overcoat, the fringe at his collar, pockets, shoulders, and outer seams of his britches flopped whenever his horse trotted. He wondered if his petite wife would recognize him with his cropped hair and thick, bushy dark beard completely concealing his mouth—an addition that kept his face warm on this exploratory expedition.[7] Believing his famous 'Custer's luck' had returned, he recalled the letter of redemption that had brought him to this very moment almost a year after his court-martial:

> *To General G. A. Custer, Monroe, Michigan*
> *Generals Sherman, Sully and myself, and nearly all the officers of your regiment, have asked for you, and hope the application will be successful. Can you come at once? Eleven companies of your regiment will move about the 1st of October against the hostile Indians ...*
>
> *P. H. Sheridan,*
> *Major General Commanding*[8]

This urgent request was one that exuded confidence, redemption, and forgiveness. Smarting from his recent court-martial, Custer was invited back into action with two months remaining in his military exile, by mentor Philip Sheridan. A million miles of central plains from Texas to Canada, as well as between the Mississippi River and the Rocky Mountains, fell into what was known as the Department of the Missouri, and "Little Phil" had jurisdiction over it. Custer had always enjoyed the short-statured commander's affection as a reliable subordinate in the Civil War and now was called on for duty once again.

Fig. 3. Elizabeth (Libbie) Bacon married George Armstrong Custer in 1864. They were married 12 years before he died. For nearly 57 years, as his widow, she worked tirelessly to promote his memory through her books and public speaking engagements. David & Sanford, 1906

National Park Service, Little Bighorn Battlefield National Monument, LIBI_00019_00888

Custer was literally back in the saddle after receiving Sheridan's invigorating telegram. He rode through the Indian Territory darkness searching for raiding Indians.

Seventh Cavalry Second Lieutenant Francis Gibson stated, "His [Custer's] arrival seemed to infuse new life in the command; we had unconsciously fallen into a state of inertia, and appeared to be leading an aimless sort of existence, but with his coming, action, purpose, energy, and a general strengthening of the loose joints was the order of the day; everybody was alert and ready to meet any emergency."[9]

Scout Ben Clark recalled, "[Custer] was impatient to proceed readily and harassed the men and subordinates by his arbitrary conduct. He was a hard master, but his dash and cavalier bearing held the admiration of his troopers."[10] His recent actions and strict discipline polarized his fellow officers into either dedicated friends or venomous enemies. One soldier recalled, "It is true there was more or less a division of sentiment in the regiment as to Custer."[11]

Custer could read between the lines in Sheridan's reinstatement: *Nearly all the officers of your regiment, have asked for you.*[12]

The reality was that some of the officers, who rode through the darkness, despised him.[13] One of them was Captain Frederick William Benteen, a man with a boyish grin and one of the few officers who remained clean-shaven in a period of abundant facial hair. There was something fatherly about the stare from his clear eyes, his soft-spoken manner, and the pipe that often protruded from his lips. His underlying self-righteous, bitter, backstabbing persona seemed strangely counterbalanced by bravery in the guise of a southern gentleman.[14] The Civil War veteran's dislike of his commander began during their initial meeting when Custer derided one of the captain's Civil War heroes.[15]

Hard-drinking and black-bearded Captain Robert M. West struggled through the darkness. West's alcoholism had been the cause of a harsh reprimand from Custer during the Hancock campaign, and he never forgot the episode. The men that were shot for desertion were from West's company. He believed Custer's court-martial sentence was too lenient and continued unsuccessfully to urge for a charge of murder against his commander.[16]

The commander of the 7th Cavalry might have guessed that Major Joel Elliott, who held the coveted leadership of the 7th Cavalry in Custer's absence, was not eager to relinquish that role. Custer's immediate revitalization of the 7th Cavalry implies that he was not happy with the state of affairs under Elliott's tutelage. Elliott was a curly-headed Civil War veteran with a large moustache that wrapped around his face and merged with his large sideburns. He received a political appointment to the rank of major over many seasoned officers. It initially alienated many of them, but his pleasant demeanor eventually won the officers over. Elliott was one of the officers ordered to shoot the deserters, which placed him in an unenviable position of choosing unethical behavior or disobeying orders. Originally known as a 'Custer man,' it seems that Elliott, in conjunction with West, had become resistant to Custer as their commander. [17]

Custer had his dedicated supporters with him as well, including 'Gentleman' Louis McLane Hamilton who was the grandson of American patriot Alexander Hamilton. Louis was a Civil War veteran with a large nose and moustache. He was a favorite among the entire regiment and was appreciated for his wit, sense of humor, and modesty. [18]

There was one faithful supporter who would follow Custer to the ends of the earth. Shivering in the blackness behind his older brother, Lieutenant Thomas "Tom" Ward Custer wondered if his health would hold up in the frigid temperatures, since he had just returned from a lengthy sick leave due to inflammatory rheumatism. He wore a scar on the side of the face as a red badge of honor. In March of 1865, near the Civil War's end, Tom and his horse jumped over enemy earthworks into a churning sea of gray at Sayler's Creek. He was shot through the face as he captured a Confederate flag and was forced to the rear by his brother. He captured another standard only days before. There was no doubt he was fearless, and one of the few soldiers during the Civil War to earn two Medals of Honor that George jealously called 'baubles.'[19] Now a first lieutenant, Tom had been one of the three officers ordered to shoot the deserters. On occasion, however, Tom was reminded of his official position and rank when he reaped the wrath of his older brother. On one occasion, Tom ignored a repeated admonition by brother 'Autie' to hurry and finish his breakfast. George Custer exploded and charged into the tent, kicking over the table with food and dishes flying through the air.[20]

Another Custer favorite was William Winer Cooke, nicknamed "The Queen's Own." Cooke was born into an affluent family of British Loyalists who fled to Canada during the American Revolution. The Civil War veteran was a natural organizer with lots of spit and polish and was easy to spot in a crowd as he wore long 'dundreary' side-whiskers that hung down to the middle of his chest. His trustworthy personality earned him the position of regimental adjutant and he became a trusted confidant to both Custer brothers. Cooke was also one of the officers ordered to chase down the deserters. The athletic Canadian was chosen to lead the newly created, crack sharpshooting unit after Custer's return to duty.[21]

Ironically one of the designers of this mission was named after the famous Shawnee Indian leader, *Tecumseh*. Civil War General William Tecumseh Sherman had inflicted an effective 'March to the Sea' strategy by scorching a path through Georgia, weakening the Confederacy. It was a strategy called 'total war' and meant that hardships were waged on the civilian population as well as fighting men, as the Union forces cut the South in half, destroying homes, factories, crops, railroads, and bridges.[22] Total warfare was brutal to both the psyche and the physical existence of its recipients. Sheridan, along with Custer, had used the same strategy when they torched the Shenandoah Valley, and now they found it expedient to enforce it against the southern Plains tribes.[23] However, there seemed to be a different standard of dealing with the tribesmen in 1867, when William Sherman dictated to Winfield Hancock that there should be 'no quarter' given to the warriors.[24]

The General Orders manual used by the army justified such a concept by stating: All troops of the enemy known or discovered to give no quarter in general, or to any portion of the army, receive none.[25] In a similar directive, Sheridan ordered Custer:

> . . . to proceed south, in the direction of the Antelope Hills, thence towards the Washita River, the supposed winter seat of the hostile tribes; to destroy their village and ponies; to kill or hang all warriors, and bring back all women and children.[26]

In the fall of 1868, Sheridan ordered three separate converging columns to discover and punish the southern tribes in their winter camps. Major Eugene Carr advanced from Colorado Territory, a second command under Major Andrew Evans from New Mexico Territory, and a third under General Alfred Sully departed from Fort Dodge, Kansas.

Sheridan, who accompanied the Sully command, envisioned Custer as the field commander of the third force and gave him instructions that indicated confidence, "Custer I rely in every thing upon you and shall send you on this expedition without giving you any orders leaving you to act entirely upon your judgment."[27]

The reason for this conflict with the native tribes originated long before the bloody battles fought between blue and gray clad armies. From their first contact with Europeans, the aboriginal communities unsuccessfully attempted to coexist with the new strangers. The Plains tribesmen realized that the creaking wooden wagons forming dusty trails, and the steam-belching iron horse following double rails, led to settlements, which in turn, equated to the loss of their food and land. This, of course, threatened their culture, religion, nomadic lifestyle, and very existence. Even when on friendly terms with the quickly encroaching non-Indians, the native tribes inadvertently picked up deadly diseases that decimated entire communities.[28] Dingy, canvas-topped wagons, longhorn steers, and Mormons pulling handcarts created new trails and settler complaints of Indian depredations demanded that military forts be thrown up for protection. New pathways, including the Santa Fe, Oregon, Bozeman, and Bridger Trails, penetrated lands occupied by native populations and over 300,000 emigrants, including gold seekers, settlers, and entrepreneurs moved west during the Civil War alone.[29]

After Indian raids in Colorado territory, local residents demanded revenge on those responsible and were not discriminating in their choice of tribes. On November 29, 1864, a Southern Cheyenne village under Chief Black Kettle, situated along Sand Creek, Colorado Territory, was surprised and attacked by Colonel John M. Chivington's 3rd Colorado Volunteer Cavalry along with other militia detachments. Black Kettle was a *Suhtai*, an Algonquian tribe that had merged with the Cheyenne in the early 18th century. As a youth, he was an accomplished warrior against the Kiowas, Apaches, Utes, and Comanches, until marrying into the Cheyenne tribe.[30] Cannon fire and a bloodbath ensued as Black Kettle and his wife, *Arnohowok*, also known as "Medicine Woman Later," fled during the carnage. About 30 Cheyenne with Black Kettle took refuge among some crude dirt trenches some distance from the killing. While most of this party escaped the horrors taking place in the camp, Black Kettle returned to the scene to find his wife shot nine times; she was rescued and survived.[31]

The savage mutilation by the Colorado volunteers matched or surpassed any ever inflicted by the warriors. Female private parts were hacked off as souvenirs, with women's breasts and warrior testicles fabricated into tobacco pouches. Such ghastly souvenirs, along with over 100 Cheyenne scalps, many from women and children, were exhibited at one Denver theater intermission. There would be approximately 150 Cheyenne killed, the majority being women and children.[32]

According to Cheyenne historian John Stands in Timber, whose grandmother survived the episode, "None of them forgot what they saw there. The Cheyennes were almost all in raids after that. When the chiefs tried to prevent it they [warriors] would ask, 'Do you want the white men to kill our people like that?' They wanted to get even . . . they killed people and took some white girls away. Some were turned loose again, and some were kept in the tribe. The chiefs tried to hold some of those things down, but the young men did them anyway."[33]

The Southern Cheyenne and other tribes were forced into two factions. Some sought to coexist and give up their way of life. Others, like the fearsome Dog Soldiers, raided, ransomed, resisted, raped, and killed those attempting to threaten their existence.[34] Beside the Cheyenne, other tribes, including the Kiowa, Comanche, and Arapaho, were threatened in Texas, Colorado, Oklahoma, Kansas, and Nebraska. During the years 1867 to 1869, some of the bloodiest conflicts occurred in Kansas when the Union Pacific Railroad cut through the prime hunting grounds of the Cheyenne, whose most ferocious defenders were the Dog Soldiers.

The fervor of Manifest Destiny—self-proclaimed God-given rights to conquer and tame the land as well as its inhabitants—perpetuated the minds of many encroaching settlers and miners. Government land grabs, poorly worded treaties, broken promises, as well as murder by both sides intensified the situation.

In 1867 Congress sent a peace commission to convene with the Cheyenne, Kiowa, Comanche, and Arapaho nations along a small creek in southern Kansas, where the Kiowa were known to conduct their medicine dances.[35] The Medicine Lodge Creek Treaty, which was actually two separate treaties, was similar to many other pieces of paper created between the two opposing sides that did not understand the complex culture of the other. Both participating parties would quickly break it.[36]

One of the signers willing to make an 'X' by his name was the moderate and astute Black Kettle, who through most of his career as a chief yearned to avoid clashes with the ever-encroaching tide of

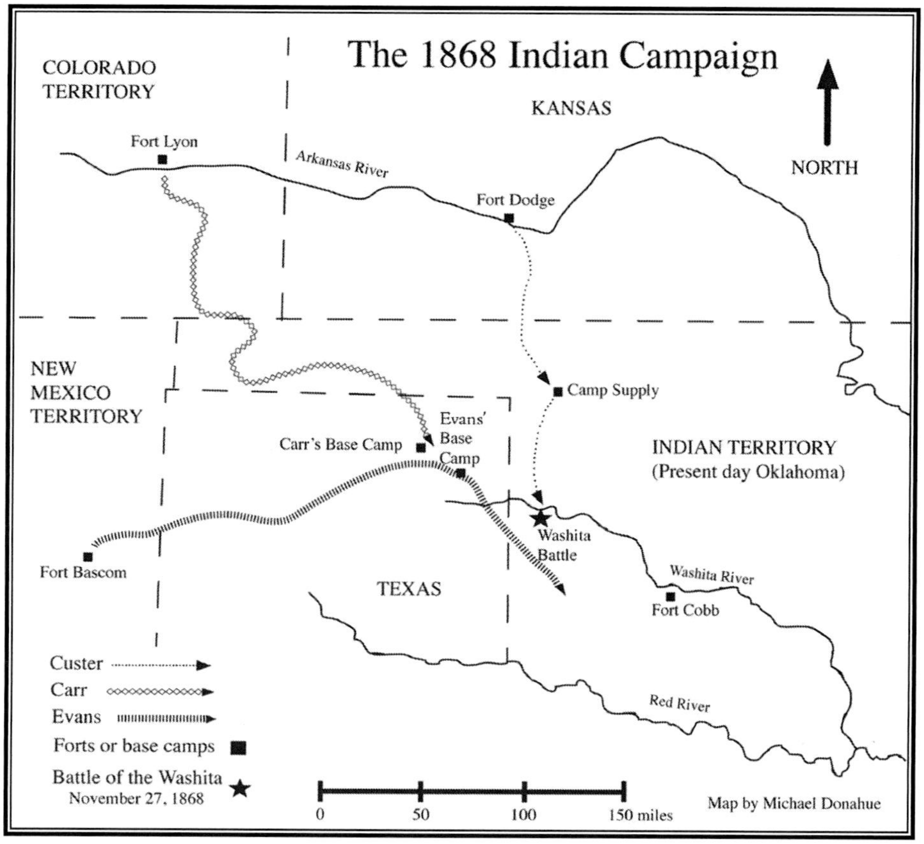

Fig. 4. Map: The 1868 Indian Campaign.

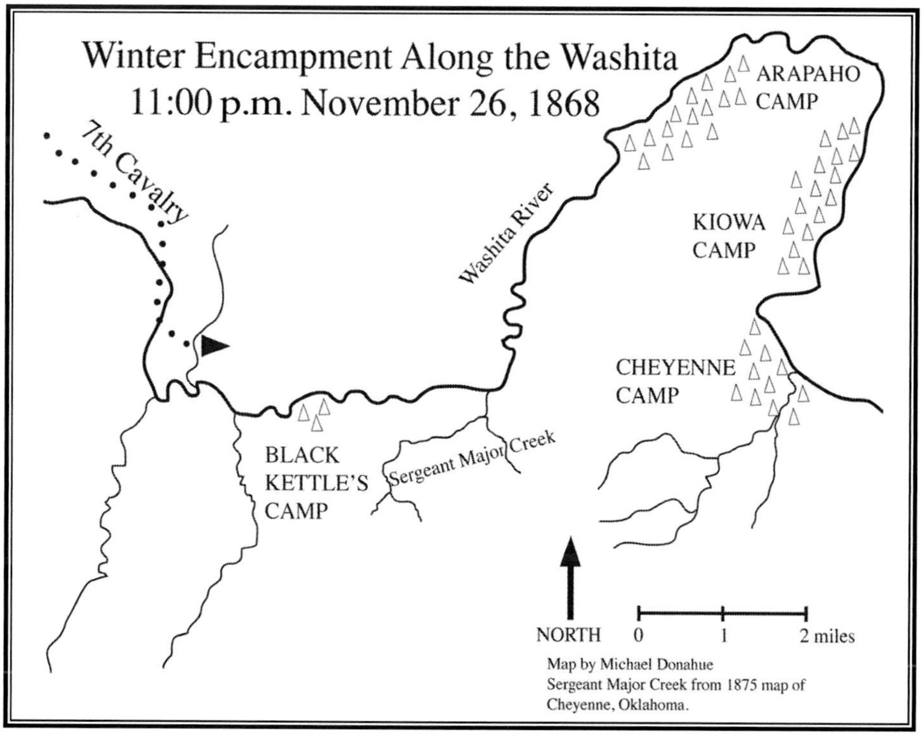

Fig. 5. Map: Winter Encampment Along the Washita. 11:00 p.m. November 26, 1868.

Fig. 6. Black Kettle (2nd from left, front row, holding the pipe.) One of only two known photographs of him, both taken at a meeting at Camp Weld, Colorado Territory in 1864. The chief unsuccessfully attempted to please both his tribe and the army, but his efforts ended in personal disaster and death.
The Denver Public Library, Western History Collection [X-32364]

settlers. The stocky man with the squinty eyes was a judicious and compromising politician. He was well aware of the attacks, rapes, and hostage-taking by the Cheyenne, which sometimes included some of his own tribal members. Even though his efforts at peace were treated with betrayal by the whites, and sometimes disapproval by the Cheyenne, a few perceptive leaders on both sides admired his powerful influence and patient wisdom.

Black Kettle was caught in a cultural dilemma that was inherent to his societal and cultural values. While the elders of the tribe had accomplished their high rank, based on past fighting and brave war deeds, they implored the young men to restrain from the same activities. The young warriors ignored the tribal elders in order to advance in status through a risky cycle of bravery, revenge, prestige, and often death. The chief admitted, "Some will not listen, and since the fighting began, I have not been able to keep them at home. But we all want peace, and I would be glad to move all my people down this way." [37]

By late summer of 1868, the settlements along the Saline and Solomon Rivers in Kansas were a continuous war zone of retribution for the Sand Creek and Hancock debacles. The Cheyenne Dog Soldiers' actions swiftly rivaled the brutality of the Sand Creek massacre.[38]

Only a year later, after signing the treaty, Black Kettle's village of 52 lodges camped along a crooked waterway called the *Hooeeohe* or 'Lodgepole River' in current day western Oklahoma. The whites called it the Washita, and the Choctaw word translation meant the 'big hunt' for good reason.[39] The valley was full of deer, buffalo, antelope, wild turkey, grouse, and quail. The water from the Washita was of excellent quality, and the valley was perfect for a winter camp.[40] Black Kettle's band was only one of several camps extending for almost 10 miles, where approximately 6,000 Kiowa, Arapaho, Comanche, and Apache had gathered.[41]

After the terrible massacre along Sand Creek, Black Kettle decided to move his band along a big bend in the river several miles from the other bands.[42] His village stretched for about a half-mile along the gray-wooded bottom that was surrounded by red-orange rolling hills that protruded from the glazed snow. While the population of his camp was changing daily, there were perhaps as many as 250 to 300 people there. Dominating the beige tipis at the western side of the camp was a large black lodge.[43] Inside sat Black Kettle's family with at least two of his wives, Sioux Woman and Medicine Woman Later, warming themselves around glowing embers.[44]

Nearby on the east side of the camp was another impressive lodge covered with beautifully painted warriors depicting brave war deeds and other exploits. It belonged to the elderly chief Little Rock, who in his early 60s, had little experience as a warrior but was respected as a holy man. The moderate Little Rock was well liked by the whites around the army posts and had signed the Medicine Lodge Treaty. He had daubed the constellation Pleiades on his hide shield that flickered from the light of his lodge fire.

Little Rock's wife, Skunk Woman, attempted to stay warm near their beautiful daughter 18-year-old *Meotzi*, or 'Dawn Walker' in Cheyenne. She would later be known as Monahsetah to the white world. The teenager had been married to a warrior named Little Eagle. During a fight she had seized a gun and shot him in the knee before leaving the union. Now she found herself pregnant by him, but living with her large family that consisted of a sister and three younger brothers including one named Hawk. Also in the camp were Little Rock's 80-year-old mother, *Tovish*, and his sister, White Buffalo Woman.[45]

The Cheyenne warrior Wolf Looking Back rested with his wife, *Mahwissa*, whose name meant "Red Dress." In her mid-30s, she was prominent among the Cheyenne females in the village and had rescued a child and her mother during the Sand Creek tragedy. Her 12-year-old son, Little Beaver, also resided in the tipi. She would soon find herself in another important role among her tribal members.[46]

During the frigid evening of November 26th, Black Kettle returned from a meeting with Colonel William Hazen at Fort Cobb. The chief asked for military protection for his band, but was told by Hazen to get authorization from Phil Sheridan. Hazen said he could not protect Black Kettle's tribe due to the fact that some of his warriors had been raiding. Phil Sheridan, however, was already in the field. Hazen also warned the Cheyenne leader that soldiers were in the field looking for them and that they should beware.

Upon his arrival at the camp, several Cheyenne proposed to Black Kettle that they move their camp closer to the other tribes camped downstream along the Washita River. His wife, Medicine Woman Later, exclaimed, "I don't like this delay, we could have moved long ago. The Agent sent word for us to leave at once. It seems we are crazy or deaf, and cannot hear." [47]

Another family with a premonition of disaster was a Lakota clan led by Bad Bank who had decided to spend the night with Black Kettle's Cheyenne. Bad Bank's wife urged that they move immediately to one of the larger camps, but her husband said it was too late in the day to move then and that they would do so the next day.[48]

Earlier in the day, a Kiowa war party passed Black Kettle's encampment and warned them of an unusual trail they had seen. Few of the tribesmen thought it could be bluecoats, because it was not customary for the military to move in the dead of winter. One Kiowa in the group, Trailing the Enemy, decided to spend the night in Black Kettle's camp to watch a dance, while his friends continued to their families camped downstream. He planned to return to his own village the next morning.[49] A sentry named Double Wolf was stationed to watch for trouble, but seeking warmth around a fire, quickly fell asleep leaving the camp unguarded.[50] Black Kettle believed that the frigid weather would make such an army movement improbable. Not far away in the dark the 7th Cavalry had been ordered to punish any Indians they could find; the trail they were following led straight to the Washita River.

Chapter 2

Washita: From the Ashes the Phoenix Rises

On November 23, 1868, Custer, along with approximately 850 heavily bundled soldiers of the 7th Cavalry, and 37 wagons equipped with supplies sufficient for 30 days, had left their field depot aptly named Camp Supply.[1] "Everybody was ready and eager for action," remembered First Lieutenant Francis Marion Gibson. "All were in prime condition as regards to health and spirits, and the whole outfit was in for it, whether it turned out to be a fight, a fluke, or a frolic."[2]

The army hired well-seasoned scouts such as the dependable and dark-mustached Ben Clark, known as "Red Neck" by the Indians because of his tendency to sunburn. The handsome, unassuming Civil War veteran was fluent in several Indian languages as well as sign language.[3] Another scout, blue-eyed Moses "California Joe" Milner was described as:

> Six-foot 1-inch, [he] wore a ponderous mat of flaming red whiskers and a growth of long knotty hair of the same description . . . his hair and whiskers were well powdered with a series of layers of dust, intermingled with stray blades of dry grass, leaves, and chips, reminding the beholder very forcibly of the old fellow's precious night's slumber on the bosom of mother earth. Joe often wiped his runny nose with a gunnysack in lieu of a handkerchief as a gentleman of the plains.[4]

From their Kansas reservation, Hard Rope and Little Beaver led ten of their fellow Osage scouts. They were traditional enemies of the tribes that the army was pursuing and, according to Custer, "were a splendid-looking set of warriors . . . they are painted and dressed for the war path, and well-armed with Springfield breech-loading guns. All are superb horsemen."[5]

The departing procession included the band mounted on gray horses playing *The Girl I Left Behind Me*.[6] During the first day of the march, one poetic soldier recalled the beauty of the landscape transformed by the falling snow, but the gray skies soon unleashed whiteout conditions. The confused Osage scouts could see nothing, so Custer took over using his compass to direct the slow-moving column south.[7]

Among the officer corps was youthful Edward Settle Godfrey who would become a dependable workhorse on the campaign. The lieutenant was easily recognized for something he would sport throughout his long military career—a moustache that hung down like the massive tusks of a walrus. The West Point graduate, nicknamed "God" by his friends, initially found cavalry service somewhat disagreeable, but

Fig. 7. This 1869 unpublished photo of First Lieutenant Edward S. Godfrey shows how he looked about the time of the Washita battle. The cool and reliable officer impressed Custer. Godfrey was ordered to capture the pony herd, destroy village contents, kill the ponies, and care for the captured hostages.
Courtesy of John W. Turner

quickly adapted.[8] Another young lieutenant struggling through the snow was New Yorker Owen Hale, a Civil War veteran with a proud heritage as a descendant of American patriot, Nathan Hale. He was considered a true gentleman who exemplified the ideals of duty and honor.[9]

The ground was blanketed with a foot of snow, but the sky was crystal clear and blue on November 24 as soldiers plodded through miles of glaring whiteness that caused snow blindness in some troopers.[10] Two days later, Major Elliott with his three-company battalion discovered two Indian trails, one made by approximately 150 warriors that was less than 24 hours old. Ruts from tipi poles indicated one trail was not from a war party and probably lead to a village camp. A scout took the intelligence to Custer while Elliott took his party over a divide into the drainages of the Washita River. The messenger was ordered to return to Elliott with instructions to follow the trail until 8:00 p.m. and if Custer had not reached him by that time, he was to halt.[11]

An impatient Custer, realizing the wagons were slowing his progress, established a new plan. He gave the regiment just 20 minutes to prepare and left 80 men with the poorest horses to guard the wagon train. He started with the rest of the regiment and with only the supplies that could be carried on horses to intercept Major Elliott's party. He ordered the cumbersome wagon train to follow the trail of the regiment.[12]

Louis Hamilton was expected to remain behind with the main pack train as officer of the day. The proud officer pleaded not to be left behind and, in a sense of fair play, Custer told him he could come if he found a replacement.[13] First Lieutenant Edward Gustave Mathey was suffering from snow blindness and agreed to guard the slow wagons.[14] The profane and swarthy Mathey was humorously nicknamed the "Bible-Thumper" since he did not smoke or drink. Having once studied for the priesthood, he eventually worked as a saloonkeeper, farmhand, and a store clerk who had a proclivity to locate hard-to-find items.[15] Benteen disliked Mathey and declared, "Of all the nonentities with which a troop of cavalry could be damned, as its head and front (Mathey) fits the bill."[16] In a later correspondence, the artistic Benteen called him a 'damned sneak' and labeled him "P. P. [piss pot] Mathey," which he thereby illustrated with two penises urinating into an initialed chamber pot.[17]

Each soldier was ordered to bring along 100 rounds of ammunition for their Spencer carbines and revolvers. They were ordered to leave sabers as well as dogs behind.[18] For unknown reasons, some officers ignored or disobeyed the last two orders.[19] First Lieutenant James Bell was placed in charge of a smaller supply train of about six wagons supplied with extra ammunition, as well as rations, and ordered to keep up with Custer.[20]

Elliott had halted and the rest of the regiment joined him around 9:00 p.m. The Osage believed that Indians were nearby, but a hesitant Little Beaver recommended that they not attack until daylight. A doubting Custer soon realized that the scouts had a fear of attacking an unseen foe in the darkness.[21] One officer recorded:

> When Custer went after Indians, he himself was always in the advance and looking for the enemy. On the Washita, we have found him with the advance scouts on all occasions and watching the enemy himself.[22]

Fig. 8. As Major Joel H. Elliott chased fleeing villagers, he was killed with his entire detachment when they encountered warriors from downstream villages. These casualties along the Washita would create controversy within the 7th Cavalry officer corps.

Courtesy of Sandy Barnard

At 10:00 p.m. the command advanced through the darkness with the order that no matches should be struck, and nothing above a whisper was to be uttered. To the soldiers who enjoyed tobacco, the loss of this single indulgence made the night seem colder and longer. One soldier recollected:

> Seen from an elevation as night approached, what a weird, serpentine specter was this body, outlined upon the white snow. As it wound around the tortuous valley, it had the semblance of a huge reptile, stealthily creeping to destroy its victim or foe. The silence was oppressive . . . the moon was not quite visible, as the sky was overcast with fleecy clouds.[23]

Coming upon a smoldering campfire and an overlook, scout Little Beaver whispered, "Heaps Injuns down there." The clear, frigid atmosphere amplified the sound of a dog barking and soon the cry of an infant echoed across the valley.[24] The village was situated in a horseshoe-shaped bend in the Washita River and was composed of "52 big lodges and almost an equal number of temporary shelters, those mostly being used by the raiders we had been following and the young boys charged with watching over the band's ponies."[25]

James Bell, Frederick Benteen and the small contingency of wagons halted several miles away. Officers were ordered to come forward, fearing noise from the command might warn the village. Excited pets belonging to Custer and his troops barked at the villagers' dogs and had to be killed or muzzled.[26] The group of officers peered into the blackness below and the plan of attack was to divide the approximately 690 men into four detachments of almost equal strength, surround the village, and simultaneously attack at daybreak. Custer realized Benteen's company would be needed in the attack and ordered him forward.[27]

Joel Elliott, Frederick Benteen, and Owen Hale were ordered to attack the opposite end of the village from the east. The oldest officer in the 7th Cavalry, Captain William Thompson—a former lawyer and newspaper editor—squatted next to Captain George Yates. Yates had survived a wound at Fredericksburg during the Civil War and met Custer while convalescing. A close bond was established that lasted until their deaths. Both Thompson and Yates were to attack from the south.[28]

During the conference with officers, Thompson asked, "General [Custer], suppose we find more Indians there than we can handle?"

Custer replied gruffly, "Huh, all I am afraid of we won't find half enough. There are not Indians enough in the country to whip the 7th Cavalry."[29] Eight years later, Custer would make a similar statement of regimental bravado as he marched toward another Indian village along the Little Bighorn River. Ironically, both the Washita and Little Bighorn villages would be similar in population, but Custer would find enough warriors at the latter site to prove him wrong.

Captain Edward Myers, an ill-tempered German who was promoted for meritorious service during the Civil War, prepared his company to attack from the west.[30] Lieutenant Charles Brewster's company accompanied Myers as they stumbled through the darkness. A former lawyer from New York, Brewster was a dedicated Custer supporter who fought with him in the Civil War.[31]

The largest contingency of troops that included four companies would attack from the northern hill, where the plan was formulated. Irishman and First Lieutenant Matthew Berry, who had an enviable Civil War record, was to lead one company along with Louis Hamilton's men.[32] Custer's nemesis Robert West and handsome Thomas Weir were ordered to charge from the same location. Weir was well educated and had been a Confederate captive during the Civil War. He had served with Custer during Reconstruction duty in Texas, and Libbie Custer had a fondness for the officer, even though he loved alcohol.[33] Custer chose to accompany Cooke's elite group of newly formed sharpshooters beside West and Weir.

The officers were commanded to prevent the killing of any but the fighting strength of the village. After the battle, Custer apologetically penned:

> In a struggle of this character it is impossible at all times to discriminate, particularly when, in a hand-to-hand conflict, such as the one the troops were then engaged in, the squaws are as dangerous adversaries as the warriors, while Indian boys between 10 and 15 years of age were found as expert and determined in the use of the pistol and bow and arrow as the older warriors.[34]

Custer ordered that the attack take place when bandmaster Sergeant Albert Piedfort struck up a song of his choice. The band had struggled all the way there with their cold, bent, and punctured brass instruments, and finally there was a chance to use them.

Washita River, Oklahoma Territory
Friday, November 27, 1868, 3:30 a.m.

The temperature plummeted, making the situation almost unbearable for man and beast. The four companies with Custer lay hidden behind a ridge, while the remainder of the 7th stumbled and cursed in the blackness to their designated positions. The cold and tired officers near Custer slumbered on the ground with the capes of their greatcoats pulled over their faces, using one another's bodies for pillows. Others could not sleep, as their minds raced with the knowledge that death might claim them at sunrise.

Along the journey some of the 14 musicians had punched holes in their instruments thinking that if they were defective, they would not have to use them. Custer had discovered the ploy, forcing them to walk in the snow or do fatigue duty. The errant band members quickly patched the holes in their instruments with soap, some likely wondering if they could play a note, or whether bubbles would float in the frost during the predawn concert.[35]

The Osage scouts huddled in a circle under a tree, and some applied paint to their faces with yellow, black, and vermilion pigment, creating mask-like visages that would strike fear into any enemy. They feared Custer might negotiate with their Indian foes and turn them over to the enemy, so they decided to ride next to his personal standard-bearer to monitor the situation and flee if necessary. Trepidation seemed to be running rampant through the ranks since this was to be a blind attack on an enemy of unknown size on unfamiliar terrain. Mistrust and uncertainty permeated the frozen November dawn.[36]

Hidden below in the blackness next to the river was a large lodge 20 feet in diameter and height. Black Kettle's lodge was painted 'black as the night sky and decorated with every manner of celestial forms'.[37] As he walked outside, the stars on his tipi seemed to merge with those in the heavens above. A strange extremely bright, flickering light began to rise over the hills. Custer remembered:

> The strange apparition in the heavens maintained its steady course upward . . . it had risen perhaps to the height of half a degree above the horizon as observed from our positions, when, lo! The mystery was dispelled. Rising above the mystifying influences of the atmosphere, which had appeared so suddenly before us and excited our greatest apprehensions developed into the brightest and most beautiful of all morning stars. Often since that memorable morning have I heard officers remind each other of the strange appearance which had so excited our anxiety and alarm.[38]

Soon the sun rose brightly over the snowy hills and valley below. A mist enveloped the village.[39] The sharpshooters removed their heavy Civil War surplus overcoats, along with their black haversacks and bulls-eye canteens that contained their remaining provisions, to allow for increased mobility during the fight. In order to fire their Spencer repeating carbines, the soldiers had to remove their clumsy woolen mittens, and some men were ordered to stay behind to guard personal items. The freezing sharpshooters began to curse and grumble after removing their coats. Custer said, "Stop that noise. It will be hot enough for you inside of an hour."[40]

Not far away, many villagers were sleeping soundly, because they had stayed up late watching a dance and sharing a feast of agency hardtack crackers, sugar, and coffee brought back by Black Kettle.[41] Ironically the U. S. Government, like an executioner, had provided many of the Cheyenne with their last meal. A 14-year-old girl, Moving Behind, was sleeping in the lodge of a young female friend. One of the subjects discussed among the girls was Moving Behind's new sweetheart named Crane.

"I began to know a little about love at that time," she recalled. Moving Behind had heard Black Kettle's wife's angry admonition for the camp to move, and the young girl "felt strange late that evening" in an eerie premonition.[42] The uneasiness was strong enough for some villagers to leave, and "some got away before daylight, but no one saw who they were," remembered the teenager.[43] One young boy, upon hearing a dog barking furiously outside his tipi on the eastern edge of the village, went outside to discover the cause. Seeing a soldier peering over the bank of the river, the youth shot into the air to warn the other villagers as the life and death struggle commenced.[44]

There was no need for rope to hang warriors; they would be shot on sight. Wounded warriors would be executed as commanded by Sheridan. On the opposite side of the village, Custer heard the single gunshot and realized that the time had come to charge. Bandleader Albert Piedfort, who also served as the chief trumpeter, sat on his gray horse and by chance selected an Irish pub tune to play that morning at 6:30 a.m.[45]

Godfrey recalled, "Gen. Custer ordered the chief trumpeter to sound the charge and turned to the bandleader and ordered him to play something lively. The bandleader [a little German] afterwards told me he was thinking of his revolver and not of his cornet and music. He quickly pulled out his music book and frustratingly opened it up to *Garryowen.*" Piedfort called out the number of the piece and began to toot his cornet and the rest soon chimed in. Such was the origin of *Garryowen* as the regimental battle tune.[46] After the battle, one of the more critical soldiers recalled how ridiculous the music sounded in the freezing and dangerous situation.[47]

> Let Bacchus' sons be not dismayed,
> But join with me, each jovial blade;
> Come booze and sing, and lend your aid,
> To help me with the chorus.
>
> Instead of spa we'll drink down ale,
> And pay the reck'ning on the nail;
> No man for debt shall go to jail
> From Garryowen in glory.

The large Irish contingency appreciated the stirring tune's subject—drinking, fighting, and resistance to paying taxes to the British government.[48] The bizarre concert ended prematurely, as saliva froze in the cold brass instruments, yet the few initial notes had inspired men, and some musicians were carried away in an uncontrollable rush into the village.[49] Custer, using a phrase he would repeat at Little Bighorn, stated, "The Indians were caught napping; but realizing at once the dangers of their situation, they quickly overcame their surprise and in an instant seized their rifles, bows, and arrows, and sprang behind the nearest trees, while some leaped into the stream, nearly waist deep, and using the bank as a rifle pit began a vigorous and determined defense."[50]

Another soldier explained, "As soon as the band struck up that tune [*Garryowen*] the horses seemed to take new heart and the men, too, and when we went into the village, the Indians didn't know what it was, thinking it was some serenade, and they came running out of the tents without their arms. As soon as they saw Custer and his command, they rushed back into their tents and got their guns, commencing firing."[51]

Double Wolf, the guard who fell asleep, quickly grabbed his weapon and ran into the chaos only to be one of the first Cheyenne casualties.[52]

Accounts of Black Kettle's death are contradictory, but it is likely that he died early in the fight. Black Kettle grabbed a weapon, firing a warning shot into the air. An account by Cooke states that he began to sing a death song of his brave life deeds

Fig. 9. Black Kettle Tree. The 'Black Kettle Tree' (in the middle of the photo) marks the site where his lodge was situated at the time of the battle. Black Kettle and his wives were killed in the vicinity of the timber at left, located along the Washita River.

Black Kettle Tree at Washita Battleground, circa 1890. National Park Service, Little Bighorn Battlefield National Monument LIBI_00015_00661, Hugh L. Scott

Fig. 10. Black Kettle Tree, detail. This previously unpublished photo shows the intentional scarring on the tree to distinguish it as the 'Black Kettle Tree.' Due to flooding, age, and changes in the river course, the tree no longer exists.

Black Kettle Tree at Washita Battleground, circa 1890. National Park Service, Little Bighorn Battlefield National Monument. LIBI_00015_00660, Hugh L. Scott

Fig. 11. Custer stands behind his Osage scouts who found Black Kettle's village and also perpetrated most of the noncombatant casualties. In a surrealistic juxtaposition, Custer's pet pelican poses in front of his headquarters flag.

General George A. Custer and group of seated Osage Scouts at Fort Dodge, Kansas, circa 1868
National Park Service, Little Bighorn Battlefield National Monument, LIBI_00019_00585

before being cut down by soldier gunfire. One of the elite sharpshooters seized the revolver from the death grip of the chief and presented the trophy to the long-whiskered Cooke. Trooper Charles Brewster noted that the chief reproached his people in a loud voice for disregarding his warriors' past deeds.[53]

It is more likely that the chief and his wife attempted to escape as they had done at Sand Creek. Black Kettle and his wife, Medicine Woman Later, having survived the brutal carnage in Colorado, fell face down into the ice encrusted river and snow near their lodge according to Cheyenne oral history. A bullet penetrated the chief's stomach, and another hit him between the shoulder blades. His wife, who wore the scars of nine bullet wounds from Sand Creek, found a tenth bullet from the bluecoats deadly. Black Kettle's other wife, Sioux Woman, was killed by a bullet in the back as she fled with her children.[54]

Chief Little Rock quickly grabbed his muzzle-loading rifle, as well as his bow and quiver of arrows. He attempted to escort his elderly mother *Tovish* and about 20 women and children across the icy river, only to be intercepted by soldiers. After shooting a cavalry horse with his firearm, a soldier's bullet hit him between the eyes splattering brain matter on the pristine snow. His son, Hawk, was able to escape toward the other camps, while his daughter, *Meotzi*, and *Tovish* were captured.[55]

According to Benteen, his company arrived first and secured the women and children, as well as the pony herd.[56] As he galloped through the camp, he encountered Black Kettle's 21-year-old nephew, Blue Horse.[57] The youthful-looking warrior fired six shots from a revolver that missed the captain who charged and killed him.[58]

Custer's black stallion jumped over the narrow Washita River in one leap, carrying horse and rider into the midst of the buffalo hide lodges. He fired his revolver into one startled warrior who was about to shoot him, and then charged through the village. Followed by scout Ben Clark, both headed toward a small hill south of the camp to monitor the action.[59] Louis Hamilton, who had begged to come along, was one of the first soldier casualties as he was shot through the heart. His last words were an admonishment to his charging troops, "Now men, keep cool. Fire low, and not too rapidly." A warrior named Cranky Man evidently shot him, and the Cheyenne was killed on the spot.[60]

Soldiers did not see the panicked warriors until they were within 50 yards of the camp. The sharpshooters ran on foot, while the rest of the companies rode on horseback into the camp before dismounting. The acrid odor of gunpowder permeated the atmosphere, while a bluish haze mixed with fog drifted under the barren tree branches. One Kiowa reported, "Splashes of snow showed where the bullets were hitting the snow."[61] In animated tones, men on both sides shouted in languages unintelligible to the other. Women sang war songs and death songs, while some of their wounded men lay in the snow screaming for help.[62]

Some terrified women and children remained inside their lodges, hoping the thin buffalo skins would offer protection, while others took flight through gaps between the soldiers. Panicky villagers jumped into the frozen river to hide, and as the surface shattered, they were cut by the ragged edges of the ice.

"The river ran red with their blood," stated one Cheyenne youth.[63]

While some warriors fought bravely, others fled. One Cheyenne named Statue angrily shouted, "That is exactly how I always felt toward you chiefs, that some day you would turn out to be cowards. Leaving the poor, helpless women and children behind, and letting them suffer!"[64] Eighteen-year-old Red Bird stood his ground as his family escaped. He fought valiantly until killed by the soldiers, and his pony remained by his side even in death. A quick-thinking family member shot the horse, so the animal could join its owner in the spirit world.[65] Kiowa warrior Trailing the Enemy found it difficult to shoot the soldiers, so he directed his arrows into the soldier horses dismounting the bluecoats before hiding in the brush.[66]

Young Moving Behind and her aunt ran and hid in some short red grass near the camp. The young girl recalled, "In the grass we lay flat, our hearts beating fast; and we were afraid to move. It was now bright daylight. It frightened us to listen to the noise and the cries of the wounded."[67] A youth named Magpie, who had survived Sand Creek, fled stumbling with a bullet wound in his leg as a trooper swinging a saber bore down on him. Magpie ducked and fired his pistol, dropping the soldier. The boy quickly mounted the soldier's horse, chased by troopers, until their attention turned to a group of women and children near them.[68]

Pilan, a Mexican who had been captured as a youth by the Cheyenne and eventually married into the tribe, attempted to flee with his young daughter in his arms, but was cornered. He begged in broken English to the soldiers, "You will take the baby, won't you?" The soldiers took his child and ordered him to move off down the ravine where a barrage of bullets in the back dropped him.[69]

Captain Albert Barnitz, who had earned brevets for gallantry in the Civil War and accompanied Elliott, charged into the fleeing warriors and killed two before charging a third. As he did, both fired at almost point-blank range. The warrior's ball penetrated the officer's ribs, coming out near his spine. Hanging on to his horse's neck, Barnitz had the strength to fire one more shot, killing the warrior. Barnitz collapsed among some snow-covered boulders, and although severely wounded, he would survive.[70]

Twenty-five-year-old Private Elihu Clear, a Civil War veteran from Indiana, reportedly shot and killed Black Kettle's daughter as she ran. Godfrey saw two fleeing villagers and recognized that both were women. As one woman threw up her hands, he screamed at Clear, "Don't shoot!" At the same instant, sharpshooter Clear fired his carbine . . . killing the woman.[71] Former carpenter and Civil War veteran Sergeant John Ryan shot one warrior three times. He scalped the warrior then discarded the bloody souvenir when it soiled his overcoat.[72] In the confusion of the fight, Ryan was about to fire on a figure wrapped in a blanket only to discover it was one of his privates attempting to stay warm.

Some of these same villagers, who had survived the carnage and indiscriminate slaughter at Sand Creek, could not distinguish the difference between the Colorado Volunteers and the 7th Cavalry who were now doing the killing. As they watched their men being gunned down in the fight, many women believed that resistance was their only hope of survival. If women or youngsters resisted, they were dispatched as if they were warriors and as expected in a warrior society where surrender was an unacceptable alternative, many fought back.[73]

The Osage scouts took switches and severely whipped the escaping women.[74] According to scout Clark they did much more; allied scouts shot a majority of the women and children. He acknowledged, "I had under me 12 Osage Indian scouts, bloodthirsty wretches, who afterward disgraced the regiment."[75] He also noted that the Osage "shot down the women and mutilated their bodies, cutting off arms, legs, [and] breasts with knives."[76] Cooke recalled, "Our scouts took great delight in stripping the bodies of the dead, both male and female, of every article of clothing, no matter how dirty, and displayed fiendish pleasure in performing every manner of outrage upon the silent forms of their old enemies."[77]

Captain Thompson and his troops arrived late, creating a gap in the surrounding cavalry line with noncombatants fleeing through the opening toward the bluffs south of the camp. Clark saw some of Captain Edward Myer's men firing at them and mentioned it to Custer who ordered it stopped. The captives, escorted by Clark, were directed safely to a lodge in the village.[78] Godfrey recalled that, as they charged through the village firing through the tipis, the soldiers took no care to prevent from hitting women.[79]

Moving Behind described a tall bluecoat with blue eyes and a brown moustache that was sympathetic to the plight of the women. She remembered:

> The soldiers would pass back and forth near the spot where I lay. As I turned sideways and looked, one soldier saw us, and rode toward where we lay. He stopped his horse, and stared at us. He did not say a word, and we wondered what would happen. But he left, and no one showed up after that. I suppose he pitied us, and left us alone.[80]

Many years later the unnamed soldier would have the opportunity to meet the young girl again.

Several men, women, and children escaped to the north side of the Washita River, hiding under an overhang on the riverbank. Soldiers appealed for them to surrender, but gunfire continued from the concealed space. The sharpshooters poured lead into the villagers until all were killed.[81] According to one soldier, "One of the last Indians to be hunted out was a squaw who had a white boy, and when she saw that she was taken, she quickly drew a knife and disemboweled the boy in her capture. California Joe took aim and shot her through the head, killing her instantly."[82] It was later discovered that the light-skinned child was actually hers and, rather than surrender her son to the bluecoats, she had killed him. As Cooke led his elite sharpshooters in the attack, he fell through the snowy crust and severely bruised his leg on a tree snag. He later recalled that Custer credited his unit of 28 men:

> . . . with securing much of our success as we did most of the real fighting and fighting it certainly was. Some of the warriors gave us absolute fits and filled the crisp morning air with leaden hail. They made several desperate stands in the heavy timber along the river and at times the fighting was as furious as anything I ever witnessed during the late war. In the end we were forced to hunt them out of their holes like wild beasts, but after two hours of 'close-in fighting' their fire slackened and by 8 o'clock we were in possession of the field.[83]

It was shortly before 8:00 a.m. when Major Joel Elliott, accompanied by Regimental Sergeant Major Walter Kennedy, chased some villagers attempting to escape eastward. Elliott reportedly shouted, "Here goes for a brevet or a coffin," and his bravery combined with a lack of communication, led him to nail his own coffin.[84] After riding over one mile from the camp, the soldiers saw a group of women and children running toward the larger village, of which Elliott was unaware. Little Rock's 14-year-old son Hawk and another youngster named Blind Bear attempted to distract the bluecoats and lead them away from the noncombatants. Whooping and hollering to get the attention of the soldiers, the boys ran toward some high ground in an effort to escape, but were quickly overtaken and killed.

The rest of Elliott's party captured a few of the panicked noncombatants and turned them over to Kennedy, who alone headed them back toward Black Kettle's village. *Mahwissa's* 12-year-old son Little Beaver concealed a revolver and attempted unsuccessfully to kill Sergeant Major Kennedy. Little Rock's sister, White Buffalo Woman was among the captives, and with her barefoot children's feet bleeding, she attempted to stall their return by stopping frequently to wrap their battered appendages. Her ploy worked as four Arapaho warriors from the downstream camp attacked the soldier, and Kennedy got off two rounds before his gun jammed. He was tomahawked and killed by Bobtail Bear west of the creek that bears his name today, while the prisoners escaped to their friends in the other camps.[85]

Elliott continued to chase other young warriors, including Trailing the Enemy. His soldier detachment retreated back on the south side of Sergeant Major Creek until stopped by nine Arapaho warriors firing from the western bank.[86] Sometime around 9:00 a.m., the skirmishing commenced, with neither side doing much damage. Soon, other Cheyenne, Arapaho, and Kiowa from the downstream campsites arrived and encircled the small band of soldiers. Taking advantage of the tall grass, the soldiers lay in an oval hiding and firing, until the warriors rushed the soldiers. Perhaps in an effort

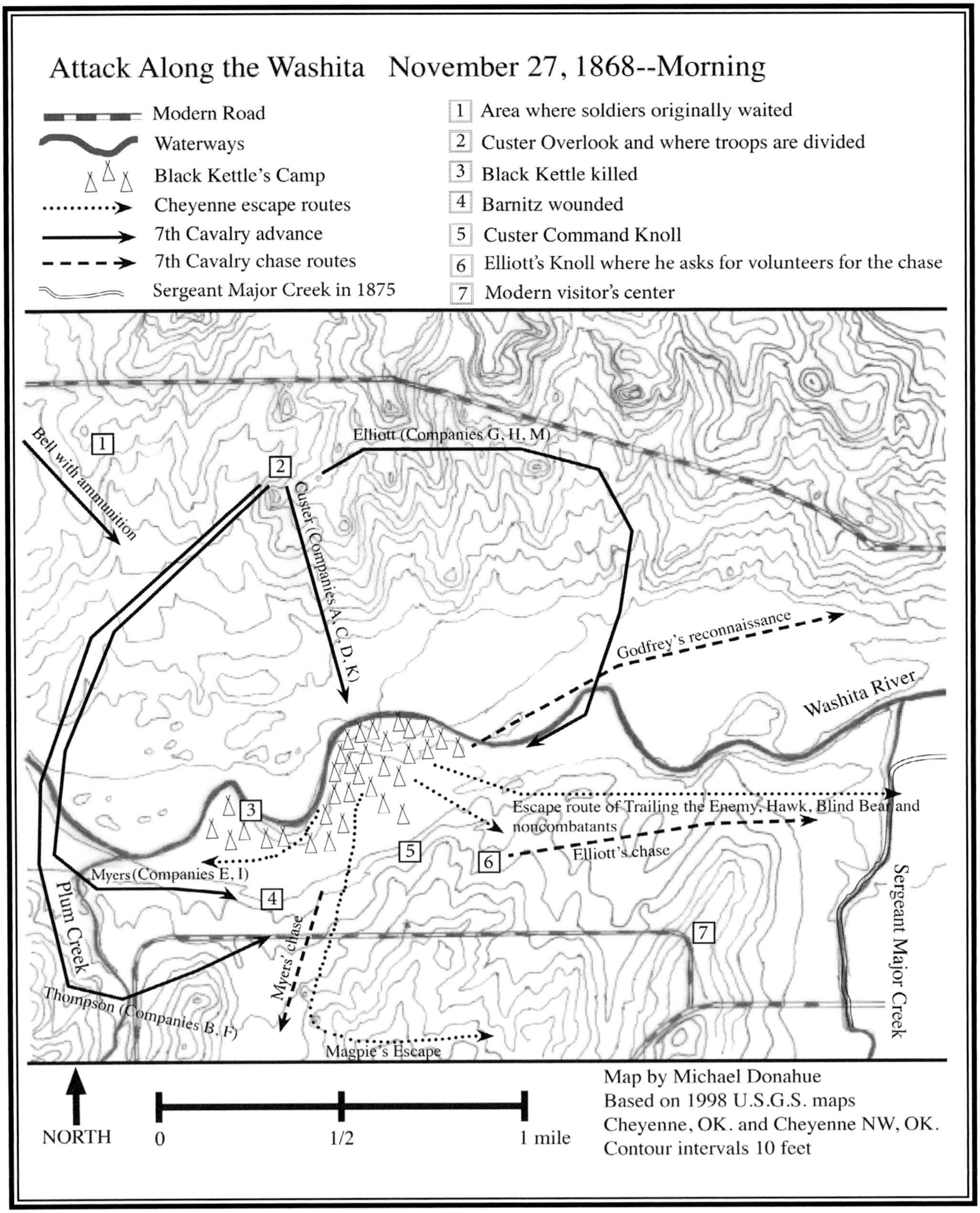

Fig. 12. Map: Attack Along the Washita. November 27, 1868—Morning.

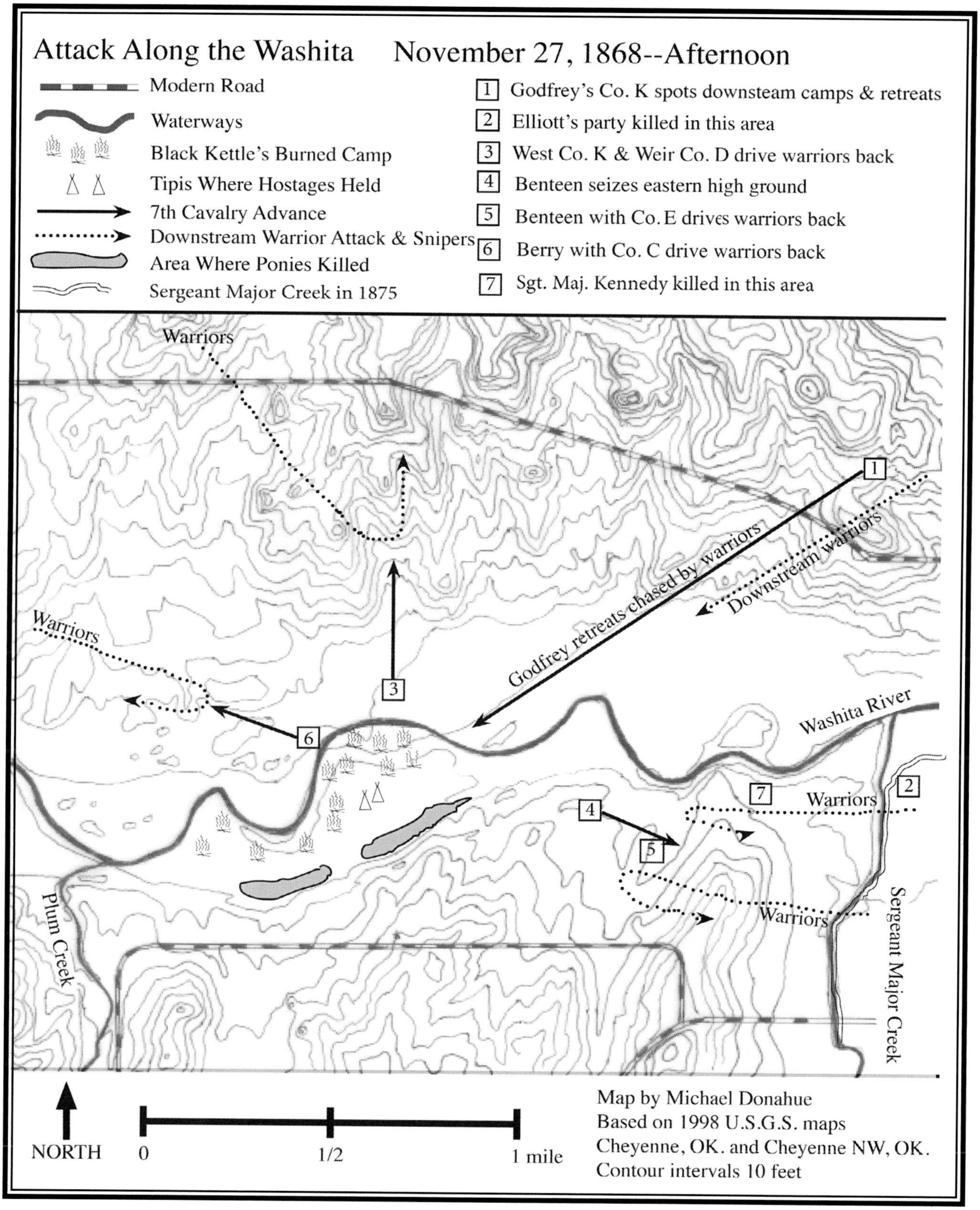

Fig. 13. Map: Attack Along the Washita. November 27, 1868—Afternoon.

to save ammunition, Elliott's men limited their firing. One warrior named Touching-the-Sky got off his horse and crawled up the ravine near the bluecoats. He and several other warriors fired into the band of soldiers, wounding several. Cheyenne Roman-Nose-Thunder broke the stalemate by galloping through Elliott's band. The incensed warriors, along with the party that had been pursued by the bluecoats, gathered to strip and mutilate the soldiers.

Two Arapaho warriors were killed in the Elliott fight, and several were wounded in the skirmish that lasted less than two hours.[87]

Godfrey and his company were north of the river, and they went a mile beyond the village in an attempt to capture ponies. A fleeing group of villagers raced eastward, and he pursued them another two miles, advancing to a nearby hill. They passed a lone tipi, a hauntingly similar image that would occur years later near the Little Bighorn River. Peering over, he was shocked by the image of a huge village with hundreds of Arapaho warriors racing straight for him. The lieutenant quickly retreated but was overtaken by the Arapaho before they reached their comrades. Dismounting with every fourth man delegated as a horse-holder, Godfrey divided his platoon and deployed it into two alternating skirmish lines. One line would withdraw as the other provided covering fire. Godfrey's pluck and the same tactical maneuver would help save fellow soldiers under similar attack at the Little Bighorn eight years later.[88]

As Godfrey withdrew, he could hear intense gunfire across the river, but he could not see the desperate fight of Elliott and his men. Around 10:00 a.m., Godfrey reported to a shocked Custer that huge numbers of warriors from other camps were heading in their direction; soldiers were repositioned to defend the seized village and their captives.[89] Custer interrogated a Cheyenne woman until she confirmed Godfrey's intelligence of a nearby massive village.[90]

Meanwhile, Lieutenant James Bell and the ammunition wagons stopped to secure the equipment left behind by the troopers, but he was attacked by the downstream warriors and was forced to abandon overcoats and haversacks left behind. Bell was able to cut his way through the amassing threat, and his timely arrival helped replenish the dwindling soldier ammunition supply.[91]

Custer soon realized that Elliott's detachment was missing and his lightly defended wagon train, under Edward Mathey, was heading straight toward possible disaster. He recalled, "On all sides of us the Indians could now be seen in considerable numbers, so that from being the surrounding party, as we had been in the morning, we now found ourselves surrounded and occupying the position of defenders of the village."[92] Scout Clark later remembered the surrounding warriors looked magnificent and gorgeous in their war bonnets and paint. Some wore the recently captured soldier overcoats and were armed with guns as well as bows and arrows. Their arm bracelets and shields glistened in the sun. Their horses were painted vermillion and were adorned with feathers.[93] The experienced Clark numbered the warriors from between 1,200 and 1,500. There may have been as many as 2,000 warriors.[94]

Custer ordered Robert West to destroy everything in the camp at noon, and Godfrey assisted in the task of demolishing the village with its contents. Heavy buffalo hide lodges were piled high and burned while soldiers hesitantly destroyed the beautiful beaded, as well as quilled, handiwork of the Cheyenne.[95] Evidence of Indian raids on white settlements and stage lines was recovered among the spoils of battle, including a pony expressman's scalp, mail taken from murdered carriers, and personal photos from the Solomon and Saline settlements.[96]

Custer packed up four captured mules with several of the finer white-skinned lodges as well as warrior saddles, a shield, bridles, and other Cheyenne handiwork as souvenirs.[97] Cooke described Black Kettle's lodge as "one of the finest examples of the red man's habitation I have seen and a masterpiece [of] Spartan construction." Custer later presented it as a gift to Sheridan as other officers retained slain Chief Little Rock's tipi.[98]

The women who had been captured expected to be put to death by the soldiers and were in a state of panicked fear. They began to scream when separated into groups to be placed into different captured lodges. *Mahwissa* seems to have become a spokeswoman for the group, and according to Custer, attempted to offer a young Cheyenne woman as bride to the commander in a ceremony he did not understand. Custer declined the relationship and assured them they were safe from harm as long as they behaved.[99]

The Cheyenne captives built a fire, sending smoke into the sky as an offering for deliverance by the Everywhere Spirit as the military created a smoke signal to the nearby warriors of destruction and forewarning. Incensed warriors from the nearby hills opened fire on the soldiers.[100] Mrs. Courtenay, an Irish laundress and cook for Custer, had arrived in an ambulance with Lieutenant Bell. During a lull in fighting, she commenced to make coffee for the officers on a stove that she had brought in the ambulance. She continued steadfastly with the task while bullets whizzed past her, when an Indian sniper killed a mule next to her. She sneered at the distant warriors, and in her strong Irish brogue, hollered, "Ye dirty devils ye."[101]

Realizing the dangerous situation he was in, Custer adopted a startling strategy to stop the warriors sniping at his troops. According to John Ryan, "General Custer took the squaw prisoners and put them on the side of the camp the firing came from. When the bullets would come close to these prisoners, they would set up an awful yell and it caused the Indians to cease shooting from that direction."[102]

As the frustrated warriors closed in from the northern hills around noon, Custer knew that the best defense was an offensive thrust. He ordered Weir along with West to drive the warriors from the area with a mere 100 soldiers, and the intimidating tribesmen withdrew as the bluecoats charged.[103]

Arapaho, Kiowa, Cheyenne, Plains Apaches, and Comanche warriors congregated in immense numbers east of the burning village as a defense between the soldiers and their escaping families.

Benteen was ordered to seize the hills while Myers was to drive the warriors away. Myers, suffering from snow blindness, could not fulfill the order, so Benteen took charge of Myers' company and drove the warriors from the ridges in the proximity of Elliott's dead. According to an accusatory claim made by Benteen:

> Custer wanted Myers and myself killed . . . Mine [Company H] were not in it all; I was ordered to take command of Myers' squadron, Myers being with it; thus disgracing Myers; the plan, we thought to get each of us killed. The balance of the Cav. [cavalry] looked on

> at our evolutions at attacking the enemy. Had Custer been the brave reckless cavalryman he had the reputation of being, there was a fine field for him; but I never saw any exhibition of bravery in him.[104]

Indian gunfire eventually threatened the soldiers from hills to the west of the smoldering camp. Matthew Berry charged this area, driving the warriors from the snowy ridges.[105] Even though outnumbered, Custer quickly learned that an offensive thrust, even with only one or two companies, was successful in driving away the warriors and better still, he suffered no casualties from these charges. Twenty-year-old Comanche Quanah Parker from the downstream camp was watching the fight from the surrounding hills. He was the son of a Comanche chief and the white captive Cynthia Ann Parker, who was taken at the age of nine. Quanah recalled:

> When we heard of the fight all of our men hurried to the scene but General Custer retreated when he saw so many of us coming. We did not get close enough to fight him. After several skirmishes without results, we returned to our camp and moved onto the plains.[106]

Two white women from Kansas, Sarah White and Anna Morgan, were being held in the larger downstream village by their Cheyenne captors. They listened to gunfire with the hope they would be rescued but were quickly ushered down the river through the snow away from possible rescue. Miss White remembered that her feet suffered frostbite during the ordeal.[107]

Custer, as part of the total war strategy, made an arduous choice for a man who treasured horses—he decided to slay over 650 of the 875 captured ponies and mules. The captives were allowed to choose mounts; then Godfrey was ordered to lead the remaining animals to a steep bank south of the camp and slaughter them.

Frustrated by the disgustingly slow pace of slitting the animal's throats, George Yates' men along with three other companies, shot what ponies remained. The detestable task took at least an hour and a half, and the enraged warriors watched from the hillsides, loudly voicing their disapproval. Moving Behind remembered, "The wounded ponies passed near our hiding place, and would moan loudly, just like human beings."[108] Another Cheyenne captive remembered, "The ponies after being shot, broke away, and ran about, bleeding until they dropped. In this way the snow on the whole bend of the river was made red with blood."[109] Godfrey recalled, "As the last of the ponies were being shot, nearly all the hostiles left. This was probably because they could see our prisoners and realized that any shooting they did might endanger them."[110]

The stiffening body of Captain Hamilton was vigilantly wrapped in army blankets and placed in a wagon. It was about an hour until darkness would once again descend, and Custer noted, "We had achieved a great and important success over the hostile tribes; the problem now was how to retain our advantage and steer safely through the difficulties which seemed to surround our position."[111] He proposed, "To guide my command safely out of the difficulties which seemed just then to beset them,

I again had recourse to that maxim of war which teaches a commander to do that which his enemy neither expects nor desires him to do."[112]

Hoping to draw the warriors away from his vulnerable supply train, with the spectacle of colors fleeing and band playing, the dog-tired troops with their despondent captives began to march in the direction of the other Indian camps, feigning another attack.[113] The troops crossed the river and marched on the north side of the river toward the hill where Godfrey had seen the other village.[114]

Clark suggested they march until darkness overtook them, then reverse course to locate the wagons. This would avoid further casualties, since the warriors would be unable to monitor the soldier recall under the cover of night. Since Custer did not know the exact location of the other camps, he nervously kept asking Godfrey about its location and the high point where the lieutenant had observed the massive camp. According to Godfrey, before they got within sight of the moving village, the troops turned and "got out of the country."[115]

The accounts of survivors Magpie and Moving Behind are clear that the warriors did not attack due to concern over the Cheyenne hostages held by the soldiers. Magpie recalled:

> The Indians were very angry and they would have attacked the soldiers had they not been afraid that such an attack would bring death to the women and children the soldiers had captured and were holding in the village. Shortly after dark, the soldiers suddenly fell into line and with the women and children in the center marched out in the direction from which they had come—to the northwest. Some of the Indians started to follow them but were called back for fear harm might come to the prisoners.[116]

Moving Behind remembered, "Someone suggested, 'They [the soldiers] are right across the river, and are going slowly; let's shoot them.' Others said not to shoot, for fear of hitting the Indian women who were being taken away by the soldiers."[117]

Custer's ruse worked, and he inadvertently learned an indispensable lesson at Washita:

> Whether the fact that they could not fire upon our advance without endangering the lives of their own people, who were our prisoners in our hands, or some other reason prevailed with them, they never offered to fire a shot or retard our movements in any manner, but instead assembled their outlying detachments as rapidly as possible, and began a precipitate movement down the valley in advance of us, fully impressed with the idea no doubt that our purpose was to overtake their fleeing people and herds and administer the same treatment to them that the occupants of the upper village had received. This is exactly the effect I desired.[118]

The soldiers did an about-face in the darkness, riding past the burning embers and mass of dead horses around 10:00 p.m. The exhausted troopers continued on until two o'clock in the morning when a halt was ordered and the officers allowed the men to rest. Uncertainty regarding the location of warriors as well as the wagons made the respite an uneasy one.[119]

After the soldiers departed, Moving Behind returned and found a family member as well as her boyfriend Crane still alive. Black Kettle and his wife were discovered lying near the river, but were too heavy to pick up, so their bodies were dragged some short distance and placed on a red and blue wool blanket that was folded over their faces. They were abandoned in the darkness. Moving Behind sadly recalled, "As we rode along westward, we would come across the bodies of men, women, and children, strewn about. We would stop and look at the bodies, and mention their names."[120]

The soldiers found the supply train at daylight, completely oblivious to the possible danger. A nervous Custer continued the march until early afternoon, when the command halted and the troops were allowed to rest. The overworked and newly proven Godfrey was given charge of the Cheyenne captives.[121] Custer boasted to Sheridan, "We have cleaned Black Kettle with his band so thoroughly that they can neither fight, dress, sleep, eat, or ride without sponging upon their friends. . . . We have lots of scalps, a sample of which I send you."[122]

As the troops neared Camp Supply on December 1, Custer sent a missive to Sheridan requesting a review in front of his diminutive commander, which was respectfully granted. Godfrey recalled that the pageantry rivaled the prototype of the modern Wild West shows. The painted Osage scouts fired their guns and were followed by the drab covered white scouts. Next, the brightly clothed Indian captives followed on ponies, trailed by Cooke's sharpshooters and the troops. The sounds of *Garryowen* greeted Sheridan as the band rode by. The 7th Cavalry was proud—it had won its first victory against the Plains warriors.[123]

Fig. 14. A portrait of some of the 53 captives that Custer used as a hostage shield during the Washita battle. Contrary to popular myth, the 7th Cavalry did not kill all the women and children, as can be seen in this photograph.

Camp (Fort) Supply, Indian Territory. Prisoners captured in the Battle of the Washita-Scout with them, circa 1869. National Park Service, Little Bighorn Battlefield National Monument, LIBI_00019_00392, W. S. Soule

Casualty figures for the 7th Cavalry included Elliott's party of 17 men that were assumed dead. Four other soldiers, including Hamilton, were killed or died from wounds in the village attack. Three officers and 11 enlisted men were wounded.[124] While the number of Indian casualties will never be certain, Custer attempted to gather reports from his officers on the second night after the fight, and they calculated 103 Indian deaths. There still is great controversy concerning the numbers of Cheyenne casualties. Trooper Henry Langley stated, "We did nothing but return the fire, and left a great many of them on the ground, between 80 and 90, I believe."[125] Cooke was more specific and insinuated that the bodies were laid out and identified by the captive women. He wrote:

> When our deadly work was finally completed, by actual count, we laid out 103 freshly minted 'good Indians' on the field of battle and I am sure that a great many more are en route to the 'happy hunting grounds.' My sharp-shooters alone accounted for 56 of the enemy, 19 in one small place, Sergeant Fay and Private Rodgers are credited with eight each. Aside from their two big chiefs, Black Kettle and Little Rock, we killed and counted for 16 of the most prominent chiefs and head men of the tribe.[126]

Fig. 15. Another photo of the Washita captives. Custer discovered their value in stopping the sniping during the Washita fight and in negotiating with their fellow tribesmen afterward.

Camp (Fort) Supply, Indian Territory. Prisoners captured in the Battle of the Washita-Scout with them, circa 1869. National Park Service, Little Bighorn Battlefield National Monument, LIBI_00019_00391, W. S. Soule

Sharpshooter Dennis Lynch, whose tally was slightly different, reinforced Cooke stating, "[We] picked up 106 dead warriors. Killed some squaws."[127]

Godfrey, who may have been unaware of Cooke's tally, disputed the adjutant's account by stating the bodies on the field were not counted.[128] Ben Clark estimated the Cheyenne losses as 75 warriors and as fully as many women and children.[129] The Cheyenne recollection of casualty numbers is equally ambiguous and contradictory with about 40 men, 12 women, and six children killed.[130]

The seemingly always critical Benteen claimed the number was inflated and stated, "Custer assembled the officers to inquire of each how many dead Indians each had seen; then what *each* had seen were added. *They all had seen the same dead Indians.* [Benteen's emphasis]"[131]

According to Custer and Sheridan's report, two white children were recaptured.[132] Perhaps it was thought that two of the captured Cheyenne boys may have been non-Indian or part white, but there is little recorded attention given to them after the troops returned to Camp Supply.[133]

Controversy, which seemed to often shadow Custer, became an integral part of the Washita story. Eastern humanists, and others isolated from the ongoing western conflict, felt the peace-loving Black Kettle and his people had been ruthlessly butchered—a repeat of the Sand Creek Massacre. Many local settlers, as well as the army, who were directly affected by Indian raids, felt the harsh action was justified. The expansionist attitude of many Euro-Americans and the military commanders concluded the action was necessary, successful and productive.[134]

The debate over whether the action was a massacre or a battle lingers to this day.

Researcher Walter Camp summed up the action accurately by stating, "Although many settlers of the west did and do yet consider this a massacre of Indians, it reflects no discredit upon Custer. He was but obeying his orders as a commander of troops. The authorities who declared war upon the Indians necessarily took the responsibility for the results of the war."[135]

As late as 1870, the House of Representatives put forth a motion for the investigation into the alleged abandonment of Elliott and his men.[136] Benteen believed that Custer had abandoned Elliott to his death and attempted to get Ben Clark as well as others to condemn their commander.[137] An angry Benteen also felt slighted that special mention was not made of his actions there. He later stated that his company was the first to break up the village and capture horses as well as women, but was upset that Custer gave the credit to California Joe. The captain also believed that Custer had sent both he and Myers downstream to be killed by the gathering warriors.[138]

An unhappy Sheridan was concerned that Custer had left soldiers on the battlefield in an unknown condition. Nonetheless, the decisive action pleased him and cemented his confidence in his aggressive subordinate. While Custer had chased, but had been outsmarted by warriors in the Hancock campaign, the fight along the Washita was his first 'real' armed conflict with the aboriginal people of the Plains. His following fights with them would be few and far between.

The Washita offered some important personal experiences and lessons that Custer would retain as well as use again in his upcoming dealings with various tribes. When aggressively pressed by only two companies, larger numbers of warriors retreated. Ordering his men into "light marching order" had eliminated excessive supplies and given his troops more mobility. The use of native scouts to find their Indian enemies had proven successful. Dismounting the tribesmen by immediately capturing their pony herd made escape difficult and placed the immobile warriors in a defensive position from the outset.

Perhaps most importantly, Custer had by accident realized the value of hostages and captives in such a situation. He had used them as a body shield in the battle and later utilized them as a negotiating tool with their fellow tribesmen. This tactic also cut down on his soldier casualties and helped save his highly outnumbered command. Dividing his troops and surprising the village from several sides, caused mayhem in the camp and favored the momentum of the attacking troops. The strategy of using the cover of darkness to get within close proximity of his adversary allowed for total surprise. Of the three commands searching for the village, Custer's command was the first to discover it. This was due to his aggressive persistence combined with luck—the same traits that would allow him to find Indians in 1876.

He would later ignore some lessons he should have learned at the Washita and commit some of the same errors in his short Indian fighting career. Seven and a half years later, he would mistrust native scout intelligence and again face their hesitancy to fight like his soldiers. He would again attack an enemy of unknown population on unfamiliar terrain in June of 1876. The exposure and outdistancing of his supply train during an attack as well as lack of knowledge about disposition of his troops would come back to haunt him.

Poorly coordinated elements of his command arrived late during the fight, causing problems. Men and animals were pushed to the breaking point in order to follow a hot trail. Such decisions would prove fatal along the Little Bighorn River. His strategy to once again take hostages along that Montana waterway would prove more difficult.

The allied Osage scouts were responsible for most of the noncombatant casualties as well as the horrific mutilation of the corpses. Ironically, eight years later at Little Bighorn, it would be the allied Arikara scouts who were the first to kill some of the women and children. From his distant vantage point, it seems likely that Custer did not condone or witness these killings and mutilations at the Washita, as recorded in Clark's observations. It was an embarrassing activity that Custer certainly did not wish to acknowledge in his official report.

Even with the pre-battle schisms between Custer and some of his subordinates, they all came together performing dutifully. Such would not be the case a little over seven years later. Without his body shield of captives, the Washita fight may have intensified or concluded as Custer's first and last stand. In only a few months, Custer had gone from an embarrassing, court-martialed exile to a publicly acknowledged Indian fighter of the plains.[139]

In ancient Egyptian mythology, the phoenix was a bird that was consumed by fire yet rose again from its own ashes. Yet another definition is a person or thing of unsurpassed excellence. Out of the smoke and ashes of Black Kettle's village, the engagement at the Washita resurrected Custer from the depths of shame that had been cast upon him by his suspension. Not unlike the morning apparition of Venus over the Washita valley on a frozen November morning, George Armstrong Custer's star was beginning to rise once again.

Chapter 3

Hostages, Threats, and Different Rules of Conduct

The Washita vaulted Custer back into the national spotlight and reassured Phil Sheridan that he was the right man for the job of subjugating the other tribes to reservation life. A pleased Sheridan viewed the Washita as only the commencement of the winter campaign, and he quickly had the 7th Cavalry back in the field looking for the other tribes that had camped along the Washita. He proclaimed, "If we can get in one or two more good blows there will be no more Indian troubles in my department."[1] Sheridan wanted a punitive action and Custer would be willing to do so if need be—yet in the upcoming campaign he would use diplomacy rather than his army fists.

The legion of warriors encountered along the banks of the Washita River compelled a wake-up call to the army concerning the prodigious numbers they could encounter. With 11 companies of the 7th Cavalry, 10 companies of the 19th Kansas Volunteer Cavalry, a detachment of black troops from the 10th Cavalry, as well as Osage and Kaw Indian scouts, Custer doubled the strength of his command to 1,500 and began the search for Indians once again in early December 1868. Custer wrote:

> Nothing short of race horses will catch Indians now—they are scared out of their boots.[2]

Accompanying the army were three captive Cheyenne women—*Mahwissa*, Sioux Woman, and *Meotzi*. A mistranslation of Sioux Woman's name is the reason why Custer believed she was a Lakota, but he acknowledged that she was "extremely communicative in regard to the winter resorts of the various tribes and other matters connected with the purposes of the expedition."[3] The stated purpose for them joining the expedition would be to render their services in communicating with the villages, if they were discovered.[4] The women seemed happy to accompany the expedition since they felt it might be the fastest way to rejoin their tribes, with the trio of women choosing an army ambulance for the journey.

George Custer described the extraordinary beauty of the young *Meotzi* or Monahsetah in his book, *My Life on the Plains*. Custer mistakenly believed her name meant 'the young grass that shoots in the spring,' and as the daughter of deceased Chief Little Rock, she was perceived as 'the cream of the aristocracy if not royalty itself.'[5] Tom Custer humorously nicknamed her 'Sallie Ann.'[6] Some have suggested that he brought the attractive Monahsetah along for sexual purposes; however, avid Custer detractors created most of the derogatory innuendo. Cheyenne oral history claims that after the Washita, the officers, as well as Custer, forced themselves on the female captives.[7] There may have been something to the accusations of infidelity since Ben Clark, Benteen, and the Cheyenne noted it.[8] Custer likely had a purpose for them, which has escaped researchers for over a century. Given the ticklish scenarios he might encounter he thought that using them as shields, as he did at the Washita, might prove valuable again.

Sheridan wished to tour the Washita battleground and determine the fate of Joel Elliott. With the bitter wind thrashing their faces, the expedition arrived at the scene of the bloody conflict as thousands of ravens and crows formed a black swirling cloud that ascended into the clear blue sky above. Feasting wolves scampered to nearby hills to watch the intruders.[9] When the troops arrived on the battlefield on Friday, December 11, they found the blanketed Indian casualties had been removed from their death sites. As many as 30 bodies were found in one location and many were buried in trees or covered with brush.[10]

Two hundred yards from the burned village rubble, the troops found the stiff, frozen bodies of the slaughtered ponies. Some of the ponies that were too badly wounded to get up had eaten all the grass within reach before dying.[11] Wolves and blackbirds had gorged for days on this wintery carrion feast. The troops followed tracks that were believed to belong to Elliott and discovered the nude body of Sergeant Major Kennedy. He was found covered with holes from withdrawn arrows as well as bullet wounds, and the top of the skull was broken into a number of pieces. His brain was lying partly in the skull and partly on the ground.

About two miles east of Black Kettle's former village and beyond Kennedy, soldiers discovered 16 naked corpses 'frozen solidly as stone.'[12] Revenge had been taken on the dead soldiers and they lay facedown, clustered on the icy turf. Almost all of Elliott's men had their throats cut and body parts were found strewn about in the dead grass. A portion of a Lancaster rifle used by a warrior was broken off and driven into the side of a soldier's torso. The arrows shot into the bodies had been removed by the warriors, leaving gaping red holes in the corpses. The angry warriors believed that in their next lives the soldiers would reappear in the physical condition in which they were left.

Along the nearby riverbank the soldiers spotted the bodies of several dead warriors from the fight. A grave was hastily prepared on a little knoll near the soldier camp and, with the exception of that of Major Elliott, whose remains were carried away for interment at Fort Arbuckle, "the bodies of the entire party, under the dim light of a few torches held by their sorrowing comrades, were consigned to one common resting place."[13]

Meanwhile the troops came upon the other deserted Indian village campsites downstream. A newspaperman recalled:

> Camp kettles, coffee pots, cups, Peace Commission and Interior Department powder kegs (empty), several philanthropic rifles, hundreds of untanned robes, hundreds of lodge poles, and a variety of other things were left behind. Orders were given to destroy everything found. It was not long before a stream of consuming smoke and fire could be described following the line of the Washita for a distance of six or seven miles all the way to our present camp. Not more than 1,000 lodges occupied this stretch of country at the time of the fight.[14]

As the military searched the refuse of the abandoned Arapaho campsite, they found the pitiful frozen remains of Clara Blinn and her son Willie. Mysteriously, a dead Indian woman was found nearby wrapped in a buffalo robe and had no marks of violence on her body except a small contusion on the back of her head. Soldiers sentimentally placed Clara on her side lying on a blanket with Willie resting on her arm. The men were ordered to march by to see if they could be recognized.[15] While the demise of Elliott's party disturbed Custer, these two deaths truly haunted him and the other soldiers. While death could often be an expected outcome of a soldier in battle, these newly discovered corpses were a disgusting revelation. The bodies of Elliott, Mrs. Blinn, and her child were respectfully taken away and buried at Fort Arbuckle.[16]

Fig. 16. Clara Blinn's death would haunt and inspire the soldiers as they pursued the escaping tribes. Custer would be more cautious in attacking the villages when he encountered them, due to the possible presence of captured white women.

Courtesy of Judith Justus

Custer could not have known that an attempt to negotiate Mrs. Blinn's release was eminent or that she was in the village. With his romantic Victorian sensibilities and love for the fairer sex, he was reminded that it could have been Libbie in Mrs. Blinn's place. Shortly before the Washita campaign, a starving white woman driven insane by Indian treatment had stumbled into his camp. He had the former captive escorted to safety.[17] In his writings, Custer seemed to downplay any personal connection to the deaths of the Blinns; however, it was something he did not forget as it altered his actions throughout the upcoming campaign. Later, when negotiating the release of two other women captives held by the Cheyenne, such was obviously on his mind.[18]

Even though he was simply following orders, Custer recognized that his actions had unwittingly caused the death of the innocent woman and her child. Such consequences must have weighed heavily on his mind, and his feelings must have matched those of the men around him, a mixture of anger and sadness.[19] One soldier noted, "They were wantonly murdered to prevent their rescue by Custer. The names of Mrs. Blinn and Major Elliott will always be synonyms for the tragic death, in the region once known as 'the plains.' "[20] The battle intensified friction within the already divided 7th Cavalry. Godfrey wrote:

> Benteen was a bitter enemy of Custer and had criticized the Washita and tried to belittle that action. He was, I feel sure, responsible for much of the adverse sentiment against Custer. While under Custer's immediate command, B [Benteen] very carefully watched his 'Ps and Qs'—but later indulged to excess. . . . He loved his J. B. [John Barleycorn was a mixture of beer and whiskey].

Fig. 17. Captain Frederick Benteen lost four of his five children while serving his country. His hatred of George Custer began with their first meeting and lasted until his own death. His reluctance to take initiative may have doomed Custer, but his bravery and control on Reno Hill helped save the remainder of the 7th Cavalry.

Courtesy of Christopher Kortlander & the Custer Battlefield Museum, Garryowen, Montana

Benteen's fondness for the drink loosened his tongue and amplified his negativity toward Custer and fellow soldiers.[21] The bitter captain later proclaimed:

> From being a participant in the Battle of the Washita, Nov. 1868, and from the manner the 7th Cav. were handled there, by Custer, I formed the opinion, that at some day, a big portion of his command would be 'scooped' if such faulty measures were persisted in. *The time came.* [Benteen's emphasis.][22]

Born in 1834 into a slave-owning family in Virginia, Benteen's early life helped to formulate his strong-willed personality. Later, living in St. Louis, Missouri, young Frederick showed a talent for art and worked as a sign and house painter with his father. In 1841, the first of many tragic events that would shape him occurred. His mother died, leaving his father to raise three young children. When the Civil War broke out, Benteen joined the Union's 10th Missouri Cavalry. A Unionist lady named Catherine Norman that he married in 1862 may have influenced Benteen's defection.[23]

Feeling betrayed by his son, his father supposedly disowned Fred, allegedly proclaiming, "I hope the first bullet gets you!"[24] An early show of acrimony occurred in 1862 when Frederick Benteen's company was part of a force that captured the Confederate supply steamer *Fair Play.* The chief engineer aboard the ship was none other than his father, Theodore. All civilians aboard were released; however, likely due to the influence of Frederick, his father was placed in a Union prison for the remainder of the war. Forgiving and forgetting was not one of his personality traits.[25]

Benteen's dedicated and loving relationship with his wife was in direct contrast to the sarcastic behavior he exhibited with others, and his private letters relate an intense and passionate love affair.[26] The Benteens conceived five children; four died in infancy with only a son surviving into adulthood. Benteen felt guilty about these deaths and attributed them to the hardships his family had to endure due the unforgiving conditions of military life.[27] Often speaking in a disparaging manner about almost every officer and soldier with whom he served, he would later claim few friends from his military service.

An episode in 1862 foreshadowed Benteen's later exhibitions of insubordination that were repeated on numerous occasions throughout his army career. When one of his corporals was ordered to serve as an orderly for the headquarters staff of Captain Stanford Ing during the Civil War, the soft-spoken Benteen felt slighted. He sent a threatening and hot-tempered response, and offered his resignation to Ing.[28] Benteen was charged with disrespectful behavior towards the commanding officer and conduct prejudicial to good order and military discipline. Luckily for Benteen, Ing withdrew the letter of condemnation within weeks and resigned for unrelated reasons sparing further contention.[29]

Benteen went on to serve with distinction at Pea Ridge, Wilson's Creek, West Glaze, Bolivar, Pleasant Hill, Vicksburg, and many other engagements. Even with his sometimes biting disposition, Benteen had a certain easy-going Southern charm about him that some of his fellow soldiers appreciated. Things took a tragic turn for the Benteens when in July of 1863, Frederick and Catherine's first daughter died unexpectedly.

In 1865 General James Wilson recommended Benteen for the brevet rank of brigadier general, but his superiors did not accept the commendation.[30] He always exhibited pluck combined with bravery under fire, and this lack of recognition was a bitter pill to swallow. By the end of the Civil War, Benteen suffered through the death of his first child and a deserved promotion had been declined.[31]

Mrs. Benteen (known as "Kate") had a second child, Frederick, born in March of 1867. While Captain Benteen was engaged in the Washita campaign, his wife delivered their third child, Kate, but the girl died sometime in mid-December at Fort Harker. Benteen returned to Fort Cobb on December 18, 1868, and reported being sick with chronic rheumatism until Christmas Day.[32] He was also suffering from snow-blindness after loaning his goggles to a doctor at the Washita.[33]

How the death of a second child affected Benteen is not known, but the combination of her unexpected death, absence from his wife, his poor health, lack of recognition at the battle of the Washita, and death of friend Elliott, all within weeks, surely must have caused Benteen intense depression and anger. It was almost a repeat of the events that had transpired at the close of the Civil War. Benteen would focus his anger and bitter sarcasm on his commander.[34] While at Fort Cobb on December 22, 1868, a sick and frustrated Benteen penned a letter condemning Custer to a friend. Benteen blamed his commander for leaving Elliott to his death and the letter was published anonymously in a St. Louis newspaper. It was then reprinted in the *New York Times*.[35]

According to Benteen in late February, "Tom Custer and Lieut. Cooke came in, Tom with a newspaper in his hand, showing it to me, saying, 'Isn't this awful?' I looked at it, reading down a line or so and said, 'Why, Tom, I wrote that myself,' and so I had."[36] George Custer had officer's call sounded and threatened to horsewhip the man responsible. It is likely that he did not expect anyone to step forward and hoped his broad threat would deter such future activities. Benteen, with revolver loaded and ready for action, stepped forward and admitted to the letter. Custer backed down, stammering, "Colonel Benteen, I'll see you again, sir!"[37]

According to Benteen's version of events, he went straight to correspondent De B. Keim who already knew about the letter, and "told him I wanted him to go with me at once to Custer's tent, taking his notes with him of all I had told him, as a whipping was due somebody, and I didn't want a word I'd said omitted. Keim went with me, and though I'd told him enough, Custer wilted like a whipped cur."[38]

The red-faced commander never mentioned the episode again and was likely surprised by Benteen's boldness and unflinching audacity. The 7th commander had hoped that the conference would halt such subversive criticism, but instead it seemed to empower and embolden his subordinate. Not only did Benteen dislike his commander, he despised his family members as well, and he labeled them 'his clan' or the 'Custer gang.'[39]

Despite Benteen, a nepotistic Custer surrounded himself with supportive family members and friends within the regiment. James "Jimmy" Calhoun served in the Civil War and had difficulties getting a regular army commission until his attachment to Custer's sister Margaret, known as "Maggie." The couple married in 1872, and George helped him with an appointment as first lieutenant in the regiment. The handsome 6-foot-tall blonde officer was nicknamed "Adonis" and had a real problem with gambling, often losing large sums of money.[40]

Another dependable relative through his marriage was Myles Moylan from Massachusetts, who married James Calhoun's sister. Moylan had a distinguished career during the Civil War, but he was dismissed from service for traveling without a pass and failing to appear for duty. Custer's recommendation helped gain his commission as first lieutenant during the regiment's formation, and he served as the regimental adjutant during the Washita campaign.[41]

New Yorker George Yates struck up a friendship with Custer during the Civil War when a shell exploded under his horse wounding Yates, and the two met during his recovery in Michigan. Custer helped him get a staff appointment under mentor General Alfred Pleasonton and later as a captain in the newly established 7th Cavalry. Custer liked the polished captain who believed appearance was important, and the captain's wife Annie and Libbie Custer became close friends.[42]

Another close member of the 'Custer clan' was New Yorker Algernon Smith, who had been severely wounded in the arm during the Civil War while attacking Fort Fisher, North Carolina. The officer had to have help putting on his coat or mounting a horse due to his injury, making him vulnerable in combat. He became a first lieutenant in the 7th, a month after the Washita. He was popular in the regiment and acted as quartermaster until relieved of that duty due to questionable practices. His wife Nettie Bowen was a close friend of Libbie, and the couple shared a living quarters in Kentucky for a time. Algernon became known as "Fresh" Smith to distinguish him from another 7th Cavalry lieutenant—merchant seaman H. W. "Salty" Smith.[43]

Many of the opposing officers that clashed with Custer slowly disappeared from the 7th Cavalry—Benteen was not one of them. He stated that he had too much pride to allow Custer and his friends to drive him from the regiment. He knew that his commander wanted him as a friend, but he would not allow it. At the end of his life, Benteen posed an unflattering synopsis, "I've been a loser in a way all my life by rubbing against the angles—or hair—of folks, instead of going with their whims; but I couldn't go otherwise—'twould be against the grain of myself.' "[44]

Custer did not normally hold grudges for long. If there was a sense of vindictiveness in his personality, it exhibited itself in his preference to keep adversarial personalities like Benteen at a distance while keeping friends and family nearby. So while there was an undercurrent of discontent by some within the regiment, the hunt for the tribes resumed. It was the Kiowa that the bluecoats first encountered on this secondary expedition in current day Oklahoma. The Kiowa, like Black Kettle's band, were a mixed lot; some wished for peace while others raided and seemed anxious to fight. The soldiers, horses, and wagons slogged through the mud and snow in search of faint Indian trails by mid-December 1868.[45] Eventually following a fresh trail, Custer arrived within striking distance of the Kiowa camp. Alongside 20 warriors, Kiowa chiefs Lone Wolf and Satanta met Custer under a flag of truce.[46]

As both sides eyed one another and prepared for possible deception, Sheridan's Adjutant Lieutenant Colonel J. Schuyler Crosby, *New York Herald* correspondent De B. Keim, and Custer advanced toward the Kiowa chiefs. The two officers refused to shake hands when Satanta offered his with Custer proclaiming, "I never shake hands with any one unless I know him to be a friend."[47] The Kiowa had recently made a hostage of an army courier generating a tense atmosphere, and the chiefs agreed to release him.

One Kansas volunteer recalled, "Satanta is quite large and very strongly built, much more noble looking than the others who are darker and just ordinary looking Indians, only finer dressed."[48] The chief was in his early 50s at the time of this meeting. Satanta, an English bastardization of his Kiowa name, *Set-t'ainte,* meant 'White Bear,' and the name belied the physical prowess of the warrior. Photographs depict a stocky man with a tapered forehead and wider than normal visage. He was a brutal warrior but was also known for his dramatic oratorical skills and attempts at diplomacy. Satanta was remembered and hated by Texans for his bloody raids into that region that eventually led to his imprisonment. The Kiowa chief was a would-be musician known to play an army brass bugle in order to advance his war parties.[49]

Sitting on his horse, the smaller Lone Wolf stared, with his dark eyes framed by dark braids on each side of his oval face. His thin lips made it appear that he wore a constant smile. Lone Wolf, with his dark

Fig. 18. Kiowa chief and warrior Satanta was taken hostage by Custer in 1868, and the threat of his hanging helped force his people to surrender to reservation life. He died tragically by jumping from a hospital landing after a second prison incarceration.
Courtesy of the Research Division of the Oklahoma Historical Society

Fig. 19. Kiowa chief Lone Wolf, who was also taken hostage and nearly hanged beside Satanta, surrendered and was exiled to Florida. Eventually allowed to return to his people, he soon died from malaria.
Courtesy of the Research Division of the Oklahoma Historical Society

complexion and wide-bridged nose, was one of the Kiowa leaders who had vied for the support of the warring factions within his tribe. However, military records indicate that Satanta and Lone Wolf had not been encamped in the villages scattered along the Washita, weeks earlier.[50]

After the parley with Custer, the Kiowa chiefs, along with about 20 warriors, agreed to accompany the army to Fort Cobb and to get their village to join them as they proceeded. As they trekked toward the fort, warriors from the original 20 began to disappear, one by one, under the pretense of taking messages to their people, until only Lone Wolf and Satanta remained with the soldiers.[51] Custer, realizing the warriors had betrayed him with no intention of bringing in their tribe, wrote:

> As soon as the scheme of the Indians was discovered, I determined to seize the most prominent chiefs as hostages for the fulfillment of their promises regarding the coming on of their villages; but as for this purpose two hostages were as valuable as twenty, I allowed all but this number to take their departure apparently unnoticed.[52]

The idea of taking chiefs hostage had been first suggested to Custer by General Sherman in 1867. Frustrated by his quickly disappearing provisions and the possibility of a long winter chase, Custer took hostages for the second time in his Indian fighting career. In a letter to his wife, he simply stated, "I hold Satanta and Lone Wolf under strong guard." Both men were placed in leg irons when dismounted.[53] The return trip was brutal on the men and horses as their uniforms began to wear out, and many were not equipped for the extreme cold temperatures. John Ryan was ordered to shoot 48 horses and 88 mules in one day. Not only was Custer guilty of killing Indian ponies, but he was also decimating his own stock in the dogged pursuit of 'total war'.[54]

Arriving at Fort Cobb with his prisoners on December 18, 1868, Custer waited for the Kiowa to surrender, and after several weeks, only a few trickled in.[55] An officer wrote:

> Satanta was put in a Sibley tent and an armed guard around it. He would wrap his blanket around himself and come out and sit down by the side of the tent, then swing back and forth [and] chant the most doleful and monotonous death song. Then stooping over, he would scoop up sand and dirt and put [them] into his mouth. Then he would go around to the southwest side of the tent and, shading his eyes with his hand, would sweep the horizon to discover if possible the approach of his people.[56]

An angry Sheridan demanded, "You can inform Lone Wolf and Satanta that we shall wait until sundown tomorrow for their tribe to come in; if by that time the village is not here, Lone Wolf and Satanta will be hung, and the troops sent in pursuit of the village."[57]

Satanta's son brought the tribe in on the next day, securing a success for the army in getting most of the Kiowa onto the reservation. Custer and Sheridan's threats to kill the hostages proved successful. Satanta and Lone Wolf were eventually released to their people.[58]

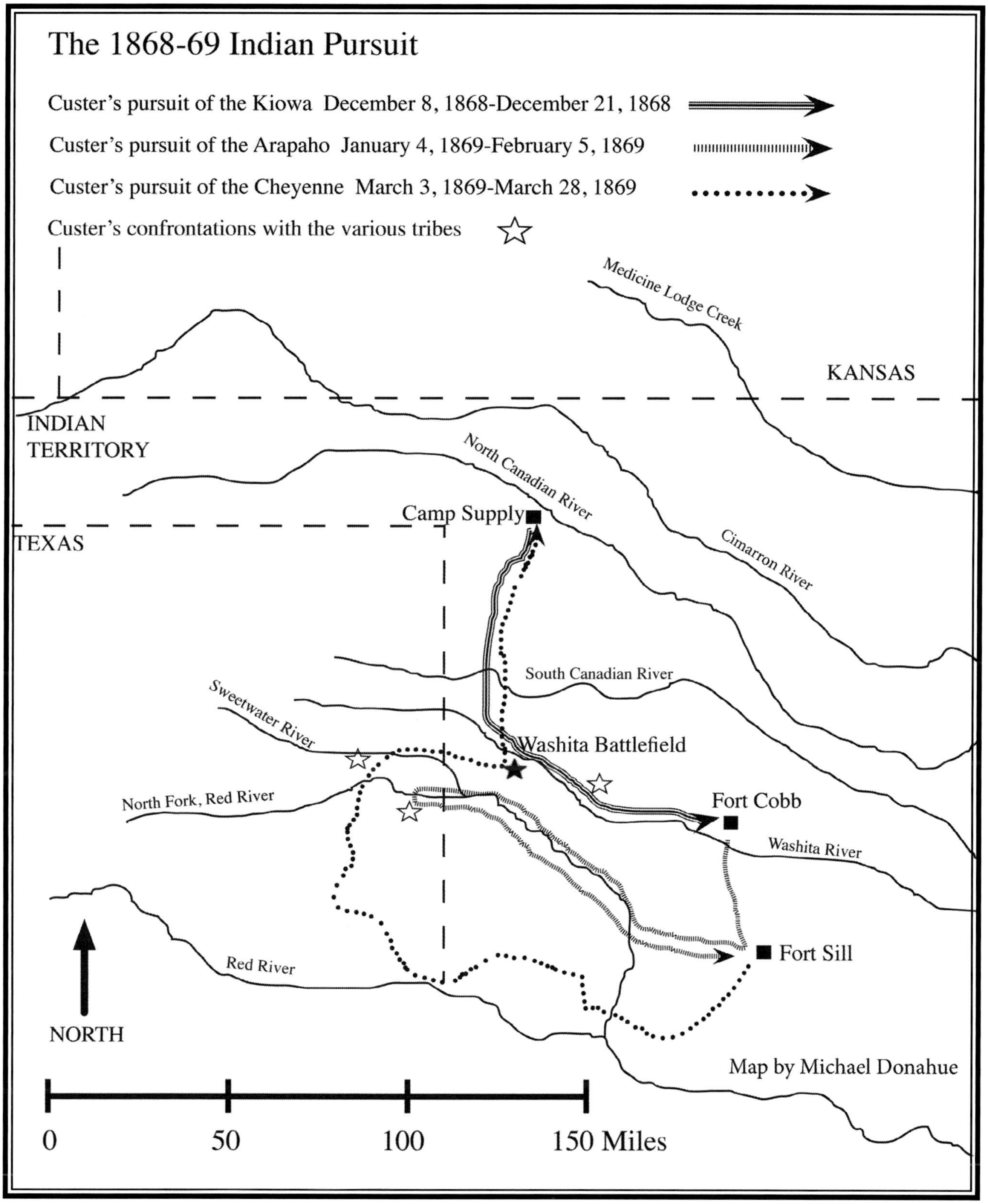

Fig. 20. Map: The 1868-69 Indian Pursuit.

On January 4, 1869, the troops left their camps near Fort Cobb in search of the remaining Arapaho and Cheyenne.[59] Cheyenne Chief Little Robe and Arapaho Yellow Bear agreed to parley with Custer who had sent *Mahwissa* as a messenger. The chiefs stated that the Arapaho were hesitant to come in. The more defiant Cheyenne kept messenger *Mahwissa* and would not allow her to return. Custer knew the tribes were monitoring his every move, and it would be difficult to sneak up on them.[60] A somewhat skeptical Custer was convinced that the Arapaho intentions were honorable and called them, "the most upright and peaceable inclined Indians I have ever known."[61]

With permission from Sheridan, Custer accompanied Yellow Bear to his village, bringing along only 40 soldiers to avoid frightening the Arapaho camp. It was a dangerous proposition due to the fact that they would be completely outnumbered. Benteen offered his pocket derringer to Custer so that his commander could take his own life rather than be captured and tortured. This may have been wishful thinking by Benteen; however, Custer had no intention of giving such satisfaction to his subordinate or the natives.[62]

After several days journey, Custer's small detachment found themselves in the midst of the Arapaho warriors who had recently surrounded them at the Washita. He decided to camp among the women and children in case it became necessary to use noncombatants as pawns once again. Custer wrote:

> Indians contemplating a battle, either offensive or defensive, are always anxious to have their women and children removed from all danger thereof. By our watchfulness we intended to let the Indians see that there would be no opportunity for them to take us by surprise, but that if fighting was intended, it should not be all on one side. For this reason I decided to locate our camp as close as convenient to the village, knowing that the close proximity of their women and children, and their necessary exposure in case of conflict, would operate as a powerful argument in favor of peace when the question of peace or war came to be discussed.[63]

Custer met with the stout Arapaho Chief Little Raven, described as 'remarkably intelligent' and about 50 years of age.[64] Through the intervention of Yellow Bear and Little Raven, Custer was able to convince the Arapaho to surrender and head to their designated reservation without firing a shot or taking hostages.[65] Custer as he had done at the Washita used noncombatants as a body shield for a second time.

The last Washita band refusing to surrender to the dreaded reservation life was Little Robe's Cheyenne, who fled west into Texas.[66] One of the additional goals of the expedition was to obtain the release of two white women who had been taken captive by the Sioux and traded to the Cheyenne: 19-year-old Mrs. Anna Morgan and the slightly younger Miss Sarah White.[67] Since both women had been seized in Kansas and one of their family had been killed, the volunteers from that state were especially anxious to punish these warriors and free the women.[68]

The soldiers surprised the Cheyenne on Sweetwater Creek, in the Texas panhandle, in early March. Chief Medicine Arrow, with about 40 warriors, notified the bluecoats that 260 lodges were directly to his front, and that Little Robe with his band were about 10 miles farther on. Custer surmised that the two white women might be in the villages of about 1,200 to 1,500 Cheyenne.[69] A cautious Custer recalled the death of the Blinns and stated:

> There was but little doubt that it was the Cheyennes, for whom we had been searching. If this should prove true, the two white girls, whose discovery and release from captivity had been one of the objects of the expedition, must be held prisoners in the village we were approaching; and to this effect their release unharmed then became my study, for I remembered the fate of the white woman and child [Mrs. Blinn and Willie] held in captive by a band of the same tribe at the battle of the Washita.[70]

Custer decided to parley with the chiefs and took a tremendous personal gamble by riding into the village accompanied by only two other men: William W. Cooke and the regiment's medical officer, Dr. Henry Lippincott.[71] Sergeant Ryan recalled, "I thought it was a very foolish thing to do, to ride right in among the Indians."[72] Most of the villagers fled as the angry soldiers watched from a distance, hoping to attack the Cheyenne. Not understanding the situation, the Kansas volunteers almost revolted against Custer. Thinking of once again using the women and children as a body shield, he projected:

> I knew, before accepting the proposal of the chief to enter the village, that he and every member of his band felt it to be in their interest not only to protect me from harm . . . knowing as they did that in case of an outbreak of any kind it would be impossible for a great portion of the village, particularly the women and children, to escape. I considered all this before proceeding to the village.[73]

As Custer proceeded through the Cheyenne crowd *Mahwissa's* familiar face appeared. Medicine Arrow ushered the bluecoat chief to the buffalo robes in his lodge as the prominent Southern Cheyenne chiefs surrounded Custer.[74] Cooke was taken into an adjacent lodge, while Dr. Lippincott rode back with an order to bring up the troops as quickly as possible.

In a strange and ironic resemblance to the series of events at Little Bighorn years later, the task of quickly reinforcing Custer fell to Benteen. As Benteen began to trot the horses toward the village, he lost about 50 animals due to their exhausted state. Benteen took his own initiative to slow the advance to a walk, aware that Custer and Cooke were in a precarious position. Benteen arrived with the troops as Custer was "undergoing their mummeries and medicine making over him."[75]

A lengthy ceremony ensued as Medicine Arrow and other chiefs smoked the pipe with the lieutenant colonel. The proceedings culminated with the chiefs symbolically tapping the ashes from the pipe onto Custer's boots as a warning that should Custer ever attack his people that he would be killed. Young Magpie peered through the tipi entrance. He had survived Sand Creek and still limped from being shot at the Washita. He could see the figures sitting cross-legged in a circle and grasped his knife in case of some betrayal.[76]

As Benteen's troops began to arrive on the scene, they formed a skirmish line, and panicked warriors, women, and children fled downstream toward the camp of Little Robe. Custer ordered his troops to hold their fire and promised the fugitives he would not destroy their camp items left behind. He also requested that Little Robe come to the camp, promising safe passage back to his people.[77] Custer unsuccessfully attempted to negotiate the release of the two women with Medicine Arrow and realized they were stalling to allow their villagers time to escape to safety.[78] Custer stated:

> My attention was devoted to the chiefs, and acting upon the principle that for the purposes desired half a dozen would be as valuable as half a hundred, I determined to seize the principal chiefs then present, and permit the others to depart.[79]

As Custer parleyed with some of the chiefs and warriors, soldiers slowly outnumbered the remaining Cheyenne. Custer arrested the struggling chiefs, placing them under armed guard.[80] Hundreds of incensed warriors raced around the camp "in the most threatening manner . . . screaming with rage and baffled hate."[81] It was a great spectacle to Kansans, who had never seen an Indian up close; one officer noted, "This is the farthest I ever walked to see a circus."[82]

If the Cheyenne decided to object to the conditions and fire into the camp, the Kansans were to open fire, but fortunate for both sides, no firing took place. Monahsetah was called on to convince the Cheyenne that it was in their best interest to surrender and return to the reservation. One of the hostages, probably Medicine Arrow, was chosen as a messenger to deliver three demands to the Cheyenne: unconditional release of the two captive white women, their return to the reservation, and an audience with Little Robe. If the first two demands were not met, the soldiers promised an immediate pursuit of the Cheyenne.[83] The three elderly chiefs that remained in the custody of the 7th were Dull Knife, also known as "Lean Face" (or "Slim Face"), Fat Bear, and Big Head also known as "Curly Hair."[84] George Bent, who lived with the Cheyenne, later recalled that the tribe considered it a treacherous act and hated Custer for it.[85]

The Cheyenne continued to delay any exchange with the soldiers. The cavalry troops were running out of provisions, and their starving livestock were dying by the score. After three days of unsuccessful negotiation for the release of the women, Custer decided to bring the issue to a head and gave the Cheyenne until sunset the next day to return the white women, or the three chiefs would be hanged.[86]

At 3:00 p.m. the next day, the Cheyenne appeared with two female figures mounted on a single gray pony. The ladies dismounted and were approached by the army interpreter who told them to walk toward their freedom. The two women hung onto each other as they slowly plodded toward the soldiers. Custer requested that since the women were from Kansas, the officers of the 19th Volunteer Cavalry should receive them.[87] Miss White and Mrs. Morgan, dressed in patched rags and Indian jewelry, were safe but had suffered so much hardship during their captivity that they did not smile when released. Mrs. Morgan simply asked, "Are we free now?"

A melodramatic Custer had the band play the tune, *Home, Sweet Home*. Daniel Brewster, the brother of Mrs. Morgan, had accompanied the troops in the hope of finding his sister and was relieved with the results. According to Benteen, both women were pregnant, and their mistreatment had ceased after the Washita fight.[88] Custer described their release, "Men whom I have seen face death without quailing found their eyes filled with tears, unable to restrain the deep emotion produced by the joyful event."[89]

While Custer was delighted with the result, he still contended that until the Cheyenne came to the reservation, the three chiefs and Washita captives would be used as bargaining chips. He recalled, "I determined to march in the morning for Camp Supply, Indian Territory, satisfied that the three chiefs in our possession, and the squaws and children captured at the Washita still held as prisoners at Fort Hays, Kansas, we could compel the Cheyennes to abandon the warpath and return to their reservation."[90] The tribesmen began to trickle into the reservation.

The soldiers made the long, miserable march back through Oklahoma territory. The troops returned to Fort Hays, Kansas with the prisoners. Upon arrival, the Kansas volunteers were mustered out

Fig. 21. The three Cheyenne captives seized by Custer are noted here as Fat Bear, Big Head and Dull Knife. While imprisoned at Fort Hays, there was a tragic misunderstanding in which they believed they were to be hanged. Attempting a breakout, they pulled knives and killed a guard. Both Dull Knife and Big Head were killed with only Fat Bear surviving.

Three Cheyenne Indian Chiefs: Fat Bear, Big Head, Dull Knife, circa 1869.
National Park Service, Little Bighorn Battlefield National Monument, LIBI_00019_00395, W. S. Soule

of service.[91] The Cheyenne chiefs, women, and children were placed in a wooden stockade equipped with tents at Fort Hays.

Some of the prisoners secretly honed dinner knives into lethal daggers and decided on an escape. In a tragic scene foreshadowing the death of Crazy Horse in 1877, the three chiefs were moved to the guardhouse for security and mistakenly thought they were being led to their deaths—or they may have seen an opportunity to escape. Fat Bear stabbed and mortally wounded the sergeant of the guard. A soldier's gunshot quickly killed Big Head, while a bayonet mortally wounded Dull Knife. Fat Bear was knocked unconscious but survived. Weeks later, Fat Bear, along with Monahsetah and the hostages from the Washita, were released.[92]

In only a matter of months, Custer had four separate encounters with the Plains warriors and had been successful in achieving the goals set forth by Sheridan. His department commander was happy with his performance, stating, "I am very much rejoiced at the success of your expedition, and feel very proud of our winter operations, and of the officers and men who bore privations so manfully."[93] Sheridan complimented Custer with some leave time and promised to look into a promotion for his subordinate, but it never developed.[94]

Custer's conflict with Plains Indians was quickly replaced with one involving Benteen upon his return. It seems likely that after the accusatory Washita letter, the lieutenant colonel wanted Benteen away from him. Benteen was ordered to report to Fort Dodge to serve under an infantry officer. According to Benteen, "I did not want it; this Custer knew." Not happy with the assignment, Benteen asked a higher-ranking officer to intervene on his behalf, and within three months, he returned to Fort Hayes where the 7th Cavalry was stationed. However, Custer forced Benteen to remain outside the fort proper and in the temporary cavalry camp for three more months, before being brought into the post for the harsh winter.[95]

During his military career, Custer would have only one more occasion to take hostages before his final campaign: an expedition into *Paha Sapa*, the sacred Black Hills, in July 1874. He led a reconnaissance force into the area in search of a site for a future army fort, as well as to map and collect information about this largely unknown terrain. During the trip, the 7th Cavalry came upon a small Oglala camp in the hills where they captured an old man named One Stab.[96] A newspaper reporter with the soldiers noted, "He would be kept a hostage until we were out of the [Black] Hills and must show us a good road."[97] Although a captive, One Stab was not harmed, and he led the soldiers through the hills until he was eventually loaded with provisions and released. One wonders if Custer would have used the captive as a defensive option if attacked by other warriors.

Sherman had suggested taking chiefs as hostages in the 1867 campaign, and two years later, Sheridan initiated the idea of hanging chiefs to achieve results. Custer embraced both strategies, with hostage-taking becoming his *modus operandi* in dealing with the guerilla fighters of the plains. A perceptive Sheridan knew that the act of hanging reached deep into the psyche and superstitions of native peoples—a dreaded and shameful manner of death. The slow strangulation by choking to death was as fearful for Indians as the equal dread by whites of being slowly tortured. Most Plains people believed that the spirit escaped the body via the mouth at the moment of death. Hanging or strangling a person by crimping the windpipe would trap the spirit inside a decaying corpse for eternity.[98]

In its behavioral bible known as *General Orders No. 100* that dictated military conduct during wartime, and of which every officer was familiar, the U.S. Government realized the slippery slope of hostage taking and hanging. It stipulated:

> A hostage is a person accepted as a pledge for the fulfillment of an agreement concluded between belligerents during the war, or in consequence of a war. Hostages are rare in the present age. . . . If a hostage is accepted, he is treated like a prisoner of war, according to rank and condition, as circumstances may admit. . . . A prisoner of war is subject to no punishment for being a public enemy, nor is any revenge wreaked upon him by intentional infliction of any suffering, or disgrace, by cruel imprisonment, want of food, by mutilation, death, or any other barbarity.[99]

From Custer's comments, it is clear that he intended to use the chiefs and other prisoners to fulfill the one-sided agreement of surrender by the tribes. Semantically, whether the term 'hostage' or 'prisoner of war' applied, Grant, Sherman, Sheridan, and Custer were clearly willing to turn a blind eye to some rules when deciding to hang the enemy. Fortunately, in the preceding scenarios, no chiefs were hung, and it will forever remain a mystery whether the commanders would have carried out their threats, and if there would have been legal repercussions. There was, and still is, a stigma about hostage-taking which in the eyes of many seems underhanded and treacherous.[100]

The taking of hostages by the tribes is often ignored, and the Indian ransoming of hostages was a lucrative and commonplace business, whether it was a white or non-white captive. Ironically, most warriors did not take hostages or prisoners for political reasons, but for personal or financial gain. While such activity was planned to deter the invasion of non-Indians, large amounts of money and goods were paid for the return of white captives. Female captives and young children could be bartered within the tribe or with captives' families like money as a spoil of war.[101] "Stealing white women is a more lucrative business than stealing horses," was the conclusion of an 1866 report from the commissioner of Indian Affairs.[102] More empathetic frontier army officers often found themselves at odds with the policy of non-negotiation when given a chance to free captured women and children. Sherman and Sheridan were opposed to such ransom payments as early as the Hancock campaign.[103]

Raising young captive boys was a way of replenishing the village's warrior population. While the U. S. Army could find endless recruits and volunteers, the Indian village had to grow its own. In cases where the captive became a liability or was unmanageable, they were sometimes traded or killed. In other cases, captives were not subjected to outrages or mistreatment and recalled being well treated.

Many contemporary writers and historians have avoided scrutinizing Custer's hostage-taking strategy. Rather than looking at it as a functioning and fascinating stratagem in dealing with the southern tribes, it seems to receive only cursory notice. Few authorities search for clues into his initial mindset, and how it may have affected his actions throughout the Indian wars. Custer's own words, and more importantly, his actions, are extremely revealing to anyone who reads the lines he wrote as well as in between them.

Custer defenders would rather not talk about this aspect of his manipulation of Indians because of the stigma mentioned, while his detractors emphasize it as the pinnacle of deceit, dishonesty, treachery, and ultimate character flaw.

This historian takes it for what it was—a strategy that was effective and cost few soldier lives. It occurred no matter how one wishes to interpret, ignore, rationalize, or re-label it.[104]

War by nature is brutal, and Custer, his superiors, and the tribes themselves reflected some of that brutality in their actions. The frustrating conditions of fighting the southern warriors called for different strategies since these officers were trained to fight using European tradition and tactics. Custer noted that Indian warfare was "a distinct and separate species of hostilities, requiring different talent, different material, as well as personnel, and different rules of conduct."[105]

One of the most frustrating episodes for Custer in the Civil War was fighting the guerilla forces under John Mosby, because his tactics were outside the mainstream of traditional warfare.[106] Custer found similar warfare on the plains and adapted quickly to body shields, hostage taking, and the threat of hanging, all of which garnered positive results. The negative media image that resulted from the Washita, the death of Elliott and his men, as well as the resulting death of Clara and Willie Blinn, caused the 7th Cavalry commander to ponder other strategies in 1869.

He would have plenty of time to think about his Indian fighting experience before his next confrontation with warriors. In his next meeting, the warriors would be the aggressors, noncombatants would be nearby, and Custer would once again chase the tribes, and it would be on the northern plains.

Chapter 4

Decoy Along the Yellowstone

The campaign against the southern tribes only reinforced Custer's attitude that he could not fight the warriors the way he had fought the rebels during the Civil War. He had used hostages in three of four situations to manipulate the tribes, and in two situations, the commander of the 7th had moved next to their noncombatants, believing that there was personal safety in their proximity. In reality and retrospect, Custer had done very little fighting and had imparted a great deal of psychological manipulation.

While Custer was saving white women on the plains, new Republican president Ulysses S. Grant took office in March of 1869. Even though Custer remained in the army, his social life and private business ventures occupied much of his time during the next three years. He entertained the Russian Czar's son, the Grand Duke Alexis, as well as other dignitaries in 1872, and spent much of his time writing articles for *Galaxy*, which eventually became his biographical volume, *My Life on the Plains*. It would not be until the summer of 1873 that Custer would have an opportunity to do what made him feel most alive—fight.[1]

In the winter of 1869, a new officer named Marcus Albert Reno arrived in the 7th Cavalry. Born in 1834 in Illinois, his West Point stint was memorable for insubordination rather than excellence—strangely similar to his commanding lieutenant colonel. Reno excelled in French as well as demerits and graduated 20th in a class of 38 cadets.

However, there was a difference in the two personalities that was quickly apparent to West Point classmates. While Custer's shenanigans gained him popularity among other cadets, Reno's misdeeds seemed to make him less likable, and as one fellow cadet stated, "Reno is not very popular."[2] Ironically, this sentiment did not change with the passing of 20 years, as 7th Cavalry Lieutenant Francis Gibson noted, "Reno was not at all a popular officer."[3]

There was a strangeness about his dour personality that did not endear him to cadets, enlisted men, or officers. Even the allied Arikara scouts saw something iniquitous about the major, naming him "The Man with the Dark Face."[4] Reno was a man who seemed to have a chip on his shoulder, and with a little alcohol, was in a constant search for someone to challenge him. His inflated sense of self-importance combined with whiskey caused him to initiate fights on several occasions.

Commissioned in the 1st Dragoons, Reno was stationed at Fort Walla Walla, Washington Territory, where he spent more time with commissary duties and court-martial boards than fighting the

Fig. 22. Major Marcus Reno seemed to have a dark shadow that followed him throughout his life. Thrown into an attack on the massive Indian village along the Little Bighorn River, he probably suffered from battle shock when scout Bloody Knife's brains were splattered across his face.

Courtesy of John Doerner

Snake Indians in the region. During the Civil War he quickly climbed in rank, and in 1863, he was brevetted the rank of major after a gallant charge that left him pinned under his horse, causing a hernia. His recovery allowed time to court Mary Hannah Ross, the daughter of a wealthy banker and industrialist, and he married her two months later.[5]

In 1865, he found himself brevetted brigadier general of volunteers, and after the war, he served short stints as an infantry instructor at West Point and a provost marshal in New Orleans. In 1869, he was appointed major in the 7th Cavalry and was stationed at Fort Hays, Kansas. In 1873, he was ordered to accompany the Northern Boundary Survey to aid in establishing the line along the Canadian/U. S. border.[6] Reno was not a fan of Custer, admitting, "I had known General Custer a long time and had no confidence in his ability as a soldier. I had known him all through the war."[7]

◆

In the late 1860s and early 1870s, the changing social and political climate following the Civil War created great unrest, as well as the backlash of the Ku Klux Klan in many of the southern states. Defeat and reconstruction were bitter pills to swallow along with southern pride, and several of the 7th Cavalry companies were posted throughout the South as a police force to monitor the tense situation.

Benteen was stationed in Louisville, Kentucky, where Cooke, George and Tom Custer visited him, with the hope that he would temporarily swap an upcoming posting in Nashville, Tennessee, for one in South Carolina. Tom, who evidently did not want the assignment in South Carolina, hoped that Benteen would make the switch. Benteen, in his typical adversarial manner, refused. One biographer recalled, "Custer told Kate [Benteen's wife] that whenever he recommended anything to Benteen, he could be sure that Benteen would do just the opposite." Kate agreed with Custer and told her husband that he was "dead wrong." [8] It is intriguing to contemplate, when put in a no-win situation between her husband and his commander, whether her conciliatory statement was truly heartfelt or an attempt to avoid conflict. In April of 1872, Kate gave birth to a girl they named Fannie, and by 1873 the Benteens were stationed along the Missouri River in Dakota Territory at desolate Fort Rice.[9]

As the deep wounds of the Civil War slowly closed and reconstruction began to wane in the consciousness of many northern Americans, many turned westward in a spirit of expansionism and entrepreneurism.[10] More railroad lines were being planned in order to join the eastern and western states that were separated by territories primarily dominated by native tribes. Jay Cooke, a banker and financier in one of the most powerful banking houses in America, had a vision of a second transcontinental railroad that would closely follow the northern route as explored by Lewis and Clark decades earlier. The millionaire's hope was that the Northern Pacific Railroad could push its way along the Yellowstone or Musselshell rivers opening the West. The idea was a risky one, since the Lakota and the Northern Cheyenne traditionally occupied the region.[11]

In 1871 and 1872, Northern Pacific Railroad surveying parties probed the region for the best route to lay the tracks for the iron horse, but both surveys were incomplete, and in 1872, the surveyors were attacked by superior numbers of Teton Lakota. The tribes recognized this as prime bison hunting territory and claimed it as their own. Such resistance induced growing fears back east about the feasibility of such a venture.[12]

By 1873, one of the railroad company's main priorities was to extend the track westward beyond Bismarck, Dakota Territory, while the army focused on placing military posts along the new route.[13] Sheridan realized that the establishment of railroads, forts, and settlements, as well as the loss of native food supply, would repeat the success that he had on the southern plains. Another expedition was needed to complete the work and to explore the area that had not been surveyed between Pompey's Pillar and the mouth of the Powder River.[14]

The army was to provide protection for the Northern Pacific Railroad surveyors but was also ordered to 'subdue and intimidate' the Lakota, bringing them to battle if the opportunity presented such.[15] The military would depart from two different forts; most of the large protective army and supply train would leave from Fort Rice, while the surveyors with a smaller soldier escort would leave from the under-construction Fort Lincoln. The plan was for both parties to merge in the early stages of the westward march.

Fig. 23. Major General David S. Stanley was an able veteran of the Civil War and the leader of the 1873 Yellowstone railroad survey. His occasional drinking sprees placed him at odds with the teetotaling George Custer.

Courtesy of the National Archives

Colonel David S. Stanley was chosen to command the survey expedition. During the War Between the States, he had been offered a similar rank in the Confederate army, but instead chose to join the Union, where he became a major general of volunteers.[16] In many historic photographs taken on the campaign, he is shown as a stern-looking, silver-haired older man with a drooping moustache overlapping his beard.[17] What they don't show is Stanley's love of the bottle that became an irritant to many of the men and officers on this arduous journey.

Fig. 24. Lieutenant Charles Braden is pictured in his West Point uniform. Charged with holding a ridgetop and repulsing overwhelming numbers of warriors on August 11, 1873, his being wounded there ended his active military career.
Courtesy of the West Point Military Academy

Another Civil War veteran accompanying the expedition was the dashing Virginian, Thomas Lafayette Rosser, Custer's close friend and classmate at West Point. The 6-foot 3-inch, dark-headed Southerner had grown up in Texas, and he joined with others from his adopted state when it seceded from the Union. He distinguished himself commanding artillery batteries before being promoted to colonel of the 5th Virginia Cavalry. After the war, the talented Virginian studied law and became an assistant engineer for the Pittsburg and Connellsville Railroad before becoming the chief engineer and survey chief for the Northern Pacific Railroad.[18]

7th Cavalry Second Lieutenant Charles Braden, a 25-year-old from Detroit joined the military forces. He was stationed in Louisiana until called to duty along the Yellowstone. The reliable and well-built youth was nicknamed "Old Gothic" because of his solid physical stature.[19]

Another soldier who joined the adventure was John Charles "Chip" Creighton of Ohio. He spent his first summer of cavalry service in the sweltering, muggy states of South Carolina, North Carolina, and Alabama. He was ordered north to Dakota Territory just in time to be 'baptized in the big blizzard' of Easter Sunday, 1873.[20]

Perhaps one of the youngest participants of the expedition was an underage boy who had left home at age 14 to drive a wagon team across his home state of Wisconsin. Louis E. Hills, nursing the soul of an adventurer, soon jumped onto a steamboat that ended up in Memphis, Tennessee. Impressed by the Custer persona, he decided to enlist in the cavalry and was issued a gray horse as well as a double-coiled brass F trumpet. A photograph of a boyish trumpeter shows Hills wearing a kepi and a huge overcoat, with the instrument in his hand.[21]

The massive expeditionary force included five companies of the 22nd Infantry, three companies of the 17th Infantry, one company of 6th Infantry, as well as 10 companies of the 7th Cavalry. A soon-to-be Custer favorite, Bloody Knife, was one of 27 Indian scouts that acted as guides and interpreters. One of the infantry companies acted as 'pioneers' and artillery support for the two 3-inch Rodman guns. The massive entourage included 275 wagons and ambulances, 353 contracted civilian employees, and 2,321 horses and mules. Seven hundred beef were herded to provide food for the men, and at times, the expedition looked more like a cattle drive than a railroad survey. The entire military force consisted of 79 officers and 1,451 men.[22]

It was decided that the expedition would contain a party termed the 'Scientific Corps' that included a zoologist, a geologist, an artist, a writer, and a photographer. Also accompanying the expedition were three 'casuals,' as labeled by the troops, which included the son of an English Lord, the son of an English baronet, and a London University student. Lieutenant Fred Grant, son of President U. S. Grant, also joined the expedition.[23] Among several correspondents was Samuel Barrows, who was sending back dispatches to the *New York Tribune*.[24]

The huge supply train from Fort Rice departed on June 20, 1873 with Custer's band playing the popular tune, *Arkansas Traveller*.[25] Four flat-bottomed steamboats named *Far West*, *Key West*, *Peninah*, and *Josephine* carried additional supplies for the expedition and ferried troops across the rivers. They bellowed black smoke as they navigated the snake-like Missouri and its tributary, the Yellowstone.[26]

It was a strange symphony of music, creaking wooden wagons, braying mules, bellowing cattle, neighing horses, cursing teamsters, plodding infantry, and clattering metal against the McClellan saddles of the cavalry troops. Stanley's main concern was the strict logistics and timing necessary for the small survey party to rendezvous with the main supply train. Incessant and torrential rains caused the units to mire in a muddy purgatory for fourteen of the first 17 days of the journey. An abrupt hailstorm battered the isolated engineers under Rosser and their escort, stampeding their mounts, shattering wagons, and killing nearby antelope. The separated parties finally combined on July 5.[27]

Six days into the journey, Custer had an amazingly different perspective that belied his love of adventure, "Our march has been perfectly delightful thus far. I never saw such fine hunting as we have constantly had since we left Fort Rice."[28]

Custer divided the 10 troops of the 7th Cavalry into two wings of five companies each. A Civil War veteran from Indiana named Verling Hart led one wing while Benteen commanded the other.[29] The westward journey of the combined parties began with excessive drinking by many of the officers, and hostilities between Stanley and Custer. While Custer was temperate, Stanley was the opposite. Lieutenant Charles Larned of the 7th Cavalry wrote:

> Stanley is under the whiskey curse and gets on periodic tears of two or three days duration during which he manages to disgrace himself and insult everyone who happens to displease him.[30]

Evening poker parties and the resulting late morning starts irritated the crusty Stanley. Custer brought along a cook stove that required cooling off before being loaded into a wagon, and this angered Stanley as well.[31] In a letter to his wife, Stanley made it clear he did not like the lieutenant colonel, writing:

> I have had no trouble with Custer, and will try to avoid having any; but have seen enough of him to convince me that he is a cold blooded, untruthful, and unprincipled man. He is universally despised by all the officers of the regiment excepting his relatives and one or two sycophants.[32]

At about the same time, Custer was venting to Libbie:

> General Stanley is acting very badly, drinking, and I anticipate official trouble with him. I should greatly regret this, but fear it cannot be avoided. Rosser has told me how badly General S[tanley] behaved last year, some days being so overcome the expedition could not go on.[33]

Custer, realizing the ambiguity and problems arising from verbal orders possibly given under the influence, asked Stanley, "To insure a correct understanding on my part, I respectfully request that all important orders from the Headquarters Yellowstone Expedition intended for my command be transmitted in writing."[34] Such a demand further infuriated Stanley who believed Custer was becoming more insubordinate and arrogant by the day.[35]

Always anxious to take independent command, Custer took every opportunity to separate himself from the older Stanley. Soldier John Ryan recalled, "Custer was a man who was used to traveling on his own hook, and he did not care to be tied down under the command of an infantry officer."[36]

On July 6, some of Stanley's officers were arrested for drunkenness, causing James Calhoun to remark, "The Commanding Officer of the Expedition sets the example, his officers follow him, and while they are placed in arrest he goes unpunished."[37] With alcohol, disparaging gossip between the cavalry and infantry, and a fight over authority creating daily tension in the upper ranks, Benteen entered the fray. Stanley had furnished a horse to civilian reporter Barrows for the journey, and taking a similar initiative, Custer furnished a military horse to civilian Fred Calhoun, who was employed by the railroad party and brother to James Calhoun.

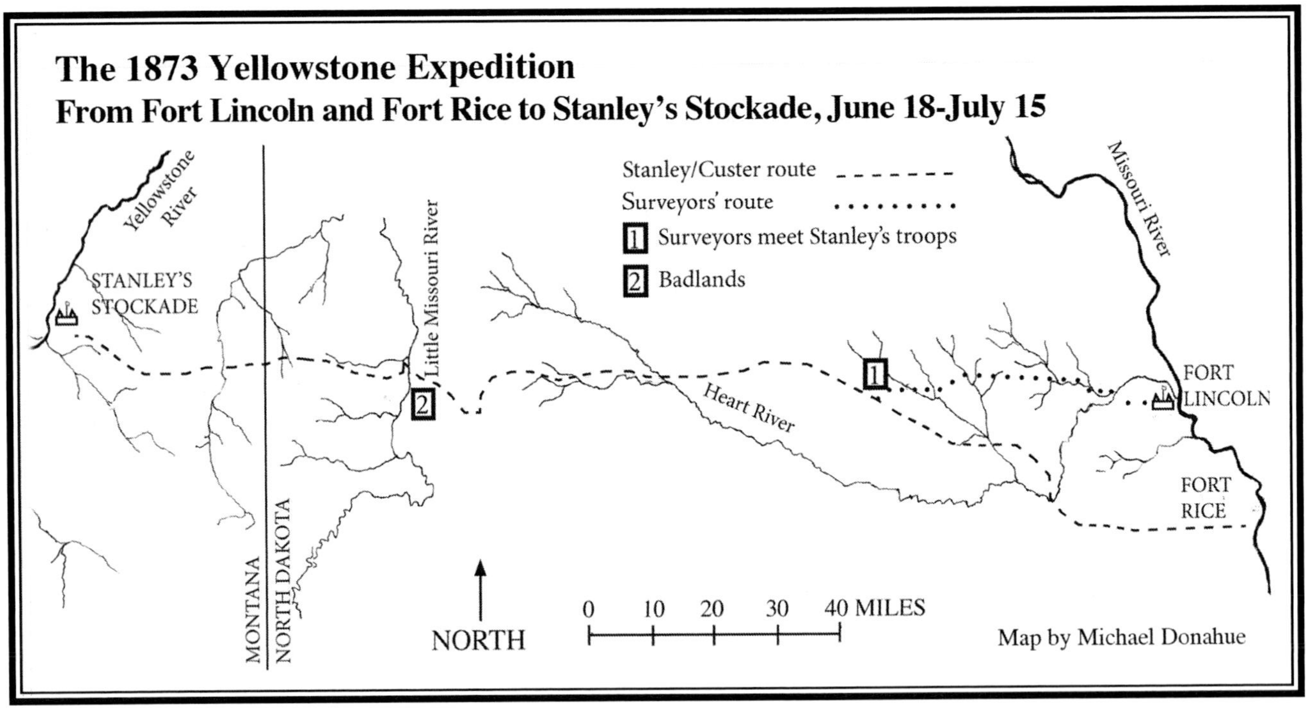

Fig. 25. Map: The 1873 Yellowstone Expedition. June 18-July 15.

Stanley's commissary chief for the expedition was poorly mounted on a mule. Benteen recalled, "H Troop was short of horses. Custer had a claybank horse, for which I was responsible and which I had assiduously endeavored to have returned to the troop, without avail."[38] Benteen, upset that Custer kept the horse for his use, told the commissary officer that the 7th Cavalry commander had provided a military mount to a civilian.[39] Stanley brought up that such an act was in violation of law and regulations, and ordered the officer be given a proper mount, which he received. When Custer noted that he was only following Stanley's example, his additional commentary only exacerbated the tense situation.[40]

The cumulative actions of disrespect, from Stanley's point of view, finally caused him to place Custer under arrest and force him to ride at the rear of the column. The humbled subordinate was released the next day as, according to Benteen, "Stanley got drunk and let Custer off the hook."[41] Custer attempted to save face as well. He later recalled, "Since my arrest, complete harmony exists between Genl. Stanley and myself. He frequently drops in at my headquarters, and adopts every suggestion I make."[42] Stanley had made his point.

The wagon train traversed grassy rolling hills westward with marches limited by the amount of water and wood needed for a camp. Sometimes wood was strapped to the sides of wagons if the supply looked sparse, and successful hunters tied deer and elk horns to the vehicles as souvenirs. Soldiers took charcoal from campfires and scribbled "Yellowstone or Bust" on the sides of wagons to be replaced on the return trip with the notation "Busted, By God."[43] Rosser and Custer, reunited after almost eight years, quickly renewed their old friendship. The two enjoyed revisiting the battles and movements against each other from the Civil War.[44]

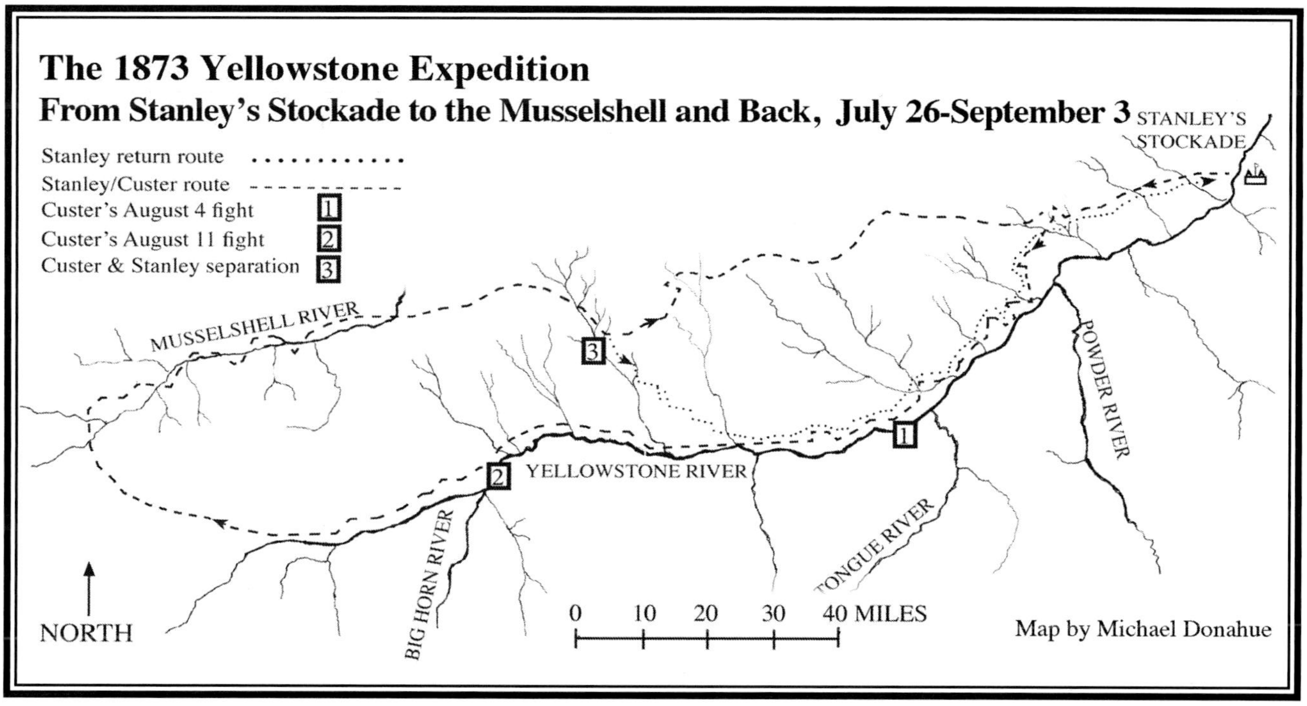

Fig. 26. Map: The 1873 Yellowstone Expedition. July 26-September 3.

Custer established a temporary supply depot christened 'Stanley's Stockade' along the south bank of the Yellowstone River, eight miles south of Glendive Creek, where the *Key West* was awaiting the troops with supplies.[45] A week later, it was decided to ferry men and horses from the south bank of the Yellowstone River to the north side, which was no easy task. It was at this point in the journey that two important individuals departed from the expedition. A homesick Rosser missed his wife and children. He had not been enthusiastic from the start on his mission and returned via the *Key West*. The 'perfectly gentlemanly' Fred Grant had a more exciting return down the river in a small open rowboat. Hearing of his grandfather's death via mail, he decided to return home. To Stanley's consternation, the president's son disappeared without his knowledge.[46]

Whiskey, which was continually being dropped off by steamers, did not help motivate Stanley or his troops to move. The boats did not arrive as scheduled, and officers took advantage of the delay to imbibe. Ryan reported, "I think that if there was not quite so much whiskey the expedition would probably have got across the river sooner than it did."[47] The quandary was so serious that Custer and other officers felt that something had to be done and convinced Stanley that the men were becoming unmanageable due to the influence of the intoxicants. It was decided that all alcohol should be destroyed. Officers quickly seized some spirits upon hearing of the plan, while enlisted men lapped it up from the puddles as it was spilled upon the ground.

By the evening of July 26, the expedition was slowly ferried to the northern bank of the Yellowstone River.[48] Up to this point in the expedition, most of the fighting had not been against Indians, but within the officer corps.[49] Even though Benteen held enough seniority to command one of the two wings, Custer made the decision to separate him from the rest of the command, leaving him at Stanley's Stockade.[50] A lonely Benteen took the occasion to fish the Yellowstone and write long letters to his wife.[51]

Benteen was concerned about the recent illness of his youngest daughter, and at some point during the journey, he requested leave to return to Fort Rice to be with his wife and sick child.[52] Custer denied his request and this only exacerbated Benteen's hatred of his superior. Also unbeknownst to Benteen, his daughter Fannie died shortly after his departure from the fort, and he had to deal with the death of his third child while away on campaign.[53] Custer's rather cold comment of the event seemed void of feeling, "Capt. Yates brought Col. Benteen word of the death of his little girl at Fort Rice, 15 days before."[54]

The expedition continued westward along the wide Yellowstone River, traversing territory that was totally unexplored by the first two survey parties, and which Custer as well as Stanley found remarkable.[55] Due to the difficulty of crossing the rough topography with wagons, Custer was allowed to take two companies and ride ahead to mark the best route and to locate the best campsites with wood, water, and grazing for the horses.[56]

On Monday morning August 4, the column was marching by 5 o'clock in the morning, and the serpentine line of soldiers and civilians would again deal with scorching heat before day's end. The entire train crossed a high plateau that overlooked the Yellowstone, opposite the mouth of the Tongue River. Custer, using sticks and stones, left visible signals along the trail to help guide the cumbersome wagon train under Stanley's command along the best course. Two of Custer's favored companies, commanded by Myles Moylan and Tom Custer, accompanied him in the advance.

Scout Bloody Knife discovered the fresh hoof prints from the unshod ponies of 18 warriors and predicted a fight that day or the next.[57] The Arikara had become one of Custer's favorites and being raised a half-breed had changed Bloody Knife's life. Born of a Hunkpapa father and an Arikara mother, he was subjected to abuse and discrimination as a youth, which finally forced him to leave the Lakota tribe. At age 15, Bloody Knife returned to live with his Arikara relatives. He became a reliable mail carrier for the area forts; however, his penchant for alcohol caused him problems on occasion.[58]

Fig. 27. Trusted scout Bloody Knife showed his bravery and skills during the Yellowstone expedition and is pictured here pointing to a map held by Custer during the Black Hills expedition. The standing figure behind Custer and Bloody Knife is believed to be Custer admirer and orderly John Burkman.

Courtesy of Christopher Kortlander & the Custer Battlefield Museum, Garryowen, Montana

In mid-afternoon, 7th Cavalry veterinarian Dr. John Honsinger and sutler Augustus Baliran, who were with Stanley, decided to wander ahead exploring for curiosities and water. The veterinarian was described as 'large and fleshy' and 'good-natured' and rode his prized and pampered horse. The dark-complexioned Baliran was a French immigrant and had been personally invited by Custer to join the expedition. Baliran was popular with many officers since he had brought along an ample supply of whiskey.[59]

Scouts from a Lakota village upstream were observing the huge bluecoat column marching toward their camp. One member of the village was White Bull of the Miniconjou band, who was born in the Black Hills. Now in his 20s, he was anxious to fight the invaders and would have a long career as a warrior that included brave feats others would long envy.[60]

White Bull's uncle Sitting Bull, the leader of the village, knew and loved the verdant Yellowstone Valley. His family members say he was born around 1831 along this very river that they called the Elk River. He was known by the boyhood nickname 'Slow' since he was determined and deliberate in thought and action; these traits did not change with age.[61] As a Lakota, he belonged to the Hunkpapa tribal band. In 1845, in a fight against the Crows, he tomahawked one of his enemies, garnering him his father's name, *Tatanka-Iyotanka*, Sitting Bull. A year later, at age 15, he made a bravery run against the Flathead tribe, earning the right to wear a red feather after being wounded in the foot. In 1856, during a running duel with a Crow chief, both fired at each other with Sitting Bull killing his enemy. The Crow's bullet hit Sitting Bull's foot, ripping a hole from toe to heel, which caused him to limp for the rest of his life.

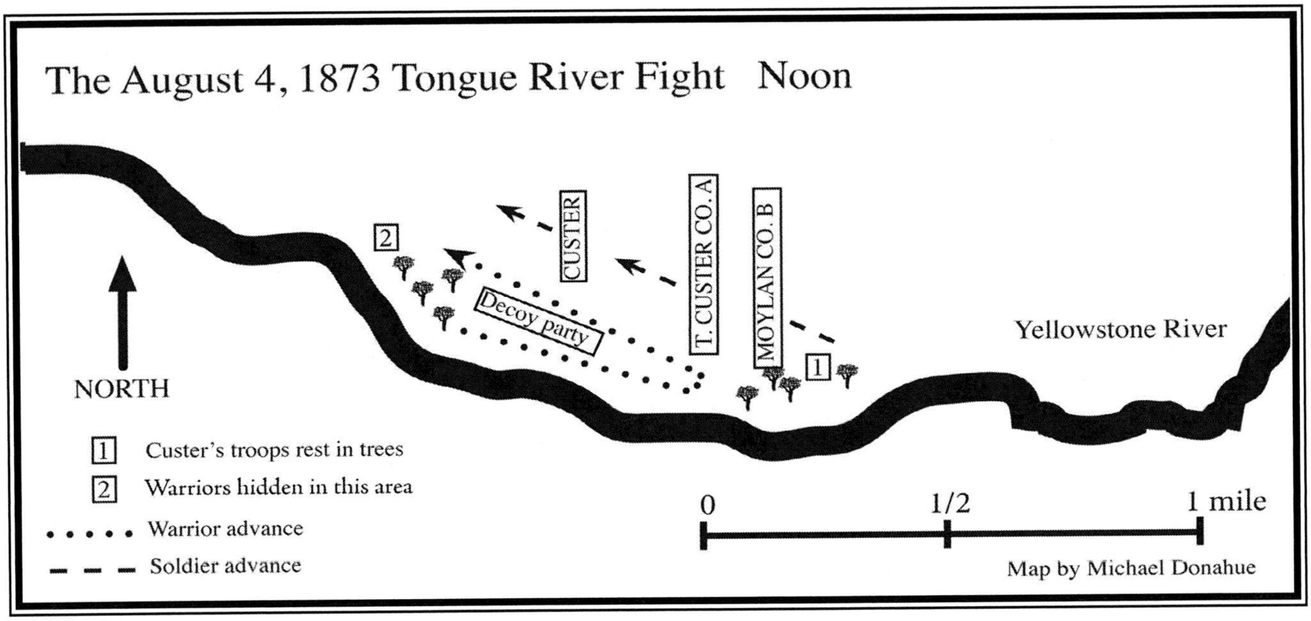

Fig. 28. Map: The August 4, 1873 Tongue River Fight. Noon.

Sitting Bull was 5-foot 10-inch with a strong muscular frame, big chest and broad head. He had deep penetrating eyes that could easily reveal his serious nature or twinkle with a sense of humor. With his warrior days behind him, he was one of the most charismatic spiritual leaders on the northern plains. Becoming a *Wichasha Wakan* or holy man, he had the gift of prophesy as well as the ability to communicate with nature and the animals. Sitting Bull's generosity and bravery, combined with fortitude and wisdom, made him a natural and respected leader among his people.[62]

Some of Custer's 'clan' and about 91 soldiers headed straight toward the unseen village. James Calhoun acted as adjutant. Lieutenant Charles Varnum also accompanied this exploratory detachment. The prematurely balding New Yorker was called "Peaked Face" by his scouts due to his unusual profile that was almost birdlike. The 24-year-old officer had graduated from West Point a year earlier.[63] Descending a high bluff that overlooked a scenic green valley, the sweaty men of two companies rode toward a welcoming grove of shade trees adjacent to the river two miles ahead as the temperature rose to triple digits.[64]

Near the Yellowstone River, Montana Territory
Monday, August 4, 1873, 11:00 a.m.

Custer's men and horses were happy to avoid the intense mid-day heat by resting under a shady grove of cottonwood trees. White cottonwood blooms fell like snow on the ground around them on this breezeless day. Next to them, the fast-flowing, greenish water of the wide Yellowstone River slid around the numerous gravel islands, only to reconnect further downstream. Unlucky soldier pickets were

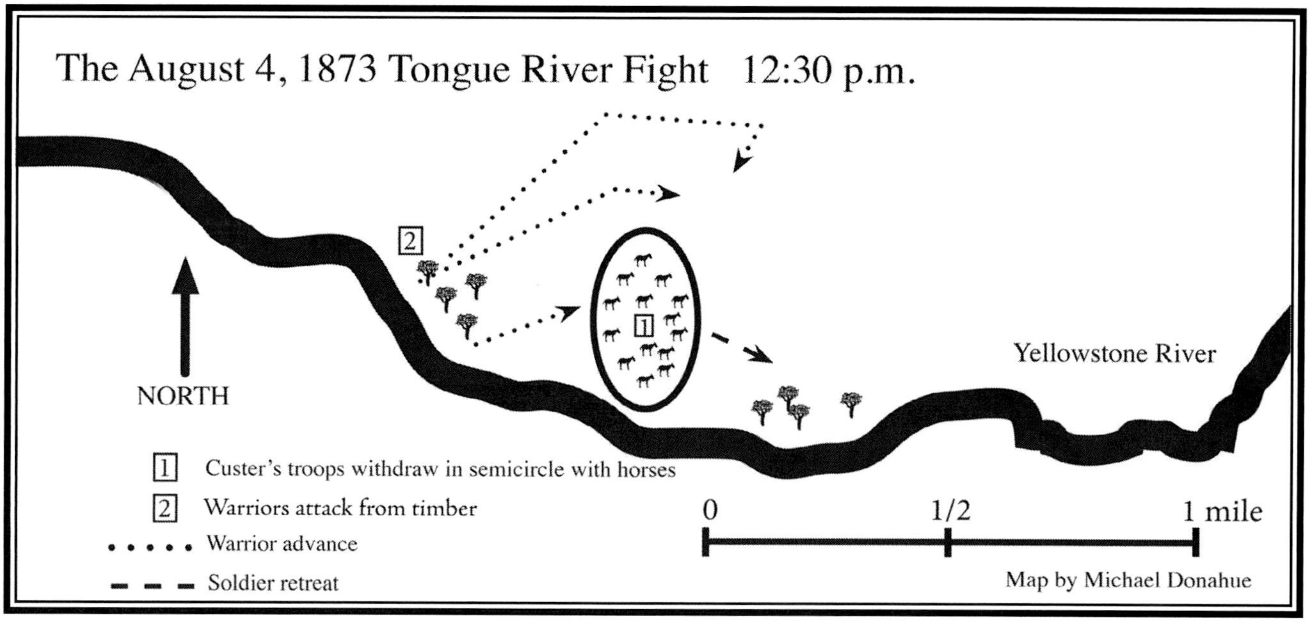

Fig. 29. Map: The August 4, 1873 Tongue River Fight. 12:30 p.m.

positioned around the shady oasis on guard duty, while the other soldiers prepared for a nap while they waited for the rest of the expedition to arrive. Some soldiers pulled their mundane diet of stale hardtack and bacon from their haversacks. Others pulled out fishing lines with hope of catching Yellowstone trout for lunch and the opportunity to have others covet their success.

Custer, wearing a scarlet shirt and knotted cravat, removed his tie and unbuttoned his collar. He pulled off his boots and used his saddle to create a pillow before falling asleep. As a safety measure, he placed his heavy 16-pound Remington rolling block rifle with octagonal barrel next to him. "I had a presentiment," stated Custer, "that we were going to have an alarm."[65] Like a mid-day slumber party, the other tired officers gathered around Custer and were resting or asleep.[66]

A corporal then noticed painted warriors galloping toward their grazing horses and discharged his weapon as a warning, shouting, "Indians are galloping this way, sir." A small party of Lakota unsuccessfully attempted to stampede the soldier mounts.[67] Custer seized his rifle and personally took a shot at the warriors, as did several other soldiers. The scream, "Bring in your horses, bring in your horses!" was heard from the officers.[68]

A trumpeter sounded *Boots and Saddles* and the two companies were formed into columns of four in the woods. Custer, with two orderlies beside him, trotted about 200 yards in advance of his troops. He quickly realized something was suspect for such a small party to attack his troop strength and believed they were decoys.[69] The six warriors rapidly turned into a dozen as they taunted the bluecoats.[70] Custer halted his favored orderly, Private John H. Tuttle, and rode around him in a circle to indicate to the Indians that he wished to parley with them; however, they were unresponsive.[71] Twenty soldiers from Company B under Tom were ordered to advance behind George Custer, leaving the rest with Moylan's Company A within supporting distance.

Fig. 30. This previously unpublished photo of Lieutenant James Calhoun (left) and Henry Harrington indicates a closeness that extended to Little Bighorn. While Harrington's body was never identified, his company was deployed and died near that of Calhoun.

Courtesy of John Doerner

As the soldiers slowed, so did the warriors, staying slightly out of reach and tempting them to follow. Custer intuitively told one orderly to return to his brother with an order to pay special attention to the timber 300 yards to the left and fall back while the lieutenant colonel rode forward. Custer was familiar with the Lakota decoy ploy that had killed William Fetterman's command only seven years earlier and believed this was another such attempt.[72] According to Charles Varnum, "about 250 warriors in line, with war-bonnets and stripped almost naked, rode up a bank out of the timber, yelling a war cry."[73]
Calhoun, in a moment of awe, described the colorful scene:

> Under any other circumstance it would have been a beautiful sight. The Indians were dressed in bright colored blankets, and as their ponies were of different hues their line presented a brilliant appearance. On they came whooping and yelling, thinking they could scatter us, and then their victory would be certain.[74]

A Miniconjou named Bull Without Hair led the attack. Another leader was an unnamed Brule warrior whose father was named Red Ears. Hunkpapa warrior Long Dog, who would fight against Custer along the Little Bighorn three years later, was also in the party that raced at full speed to cut Custer off from reaching his comrades.[75]

George Custer and his orderly were some distance ahead as the warriors angled toward him in order to cut him off from his troops. As George galloped toward his soldiers, he yelled for Tom to dismount his troops, but the order was not heard due to the excitement and distance between the two parties. Instinctively, Tom and Calhoun ordered their men to dismount, with every fourth soldier holding the horses while 15 troopers "prepared to fight on foot." The troops, mostly novices in Indian fighting, advanced several paces then dropped to one knee aiming their Sharps .50 caliber single-shot carbines at the onrushing wave of brightly colored warriors. Tom screamed, "Don't fire men, till I give the word, and when you do fire, aim low."[76]

The Hunkpapa Lakota leader Rain-in-the-Face and other warriors armed with Henry repeating rifles rode at full speed, firing at the thin blue line of men kneeling in the tall grass. George Custer and his orderly raced through the skirmishers, which had angled itself toward the onrushing warriors as 15 soldiers fired a volley of lead. The soldier firing took a toll on the warrior ponies and possibly wounded a few warriors as well. The colorful tribesmen quickly retreated out of range toward the timber that had concealed them. Moylan, along with Varnum, arrived and dismounted their men. Varnum remembered:

> Moylan handled the line and I had the left facing the timber where the bulk of the Indians were. We stood them off, but were getting short of ammunition, and I was ordered to advance, wheeling to my right and get my left on the river.[77]

In order to keep the warriors from encircling his men, Custer ordered them back toward the river and away from the open plain. While the intensity of the fight lessened, the soldiers slowly retreated on foot toward the timber where they had napped earlier. According to Custer, the retrograde movement was well organized:

> These [men] were deployed in somewhat of a circular skirmish line, of which the horses formed the center; the circle having a diameter of several hundred yards. In this order we made our way back to the timber, the Indians whooping, yelling, and firing their rifles as they dashed madly by on their fleet war ponies.[78]

Varnum was ordered to cover the retreat.[79] The troops took cover behind a semicircular natural embankment in front of the trees, while the river protected their rear. Custer relied on a successful strategy that he had used during the Civil War at Trevilian Station, when he was surrounded and outnumbered. He had his horseholders hold twice as many animals, so more soldiers could be on the firing line. This was a preventative strategy since warriors typically attempted to stampede the horses that held extra ammunition, and to dismount the bluecoats. Custer had already witnessed it minutes earlier and now reflected:

> The redskins had followed us closely . . . tempted in part by their great desire to obtain possession of our horses. If successful in this, they believed no doubt that flight on our part being no longer possible, we must be either killed or captured.[80]

Some Lakota crept through the grass to get closer to the soldiers while others bravely galloped in front of the line drawing their gunfire. Custer believed it was a ruse to exhaust his ammunition and ordered to fire only when necessary. Bloody Knife rode forward and was the first man to kill a warrior in the fight.[81] One bold warrior rode within 200 yards of the bluecoats when Bloody Knife aimed his Henry rifle at the warrior and Custer aimed at his pony. Their coordinated shots brought down the warrior along with his mount and in a characteristic show of bravery, several fellow warriors galloped on horseback rescuing their wounded friend.[82]

The great warrior Crazy Horse, who would later join Sitting Bull, was not in the fight. Sitting Bull and White Bull stayed with a reserve party of warriors near the village in case of an unexpected

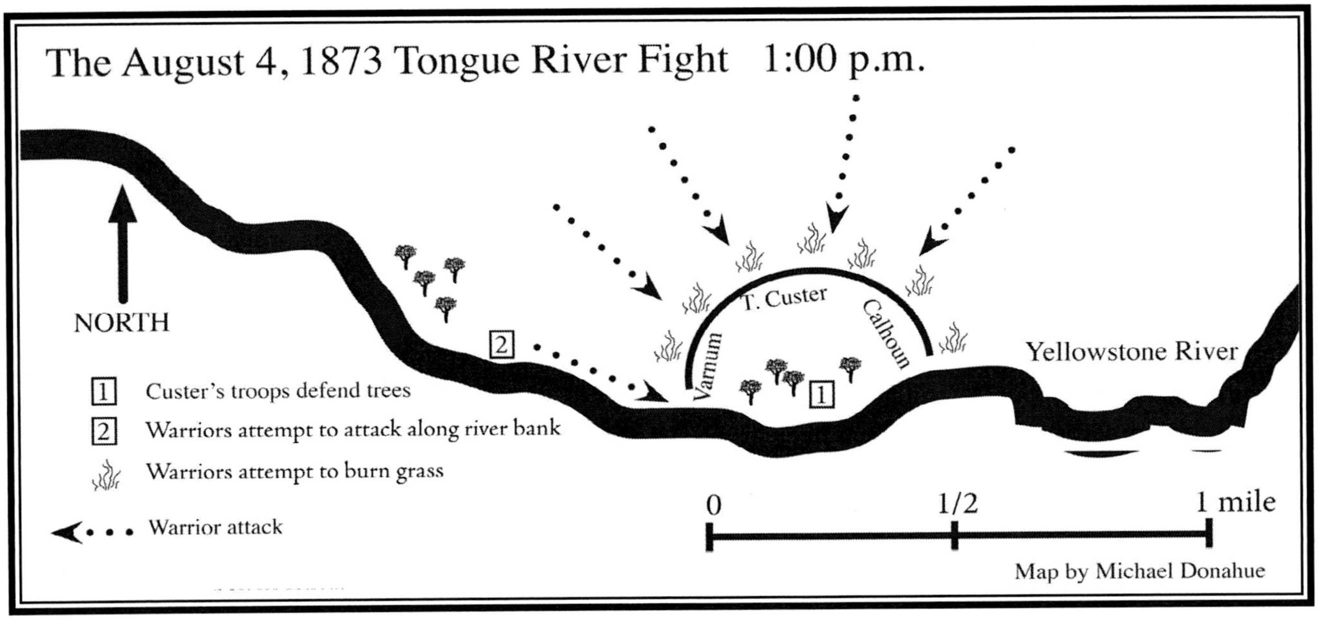

Fig. 31. Map: The August 4, 1873 Tongue River Fight. 1:00 p.m.

soldier advance.[83] During the fight, an unknown warrior was wounded and several allies collected around the man, including a war party leader carrying a lance with an attached pennant. Varnum ordered his men to fire into the party, and the man with the lance fell, shot by Private Charles Miller.[84]

Varnum protected the left edge of the timber that the Lakota seemed determined to overrun, and his duty was also to keep warriors from creeping along the river to the rear of the troops. Tom held the center of the line, while Calhoun protected the right flank, and Moylan exercised command over the entire line.[85]

Soon Custer saw smoke rising from nearby grass. He asked Bloody Knife what the Indians were doing. Bloody Knife replied, "They are setting fire to the long grass, and intend to burn us out . . . The Great Spirit will not help our enemies . . . see, the grass refuses to burn."[86]

The fire created a bluish curtain of smoke between the two opposing forces and with visibility compromised, the warriors rushed toward the soldiers opening fire. The blackened and dirty bluecoats quickly counterattacked, repeating the tactic of the Lakota, who then withdrew.

Meanwhile, there was a distraction near the bluff overlooking the battlefield, and a small party of warriors, including Rain-in-the-Face, bolted in that direction. The wagon train was closing the distance and traveling across the same high plateau that Custer had traversed earlier.[87] Unable to hear the gunfire and with the false illusion that no danger lay ahead, small parties from the wagon train went ahead to search for water in the scorching heat. Trumpeter Hills received permission to take several canteens to the river.[88] According to Private Benjamin Brown, he and John Ball had been given permission to hunt. Unaware of the danger that lay ahead, they watered their horses at a spring and separated.[89]

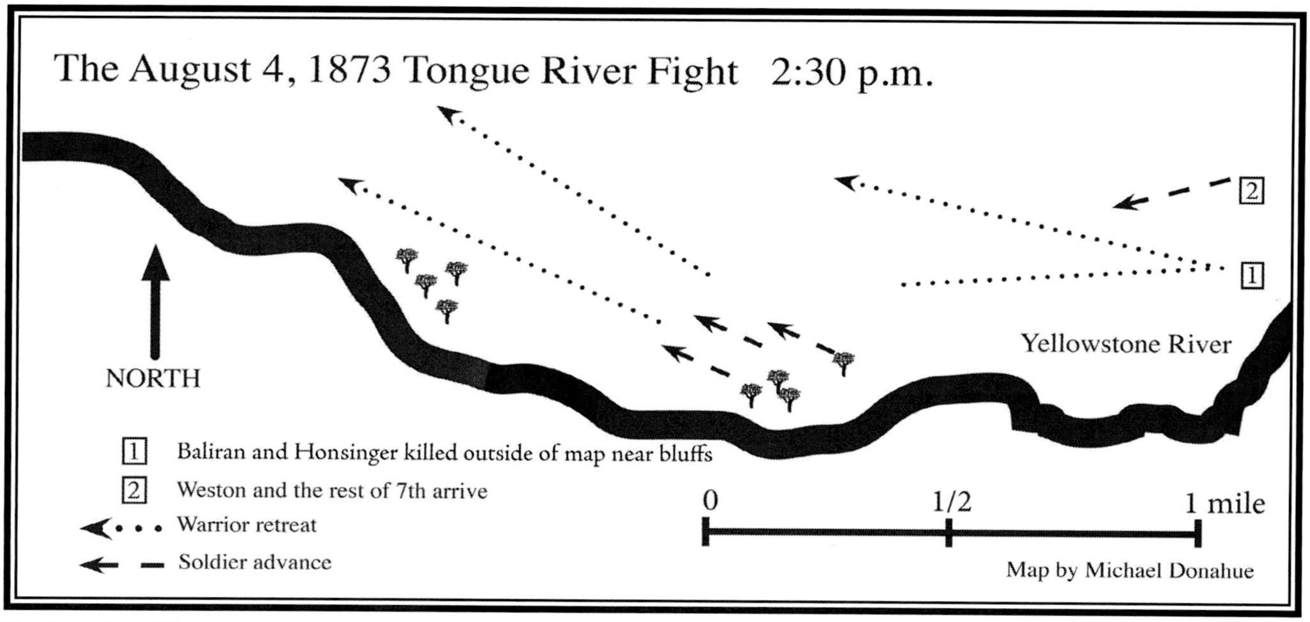

Fig. 32. Map: The August 4, 1873 Tongue River Fight. 2:30 p.m.

Around 2:00 p.m., Baliran and Honsinger also rode ahead. An alert allied Indian scout, aware of the danger ahead, attempted to warn them and grabbed the reins of their horses screaming, "Indians, Indians!" Honsinger argued, "No, no, they are cavalry."[90] The allied scout, not speaking English and unable to convince the duo, left them as they descended the bluffs. Perhaps due to the terrain and its effects on sound, none of the party seemed to hear the nearby firing or perhaps they thought the soldiers were simply hunting buffalo.

According to Brown's story, he heard gunfire and saw Baliran running on foot with two warriors shooting arrows into his back. Honsinger, also on foot, was ridden down by a big Indian and struck over the head with the stock of a gun.

The two civilians were unarmed with the exception of a single hidden derringer that was no match for warriors.[91] "When Baliran was found, there was an arrow run clear through his body and into the ground and he had hold of it with his right hand, his eyes open, dead, and the arrow still in his body," remembered Hills. "Mr. Honzinger's [sic] left hand was at his head, as he fell and was brought into camp in that way."[92] According to reporter Barrows, neither man was mutilated or scalped; he surmised that the arriving troops had hastened the departure of the warriors.[93]

Private Ball was grazing his horse nearby and was killed. Seeing Indians attacking his friend, Private Brown jumped on his bridle-less horse and galloped back to warn Stanley.[94] Ball's body lay undiscovered by the command until the return trip in September.[95]

The triple digit temperature was taking a toll on the battling soldiers nearby. A group of warriors attempted to gain access to the soldier horses by creeping along a high riverbank to the soldiers' left.

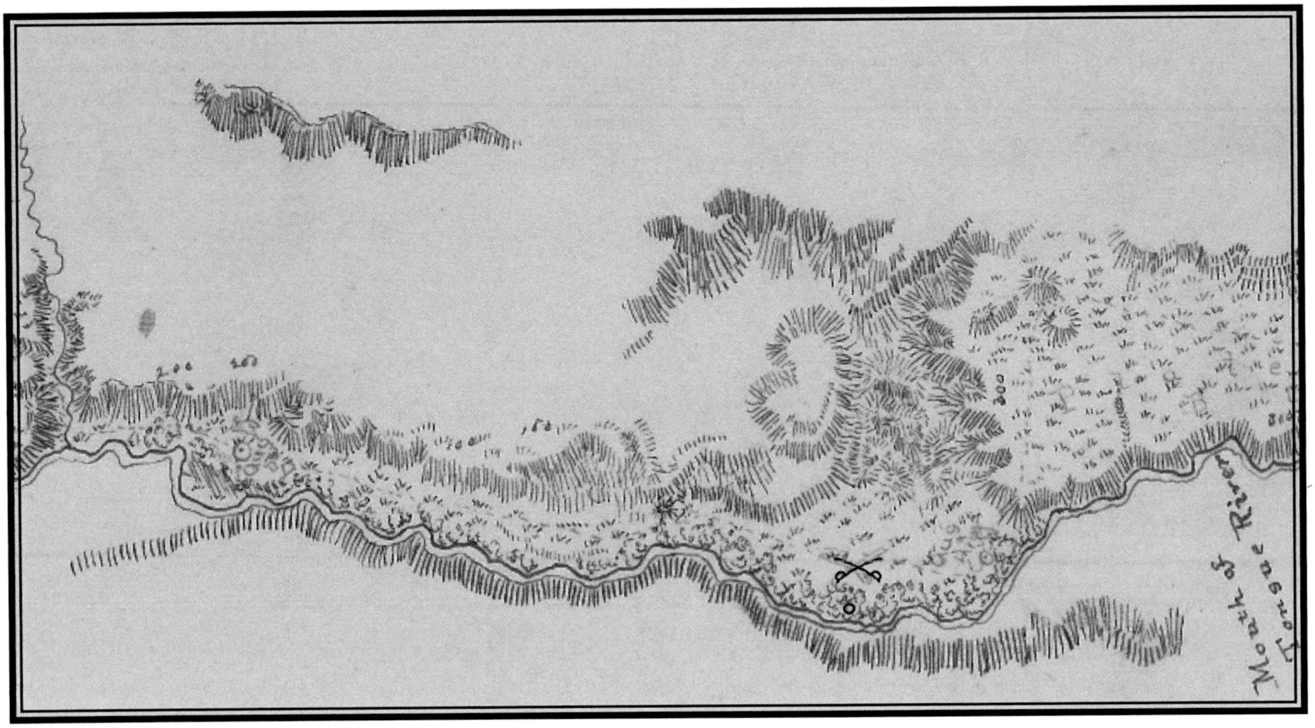

Fig. 33. Map: This is the military map made by the Stanley/Custer column noting the area near current day Miles City, Montana. It indicates the trail of the column descending the bluffs at right and the campsite of August 4, 1873. The fight site is noted by the cross sabers to slightly downstream which to date has revealed no artifacts from the battle. This could be due to changes in the river or agricultural changes. The camp site of August 5 is located at far left on the map and is the site where Honsinger and Baliran were buried.

Courtesy of the National Archives and Records Administration

A single warrior above them acted as a lookout. Custer proclaimed, "I never can see an Indian engaged in an unexplained act without conceiving of treachery or stratagem to be at the bottom of it," and in this case, he was correct.[96] Varnum drove the warrior away and stopped the advancing party. Custer, in an enlightening comment, noted, "Had they been willing, as white men would have been, to assume greater risks, their success would have been assured."[97]

The battle had raged for almost three hours, and Sitting Bull, who was apprised of the fight, told his warriors, "That is enough," and his warriors began to withdraw.[98] The likely appearance of Stanley's wagon train with fresh troops was, in reality, the impetus.
Custer recalled:

> I knew that we would soon be released from all danger by the arrival of the main body of the expedition; but a serious question presented itself. Many of our men who had been firing most incessantly now began to complain that their stock of ammunition was well-nigh exhausted . . . I sent a couple of non-commissioned officers quickly into the timber, instructing them to obtain every round remaining in the cartridge-boxes of the horse-holders and the wounded.[99]

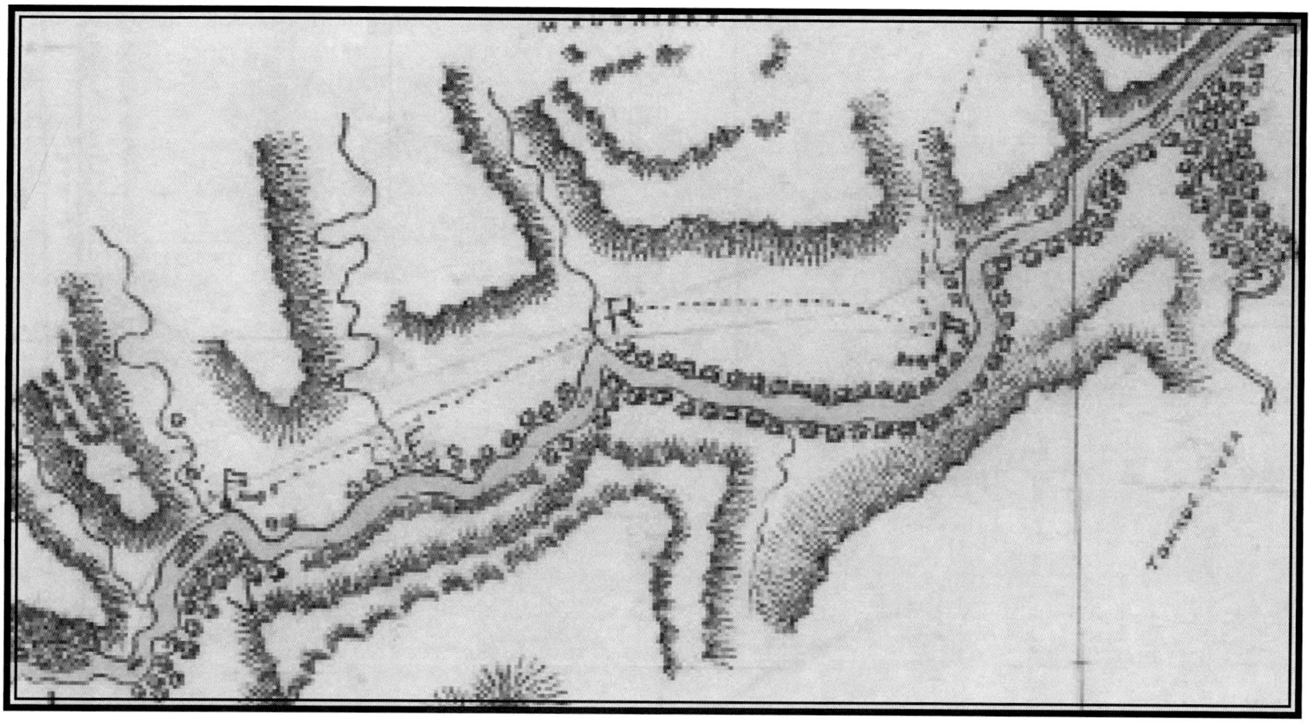

Fig. 34. Map: The Northern Pacific Railroad map that shows the same area as the map on the adjacent page. This map also indicates the campsites of August 4 and 5 in the valley near the Yellowstone River. Fort Keogh later used the area near the camp site of August 4 and many artifacts from the era have been discovered by metal detectors, but to date artifactual material from Custer's fight has not been found.

Courtesy of the National Archives and Records Administration

It was nearing 3 o'clock, and a cloud of dust was rising into the hot air above the bluff to their right. The warriors and a perceptive Custer soon realized that support from Stanley was nearby. Custer, as he had done at the Washita, decided to act on the maxim he knew worked: Do what the enemy least expects. With little ammunition, and less than 100 tired men, he decided to charge the warriors for three miles.[100] Smoke and dust stuck to their sweaty faces, making them appear as black-faced participants of a minstrel show, according to one eyewitness. With obvious contempt, Custer claimed, "They [warriors] cowardly prepared for flight."[101]

Stanley and the wagon train were several miles away on the plateau, struggling to cross a gully. Private Brown arrived screaming, "All down there are killed."[102]

Braden and Weston quickly trotted forward and found themselves on a 100-foot bluff overlooking the valley. There was no easy way down, so the troops forced the horses to slide down the incline on their haunches. Before reaching the bottom, Braden noticed five or six Indians leading several horses, including that of Balirin. "We could easily have fired on and probably killed some of the party, but we thought they belonged to the Ree scouts with us. They were in fact, as we afterwards learned, Rain-In-The-Face and a few of his warriors who had killed . . . the veterinary [surgeon] and trader and appropriated their horses and belongings," noted Braden.[103] Youthful trumpeter Hills charged down the bluffs in time to engage in the chase, recalling:

> We could see Colonel Custer and the scouting party in a fight with a large band of Indians, about 300. All the rest of the 7th Regiment came down into the river bottom about the same time we did, and when the Indians saw us coming, they fled up the valley and were soon out of sight.[104]

After the futile chase, Custer backtracked and helped the wagon train off the bluffs to a camp below. He wrote:

> My tent is pitched under the hill from which I had been so unceremoniously disturbed at the commencement of the fight; while under the wide-spreading branches of a neighboring cottonwood, guarded and watched over by sorrowing comrades who kept up their lonely vigils through the night, lay the mangled bodies of two of our companions of the march, who, although not present nor participating in the fight, had fallen victims to the cruelty of our foes.[105]

Custer exaggerated the Indian losses, noting that the number of Indians and ponies killed and wounded (about 40) were almost equaled to that of half of his force. According to a more realistic, but absent Stanley, "He [Custer] knows he killed two Indians only."[106]

The warriors were probably a mingling of bands: Brule, Hunkpapa, Miniconjou, and Cheyenne.[107] Stanley later spoke to Indian scout Antoine Clement, who told him that a Brule and son of Red Ear, as well as Miniconjou Bull Without Hair, led the attack.[108] Cheyenne Tall White Man was shot in the thigh and a warrior named Youngest Man may also have been wounded.[109] White Bull reported the three soldier/civilian deaths, but did not mention any warrior casualties.[110]

From the warriors' perspective, the event was an unsuccessful and insignificant affair. Through the historical lens, we might want to garner some great symbolic importance regarding the resistance of white invaders into sacred hunting grounds; however, this was likely not on the minds of the Lakota and Cheyenne during this coincidental meeting. The intention was to delay the army forces, give their village noncombatants time to escape, capture horses, and maybe kill or count coup on some bluecoats. The warriors were unwilling to risk a large loss of life in a situation that did not offer more lucrative results, and they quickly lost the initiative to press their advantage.

Custer's tendency to exaggerate, as well as his dramatic writing style, may have been exactly that—more drama than real danger once the soldiers were entrenched. With almost 8,000 rounds fired by the soldiers, and an unknown quantity fired by the warriors, minimal Indian and soldier casualties would indicate a long range fight with no real threat from either side. There were obviously long lulls in the fight, and it is clear that neither the soldiers nor warriors were very accurate marksmen.

The troops were armed with single-shot breech loading carbines and were fighting with warriors who carried repeating weapons as well as other cartridge firing weaponry. The organized firepower of the soldier skirmish lines was effective at keeping the warriors at a distance. Three years later, along the Little Bighorn, the soldiers would again fight some of the same warriors with single-shot weapons, but this time the warriors would have more firepower, and the terrain would be dramatically different.

The Tongue River Battle would be Custer's first fight with the northern bands of Lakota and Northern Cheyenne. He fought off a larger, well-armed force and used his knowledge of the Fetterman decoy disaster to temper his movements. Custer, with his eye for favorable terrain, masterfully used all of his skills. First, he used the river to keep his enemy from circling his troops. Second, he used the natural semi-circular rise as an embankment to protect his troops, remembering that warriors did not like to charge a position that offered protection. When needing more firepower, he ordered the number of horses held by their holders to be doubled, freeing up more men to fight. He had avoided a disaster by carefully protecting his horses and using a moving skirmish line that surrounded the animals during his retreat. The same tactic would prove less successful three years later at Little Bighorn.

Another important factor was the timely arrival of Stanley with reinforcements. Just as Bell had helped save Custer in a precarious situation along the Washita, Stanley's arrival on the Yellowstone helped to save a Custer command that was nearly out of ammunition.

The episode only reinforced Custer's negative opinion of the fighting abilities of the tribesmen. After not fighting Indians in over four years, he would have another opportunity to do so just one week later. This third experience would only reinforce his decided opinion of Indian fighting, and it would be a rematch of the very same warriors he had just chased across the northern plains.

Chapter 5

Trap Near the Bighorn River

On August 5, the expedition resumed its tiresome march; Custer had another goal besides escorting surveyors: he had permission to track the warriors who had almost killed him. There was fear the tribesmen would dig up enemy bodies and mutilate them to incapacitate the dead in the next life, so it was decided to make this prospect more difficult for Honsinger and Baliran who were wrapped in blankets and placed in an ambulance for the journey.

That evening, adjacent to the soothing sound of the Yellowstone River, in a square formed by four large cottonwood trees, a gravesite was selected and the soldiers dug a deep pit.[1] They wanted to conceal their intentions from warriors, so they waited until twilight for the funeral.[2] Luckily, a Roman Catholic priest, born in France, had journeyed alone westward and joined the expedition a few days earlier; he was called on to perform the serious task of a funeral service.[3] James Calhoun recounted:

> The priest read a chapter in Latin over the bodies, then after a short address, they were lowered into the grave and covered with earth! The band struck up a lively air, the men marched to their tents and all was over. That night the horses of a Cavalry Company were picketed over the grave, and every trace of it was thus wiped out. After this General Custer always took five companies with him, for in the event of another fight, more could be accomplished than with two companies.[4]

On August 7, one of Sitting Bull's abandoned campsites with 400 to 500 lodges was discovered near the confluence of Rosebud Creek and the Yellowstone.[5] Stanley and Custer believed the village was still in the vicinity and wanted to punish the killers of Honsinger and Baliran.[6] Sitting Bull's scouts were monitoring the soldier's movements, and the village was staying one step ahead of the slow moving blue-coat wagon train.[7]

Custer asked permission from his superior to chase the Indians, and Stanley's report bluntly stated, "I sent the Cavalry the night of the 8th to overtake and destroy them."[8] Calhoun said, "If they were overtaken, they were to be given what they so well deserved, a good thrashing."[9] Custer ordered "light marching order" with only necessary items packed on their horses because wagons would be left behind. The men were allotted 100 rounds of extra ammunition, and horses received extra forage. Three or four cattle joined the command as extra food.[10]

At 9:00 p.m. the soldiers departed under the cover of darkness led by Charley Reynolds and Bloody Knife.[11] The quiet, handsome blue-eyed scout Reynolds, sometimes known as "Lonesome Charley," was of average height and solidly built. Due to his knack of getting out of close scrapes, the Arikara called him "Lucky Man." One witness noted, "He is not bristling with knives and revolvers. His dress is such as a farmer might wear and you might pass him a dozen times on the street and never suspect he was Reynolds, the most daring, successful and popular scout of the Northwest."[12]

The Hon. Louis Clifford and Louis Molesworth, along with their university friend Robert Frost, joined the troops. Correspondent Barrows was also allowed to cover any action that might take place.[13]

A soldier remembered that 500 mounted men in two long lines followed their "eager and impetuous leader, who led us on a breathless scramble all that night down pitchy black ravines and over stony bluffs," but nature was on their side, with a full moon helping to illuminate the rough terrain.[14] The next day, the soldiers rested and remained concealed in a ravine for part of the day. The Indian trail appeared fresher with each mile they traversed. On August 9, the eight companies quickened their gait until around 7:00 p.m., when they unexpectedly discovered that the trail disappeared into the Yellowstone River. The Lakota and Cheyenne had crossed the wide river to the south bank with their families.[15]

Unable to cross the fast-running river under the cover of darkness, the soldiers decided to wait until the next morning. Many of the soldiers could not swim and "disconsolate faces could be seen everywhere. Men, who the day before were fair swimmers, now never knew how to swim."[16] An embarrassed Calhoun stated, "One would naturally suppose that where an Indian could cross a river, a white man could like wise; but it must be remembered that Indians are accustomed to swimming streams, under any and all circumstances, from the day of birth almost, and their ponies are taught to swim also."[17]

Unknown to the soldiers, the village they were chasing had been closely monitoring their movements. On the southern side of the Yellowstone River, bands of Sans Arc, Oglala, and Miniconjou had been moving slowly up Rosebud Creek from near, current day, Lame Deer, Montana. They were successfully hunting buffalo as they approached the Yellowstone River and saw their hungry friends of the Hunkpapa band on the opposite side of the river. Constructing small 'bullboats' from the hides of the bison wrapped around willow structures, they floated the meat across the river to the Hunkpapa who had been eating Honsinger's horse and could find few other animals on the northern bank. The Hunkpapa families stacked their belongings on their lodge poles and tipis, rafting the entire population across the fast-running river. It was an easy task for the experienced natives.[18]

On Sunday morning August 10, the soldiers peered across the 400-foot wide river. Two sand bars allowed the soldiers to get to the middle of the waterway. The men and horses were untrained in crossing rivers and upon entering deep water, the animals panicked and turned around. John Weston and a few brave swimmers shed clothing and swam to the other side on a log raft. They attempted to carry a long rope made of picket lines, but the swift water kept snapping the line.[19] These embarrassing attempts to get across the river took most of the day. Custer had brought cattle along and had some of them killed and skinned. Bloody Knife showed the soldiers how to bend willows and dry skins to be stretched over the basket-like frame to make 'bullboats.' It was hoped that these could be floated over the next morning with the rope cable. Warriors had been watching the soldiers from the other side of the Yellowstone and the bluecoats spotted them before sunset.

The soldiers fell into a deep slumber near the northern riverbank surrounded by 70-foot bluffs.[20] Braden placed pickets on the high ground to keep watch. At 2:00 a.m., he heard the sound of horses and surmised that some of the cavalry horses were stampeding, but one of the perceptive native scouts told him that the sound was created from Indian ponies across the river and an attack was likely at sunrise. Braden stumbled through the darkness to notify a sleeping Custer. Braden recalled, "Custer replied that he did not think the Indians were coming back. His manner indicated that he was annoyed at being awakened. I was also annoyed at the manner of my early reception, and went back to the guard."[21]

Near the Yellowstone River, Montana Territory
Monday, August 11, 1873: 4:00 a.m.

Soldiers slept under the cottonwood trees bordering the Yellowstone River in a scene similar to an afternoon nap a week earlier. The Bighorn River emptied its cool mountain waters into the wide Yellowstone only a short distance away. A mist slowly rose from the surface of the river and began to dissipate into the morning air. A few painted warriors rode up and hid in the brush on the southern riverbank. They interrupted the soldier's sleep with the 'pop-pop-pop' of gunfire. Pickets shouted, "The Indians! The Indians! They are firing at us from the other side!"[22]

The company cooks were tending their coffee pots when a sentry warned Custer there were Indians in large numbers on the opposite side of the river. The exhausted Calhoun stuck his head up, but not seeing anything, he covered up once more under his blanket, just in time to hear the 'zap' of a bullet pass right over his head. Soldiers were heard yelling as brisk firing commenced. The Battle of the Bighorn had begun.[23]

There was an air of nonchalance among the officers, perhaps knowing that the river was between them and the enemy. The bullets zipped over the heads of the soldiers for 30 minutes without reply.[24] Custer, in his official report, noted:

> No attention was paid to them, until encouraged by this, they had collected at several points in full view and within range of our rifles, when about 30 of our best marksmen, having posted along the bank, opened a well-directed fire upon the Indians, and drove them back to cover.[25]

As more warriors arrived and the gunfire continued, 30 soldier-marksmen, along with Captain Thomas French, concentrated their carbines at the warriors. Nicknamed "Tucker," French was born in Baltimore and had risen rapidly through the ranks in the Civil War as an infantry officer.[26] He was an

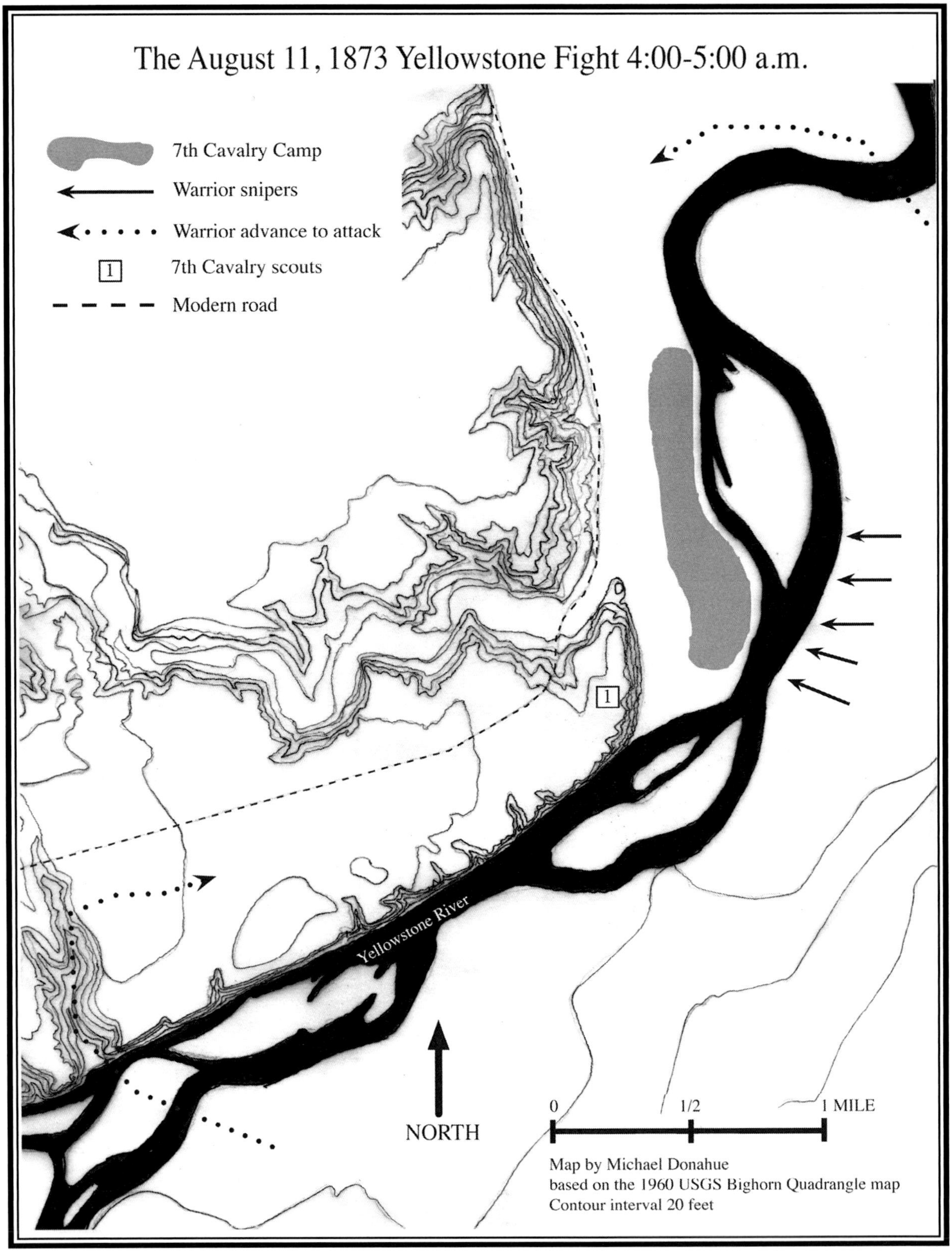

Fig. 35. Map: The August 11, 1873 Yellowstone Fight. 4:00-5:00 a.m.

incredible shot and fired a .50 caliber breech-loading Springfield rifle, hoping to cut another notch in the already scarred stock of his rifle.[27]

The Lakota, in a little psychological warfare, taunted the soldiers in Lakota and English to remember the Fetterman fight and "more Indians would be seen that day than ever before." The Lakota challenged, "We're coming over to give you hell," while the soldiers shouted back, "Shoot, you son of a bitch."[28] In a strange conversation, the inquisitive Lakota wanted to know if Honsinger and Baliran were officers. The Indians took this opportunity to describe to the soldiers the bloody fate of their missing bluecoat Private John Ball. They shouted that the soldier had fired one shot before he was killed, and then they had captured his horse and equipment.[29]

Custer's orderly John H. Tuttle was one of the best shots in the command and was firing his commander's Springfield rifle near the riverbank. Trumpeter Hills, who was watching his sharpshooting skills, laughed out loud when a warrior was hit. Custer, concerned about the well-being of the youngster, yelled for "someone to take that boy and put him behind a tree and make him stay there."[30] Tuttle hit two warriors and told a friend, "to watch him drop that Indian," and upon firing, dropped a third. Angered by this, other warriors concentrated their fire on the orderly as he was lying behind a log. A warrior bullet hit him squarely in the head, killing Tuttle instantly. The long-distance sniping back and forth across the river lasted for over two hours.[31]

The Custer brothers, Reynolds, and several other officers crawled behind another big log near French where they exchanged shots with the warriors. After the group of officers fired several shots, they moved back 50 yards and decided to have their morning coffee in the midst of the fight making for a rather surreal scene. Similar to the coffee break at the Washita, there was no reason to fight Indians on an empty stomach, and the situation seemed under control. Creighton remembered, "They commenced to fire at our cooks who were preparing our meal of hard-tack [sic] and coffee, our regular daily rations."[32] Barrows complained that he was able to grab only one piece of hardtack before he had to grab his horse and retreat. It was during breakfast when Custer was informed that Tuttle was killed. Custer later commented, "He was one of the most useful and daring soldiers who ever served under my command."[33]

The emboldened warriors found the range of the soldiers, and the bluecoats were ordered to pull back away from the riverbank. Leaves clipped by the warrior bullets fell to the ground like rain. Soldiers began to retreat slowly toward the open plain nearer the surrounding bluffs.[34]

The warrior attack was cleverly planned in three parts. They initially assembled a few warriors at the river's edge to engage the soldiers. When this did not succeed, they sent a bigger force down, hoping to draw the soldier's attention to the river. Chip Creighton wrote:

> We were compelled to move back to the bluffs and observing this move, the Indians crossed to our side of the river both above and below us, and we were finally attacked from both sides.[35]

Varnum perceptively saw the other part of the trap that was being laid. He stated, "They [warriors] sent a party to swim the river below us and attract our attention and their main body of warriors crossed the Big Horn and then the Yellowstone above and came to the attack."[36] Sitting Bull and his nephew White Bull witnessed the fight from the hills across the river.[37]

Braden and other soldiers, stationed along the southern high bluffs overlooking the Yellowstone, saw small black dots over a mile and a half downstream and signaled to Custer below. Doing what the bluecoats could not, the warriors were easily swimming their ponies across the Yellowstone.[38] Custer desired a rematch and fearlessly stated:

> It would have been possible, perhaps for us to have prevented the Indians from effecting a [river] crossing, at least when they did, but I was not only willing but anxious that as many of them should come over as were so disposed.[39]

The eight companies of the 7th Cavalry, consisting of about 500 officers and men, were divided into two wings with French in charge of the right wing that included companies under Donald McIntosh, Tom Custer, and Myles Moylan. Colonel Verling Hart commanded the left wing with companies under George Yates, John Weston, Owen Hale, and Thomas McDougall.[40]

Yates and Weston were ordered to prepare for a charge up a wide ravine that seemed to be the destination of the warriors. French was ordered to watch the smaller party of warriors crossing downstream, and he successfully shot a warrior from the saddle, adding another notch to his Springfield stock. Companies under Tom Custer and Moylan, while assigned to French's wing, remained behind in the vicinity of the headquarters group, alongside Hart's four companies.[41]

Almost simultaneously, there was an intense threat from upstream. Allied Indian scouts, who had been watching the warriors from the southern plateau overlooking the Yellowstone, came charging back toward the camp chased by large numbers of warriors.[42] Hart's wing was ordered to defend against the threat. Custer went up on the high bluff at left to monitor the situation, and using his masterful sense of terrain, made several quick decisions.[43]

The entire command below would be in a precarious position if warriors occupied the high rocky bluffs overlooking the camp. The squadron commander Yates ordered Braden to leave the guard and proceed with 20 men to take the southern bluffs. Climbing up a ravine, they reached a plateau that allowed them to view a shocking spectacle. A large party of warriors was chasing the picketed allied Indian scouts. Only 100 yards away, "Riding in our direction was what looked like the whole Sioux nation to the few of us," according to Braden. "There were probably a hundred dressed in their war toggery of paint, feathers, horns, etc."[44]

Braden was to hold the position and he quietly urged his small party to "Lie low, men . . . wait till they come close, and then let them have it."[45] The line of soldiers surprised the warriors with a volley of bullets from 30 yards away. The shocked warriors stopped as several ponies tumbled to the ground, and those dismounted jumped on the back of another's mount as they doubled up to escape. Some swerved left while the others galloped to the right. Behind the first line of warriors was another 100 who received

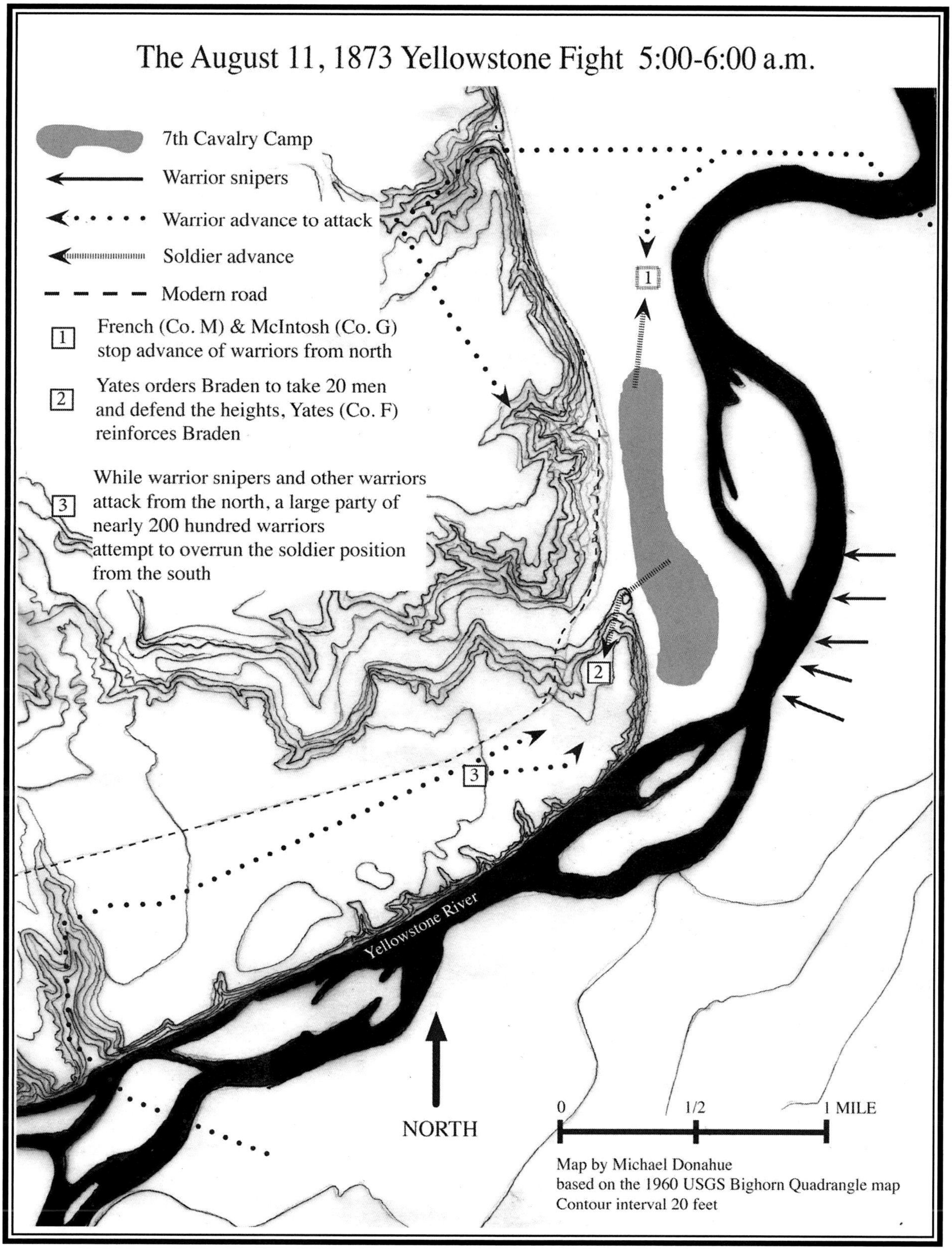

Fig. 36. Map: The August 11, 1873 Yellowstone Fight. 5:00-6:00 a.m.

the same treatment as the soldiers quickly reloaded their Sharps carbines. The bluecoats were lying down along the crest of the bluff with only their heads exposed. The returning gunfire was so intense that three or four soldiers had bullet holes in their hats, and one had the brim of his hat shot off; amazingly, no soldier was hit.[46]

Trumpeter Hills was ordered by Custer to have his Italian bandleader Felix Vinatieri move the band up on a little hill just below the point held by Braden. At some point, Braden's small party attempted to pick off warriors hiding behind rocks on the other side of the ravine near them.[47]

Braden's men fought off a third charge, and a fourth charge came closer than the first three. Concerned that the warriors had passed his left flank, he exposed himself to check out the situation. As he stood, a bullet ripped through his right leg shattering the femur and barely missing the femoral artery. He tumbled down the hill and some of his troops left the firing line to help him, but Braden screamed, "Men, I want you to hold that ridge." One of his soldiers hollered back, "You bet we will. If we move at all, Lieutenant, we'll move forward."[48]

One sergeant took off his long black silk scarf and wrapped it around Braden's bleeding leg.[49] Weston provided a blanket to four soldiers and ordered them to carry the wounded officer off the hill. As the battle raged around them, the troopers carried the poor commander across the rough terrain, but they inadvertently bounced him off a few cacti that penetrated the wounded officer's backside, causing him more grief afterwards.[50] The bluff held by Braden held a panoramic view of the entire region, and Custer decided to place his headquarters at this location near the band.[51] Yates, with his Company F, reinforced these troops and his men threw cartridges into their single shot experimental Ward-Burton bolt-action carbines.[52]

Two miles across the river, like spectators in a grandstand, women, old men, and children from the Indian village watched the spectacle and bravery of their warriors.[53] Smoke signals communicated messages to the warriors from across the river as the battle raged.[54]

Blunted by the solid defense under Yates and Braden, the warriors began to move toward the high ground overlooking the center of the soldier camp. Barrows wrote, "The second onslaught of the Indians was made on our center. Failing to dislodge Lieutenant Braden, a large party of them got into a ravine on our front and opened a galling fire upon us."[55] Yates' dismounted company was ordered to drive the sniping warriors back from the high ground overlooking the camp.[56] Other warriors returned to hidden positions as soon as soldiers withdrew.

Soldiers formed near a large ravine for a charge. The troops stood there for at least 10 minutes taking on the warrior bullets. Lieutenant Ketchum of the 22nd Infantry tightened the girth of his saddle as his horse was shot. A few feet away, Custer's horse was shot in the shoulder, resulting in spurting blood that matched his scarlet colored shirt. Ketchum screamed, "General my horse is shot." Custer excitedly shouted to Ketchum and his orderly, "So is mine, give me a horse, any horse, Dandy, the sorrel."[57] The warriors had missed Custer by inches; his luck was still intact.

Years later, correspondent Barrows wrote, "He seemed to bear a charmed life. Fear was not an element in his nature. He exposed himself freely and recklessly. In our campaign on the Yellowstone, with his bright red shirt, I protested against soliciting Indian bullets in this way."[58]

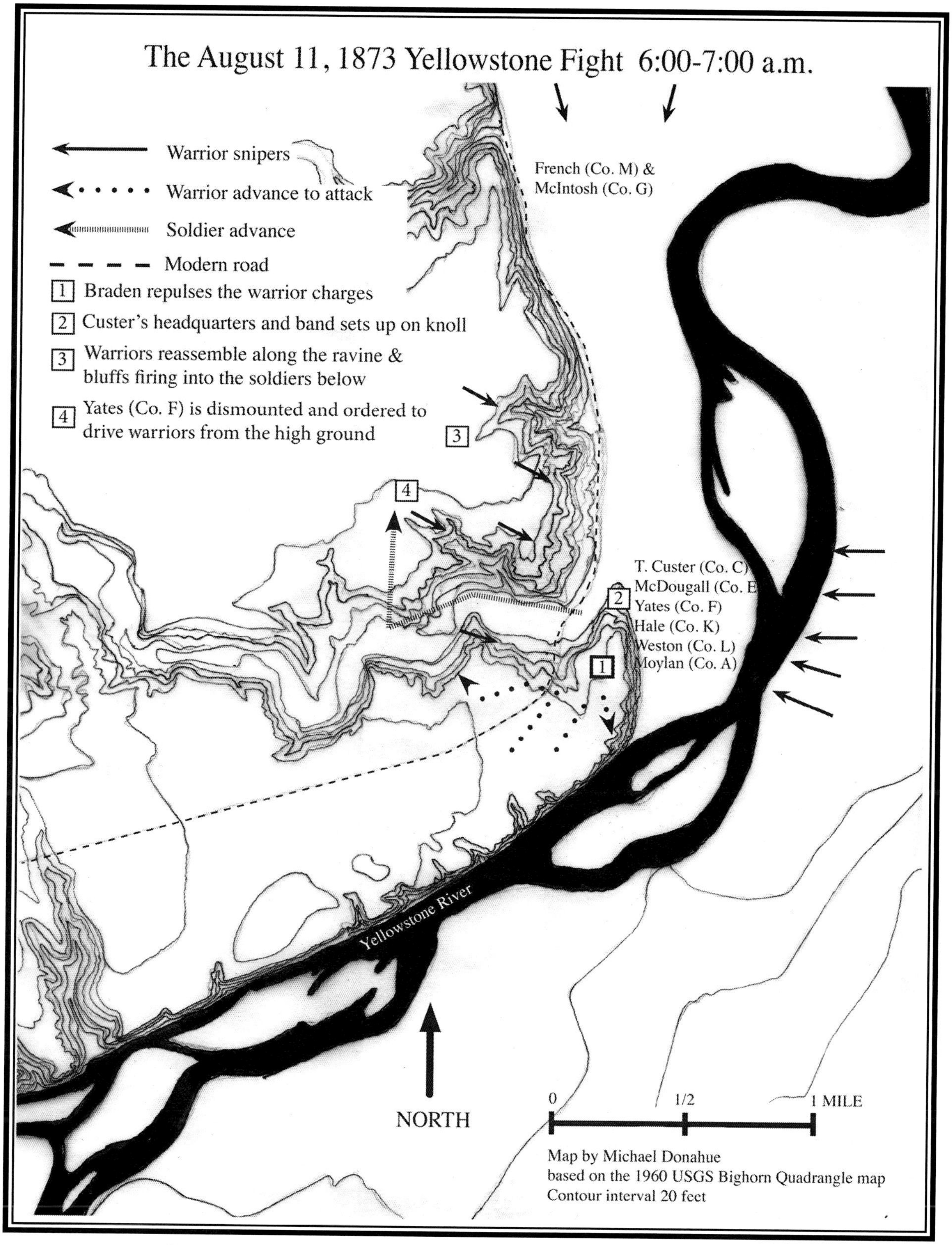

Fig. 37. Map: The August 11, 1873 Yellowstone Fight. 6:00-7:00 a.m.

Varnum's Company A formed below the bluffs, and Custer ordered him to sweep the rocky areas near the ravine. His company climbed the bluff and drove the warriors away.[59] It was now around 7:30 a.m., and the fight had been going on for over three hours. Custer felt the time had come to do what the enemy least expected: an offensive charge into the midst of a multitude of warriors. "Just before the charge was made we heard down the valley a sound, very like the report of a heavy piece of ordnance." Calhoun recalled, "and knew it must be General Stanley's command."[60] Repeating the same tactic of August 4, Custer charged the warriors, knowing that Stanley was within reinforcing distance.

Having heard the gunfire from his camp several miles downstream, around 5:00 a.m., Stanley had ordered the artillery and infantrymen forward. This was the second time in seven days they had been left out of the fight. After almost losing his artillery to a group of warriors, Stanley moved forward cautiously. The infantry arrived just in time to fire a volley into the Lakota and Cheyenne. Stanley fired several exploding shells at scattering warriors and spectators across the river around 7:30 a.m. and continued to shell them for almost an hour.[61]

Hearing Stanley's booming guns nearing him, Custer did not have to worry about warriors coming from that direction. French's troops were brought into camp to prevent an attack from across the river as the warrior snipers kept up a brisk fire.[62] As the Lakota continued firing, they were greeted with the notes of *Garryowen:*

> We'll beat the bailiffs out of fun,
> We'll make the mayors and sheriffs run;
> We are the boys no man dares dun
> If he regards his whole skin.
>
> Instead of spa we'll drink down ale,
> And pay the reck'ning on the nail;
> No man for debt shall go to jail
> From Garryowen in glory.

The cheers of excited soldiers, snorting horses, and deafening gunfire drowned out the rollicking sounds of the Irish tune as the men galloped up the gently sloping ravine. Barrows dramatically noted:

> The familiar notes of that stirring Irish air acted like magic. If the commander had had a galvanic battery connected with the solar flexus of every man on the field, he could have hardly have electrified them more thoroughly. What did it matter if the cornet played a flattering note, and the alto-horn was a little husky? There was no mistaking the tune and its meaning.[63]

One soldier had a different perspective,"Custer had the band strike up *Garry Owen [sic]*, but the noise of the firing drowned out the music, and it sounded ridiculous enough."[64]

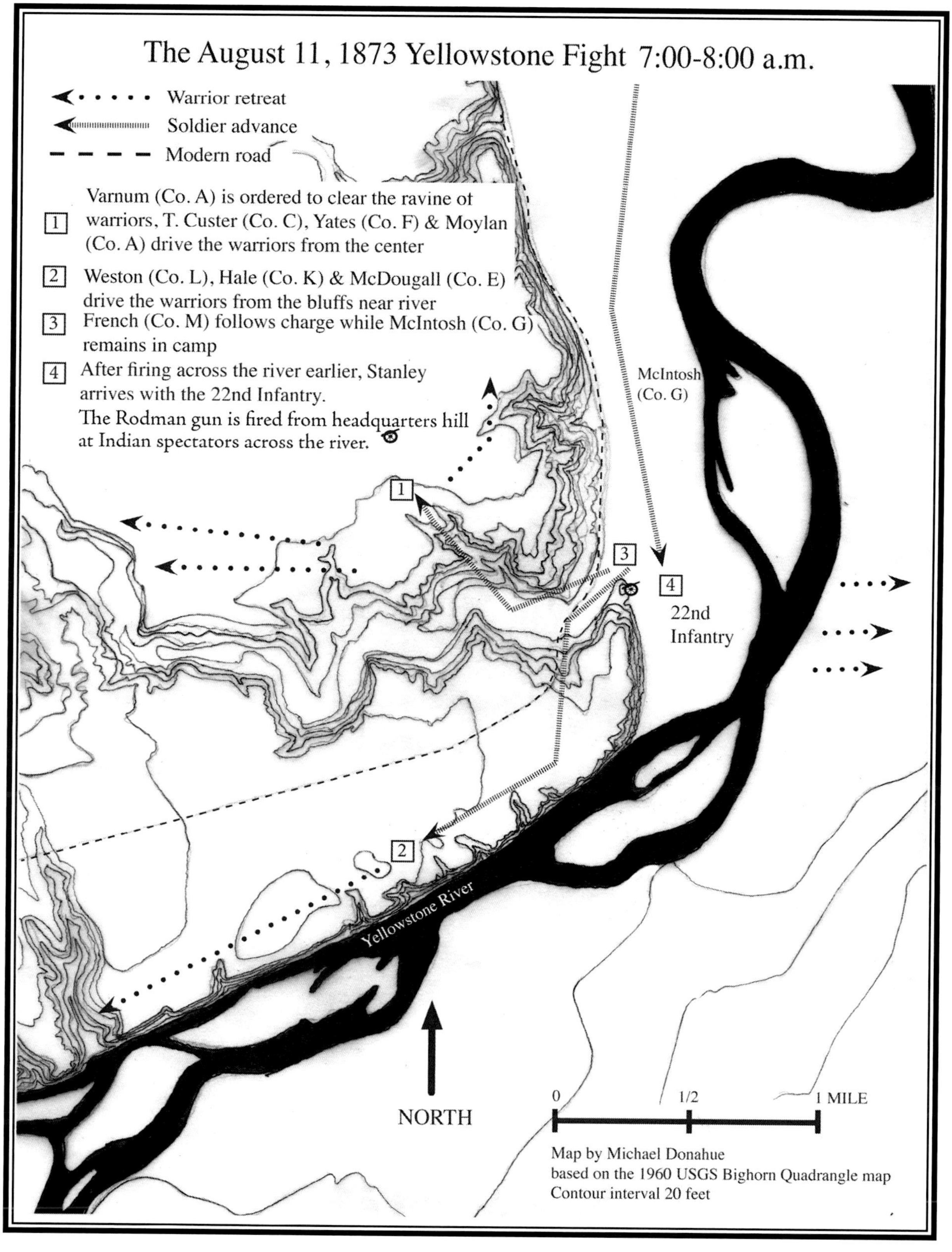

Fig. 38. Map: The August 11, 1873 Yellowstone Fight. 7:00-8:00 a.m.

Hills recalled, "While I was looking for the bandleader, my company was sent with others to charge the Indians, so I stayed near the band and watched the fight. What a sight it was!"[65] Calhoun noted, "Weston was directed to move his company up a deep ravine on our left, which would carry him to the Indian position, and as soon as the opportunity occurred, he was to charge them with all possible vigor."[66]

Trumpet calls and the scream of "Charge" were heard and each company vied for the lead position as they charged toward the painted warriors. Weston, Yates, McDougall, and Hale chased the Indians on the left and nearest the river bluffs. One warrior in a red blanket, thought to be the Hunkpapa Chief Gall, had his pony shot from under him, but quickly escaped on another horse.[67]

Tom Custer's company charged the Indians in front and center.[68] George Custer, who rarely was profane, screamed at his brother, "Charge them, Tom. Damn them. Charge them!"[69] Trumpeter Hills recalled the fight, "As the four companies charged the large band [of warriors] above us, the Indians stopped, dismounted, and fired; then mounted and retreated a short distance, then dismounted again, while the soldiers would ride up near them, then they would fire a volley into the Indians. Many an Indian was shot, and their ponies also."[70]

Moylan, Yates, and their men followed Tom as he charged toward the center. With Stanley's relief in sight downstream, Custer allowed French to join the chase as well. McIntosh also came up to make sure the snipers across the river did not get into the nearly vacant camp.[71] The warriors splintered into smaller parties and galloped across the rugged, hilly terrain. Tom's horse leapt over a ditch 15 feet wide while the soldier behind him "rolled into the gully."[72] Chasing the warriors another three miles, his company found themselves alone with a single group of warriors. The soldiers dismounted and fired as skirmishers for probably 15 minutes. Then the order to mount and charge again was given.[73]

The Lakota and Cheyenne were chased for nine miles, and the last Indian pony was killed about eight miles from the soldier camp. Once again, the warriors fled and were able to cross the river, much to the disdain of the soldiers. The two Englishmen and their friend Mr. Frost joined in the charge. Frost afterward touted, "Indian fighting was the best sport in the world."[74]

Soon after the charge began, Stanley arrived at Custer's campsite. He rolled one of the artillery pieces upon the hill where the band had played. Before sunset, Stanley fired rounds across the river into the proximity of the warriors and noncombatant spectators, causing them to flee.[75] Not to be neglected, the late arriving 22nd Infantry had missed both fights and closed the battle with several volleys from their Springfield rifles into the Indian snipers across the river.[76] Sitting Bull had not been able to stop the bluecoats as they penetrated his domain; the iron horse was coming. The Battle of the Bighorn was over.

While hundreds of bullets had been flying through the dust, there were few casualties on either side. The soldier casualties were all from the 7th. Tuttle was killed while three soldiers were wounded.[77] Custer's estimate of forty warrior casualties killed and wounded is probably high as no warriors were left on the field to verify such a number. Stanley was later told by an Indian source that they suffered four deaths and 12 wounded in the two skirmishes with Custer.[78]

Officers estimated the number of warriors to be between 500 and 1,200. Some of the warriors never crossed the river, making their numbers difficult to count. The *New York Times* correspondent noted, "some 500 Indians commenced another attack on our side of the river."[79] If this was the case, then numerically, the 7th forces were almost evenly matched with those of the warriors who crossed the river.[80] However, if all the warriors on both sides of the river are included, the 7th Cavalry was outnumbered until the arrival of Stanley's forces.

After the fight, the officers gathered around Custer's tent and were offered a nice lunch prepared by his personal cook, Mary, and all exchanged their experiences of the morning. Indeed, the atmosphere within the camp and the few casualties in the fight made the experience seem more like a hunt or the "best sport in the world." "Custer was generally regarded as very tyrannical and reckless in fighting," stated one sergeant.[81] Prophetically, later that afternoon, the same sergeant overheard three officers sitting on a log discussing the fight. One of them was Yates, who was a friend and supporter of Custer. Yates made a candid remark, "Gentlemen, it is only a question of time until Custer will get us into a hole from which we will not escape."[82]

One officer not in attendance was Braden. He was being tended to by the surgeons, who later created a contraption to keep his leg from shortening. Braden, a large man weighing over 200 pounds, would suffer in this torturous device for over 600 miles.[83] An ambulance carried the body of orderly Tuttle sewn up in canvas. He, like Honsinger and Baliran, had a delayed sunset burial under a stand of cottonwood trees. A campfire was built over the fresh grave to disguise it from any lurking warriors. The only other confrontation from the warriors would be some four or five shots fired at troops bathing near Pompey's Pillar a few days later that inflicted no casualties.[84]

Soon the expedition turned north to the Musselshell River to explore the region as another possible route for the Northern Pacific Railroad. After a brutal journey, with little water and over rugged terrain, the expedition turned again and began to make an eastward trek toward home.[85] Stanley decided that the consequences of following this path eastward were too dangerous for the wagon train and decided to retrace his route along the Yellowstone as Custer and six companies of the 7th escorted the survey party back to Stanley's Stockade.[86] On his uneventful return journey, Stanley revisited the August 4 fight site where Private Ball's corpse was discovered: a skull and partial skeleton with tattered army trousers were all that remained. The taunting warriors along the Yellowstone on August 11 had been truthful in their stories of killing the soldier.[87] The entire expedition reunited at the stockade on September 10, and Custer was allowed to ferry the 7th across the Yellowstone to the opposite bank via the steamer *Josephine*.

Custer was together again with the bitter Benteen. The white-headed captain was given the immediate task of taking the surveyors through the worst part of the journey, 'the Badlands.'[88] Benteen later bragged that "to punish me for the grand time I had at the Stockade in '73 while Custer was Mussell-Shelling [sic], I was sent thru the Bad Lands again with them on [the] return march to Lincoln. He [Custer] hadn't sense enough to know that nothing was too hard for me to do that took me away from his immediate proximity."[89] Benteen got his wish when his superior ordered him to the more desolate Fort Rice, while Custer and several of his trusted officers continued on to Fort Abraham Lincoln.

Custer arrived back at Fort Lincoln on September 21, while Stanley ferried his tired infantry troops on the *Josephine* and arrived on September 23.[90]

Upon reflection, the surveyors decided that the Musselshell route was inferior to that along the Yellowstone for laying rail. Results of the expedition rated it as a success; however, the analysis of the land's value for agriculture was mixed. Officer William Ludlow knew Custer well and recalled, "He was a natural optimist. He took rose-colored views of everything, even of the miserable lands of the Northern Pacific Railroad."[91] Stanley believed that the best location for a military post would be along the confluence of the Tongue and Yellowstone rivers.[92] The Scientific Corps could speak of little in the way of discovery.[93] The official report noted: "Gold- Some practical miners employed in the quartermaster's train frequently panned for gold, and on several occasions found decided traces of gold from the washings of sand from the small eddies of the Yellowstone above the mouth of Big Horn." However, there was no rush to the Bighorn River or national news recognition. It would be only a year later when the screams of "GOLD!" would headline newspapers and capture the imagination of the country as the same material was discovered in Custer's Black Hills expedition of 1874.[94]

The celebratory return was short lived; news of the Panic of 1873 greeted Custer as he arrived at Fort Lincoln. The fact that native tribesmen were still willing to resist encroachment by large numbers of soldiers and railroad personnel into the Yellowstone country was not encouraging to jittery investors. Custer's report of his Indian skirmishes sold more papers than reports of the natural resources and viable railroad route. The very railroad venture that they were hoping would bring prosperity to the country had instead initiated the worst financial crisis in America's history. The Stock Exchange, banks, railroad offices, and manufacturers had closed their doors. While Custer and the Railroad employees were attempting to get the trains west, overinvestment in the Northern Pacific Railway, and an inability to market several million dollars in bonds stopped it dead in its tracks.[95] Balancing the dreary newspaper headlines was the fact that Custer's old nemesis Satanta from the southern plains, had been arrested for continued raiding.[96]

The August 11th fight showed Custer's ability to use terrain effectively as he had done only days earlier. He protected his troops by using the river as a barrier and kept the warriors at a distance that affected their marksmanship. He drove warriors from the high ground above him with a small force as he had successfully done at the Washita. He was flexible in his use of troop commanders in their assignments and used them wherever they were needed to respond to the Indian thrusts from various directions. He showed patience, and like his other fight along the Yellowstone, he boldly charged when he knew Stanley was moving nearby to reinforce him.

Custer had been attacked twice during this period, rather than being the antagonist. His single-shot Sharps carbines had performed well, even against the numerous Henry repeating weapons that he encountered. He had been outnumbered both times, but his boldness and discipline had driven the warriors away. The minimal soldier casualties in the two fights made the concept of huge losses inconceivable to him. Custer believed the warriors were inferior fighters since they scampered away when he charged them both times. The hit and run tactics of the warriors frustrated the soldiers as well as their ability to do something they could not: easily cross the Yellowstone River. With the Indian families safely at a distance and across the river, the lieutenant colonel did not have an opportunity to use them as body shields or hostages. The warriors were not willing to take huge casualties but showed boldness and an attack that was thoughtfully planned.

It is with certainty that Custer recalled these Yellowstone experiences as well as the Washita fight as he rode with about 600 men of the 7th Cavalry toward Sitting Bull's camp three years later. It would be these same warriors who would not run away at their next meeting. The next time the tribesmen would be better armed, and Custer's use of terrain would be less successful. After the August 4th fight, he felt safer advancing into danger with five companies rather than two: exactly the number he would die with three years later. Custer would also attempt to sneak up on the village by traveling under the cover of darkness. While successful at the Washita, it did not work in 1873; it would again be a strategy in 1876. The village noncombatants at Little Bighorn would be in immediate danger, and the warriors would fight with a ferocity that Custer would have never witnessed during his limited Indian fighting and hostage-taking career. Most importantly, the reinforcements that arrived at the last moment to allow him to act boldly in his first three battles would evaporate along the banks of the Little Bighorn River.

Chapter 6

Paha Sapa and the Road to War

After the Tongue and Bighorn river fights, Custer's life became something of a peacetime routine in his newly-constructed home at Fort Abraham Lincoln. He was able to spend more time with wife Libbie, and from the lengthy, frequent, and loving correspondences, the marriage seemed like an ongoing love story. Several volumes have been written about their intense relationship, and it was clear that Libbie was the woman behind the man.[1]

Libbie was able to overlook his many imperfections. Sometimes she, like the rest of the Custer clan, would sometimes be the recipient of her husband's pranks and jokes.[2] That George had an eye for women cannot be ignored. His early libido was hard to hold in check. He courted several young ladies before his Civil War service. While a cadet on leave in New York, he had contracted gonorrhea from a weekend jaunt. On trips to New York, George, perhaps to make Libbie jealous, wrote about the lovely women he had met or seen. Women wanted to be photographed alongside the famous cavalier.[3] While on a political tour in 1866 with Andrew Johnson, one reporter noted that after Custer refused to give a speech, he visited the ladies and was "an especial favorite with the fair sex."[4] When he waved at a group of girls near Niagara Falls on the same tour, the women in the audience sighed loudly in response to seeing him.[5] In a revealing letter to Libbie, her husband stated, "All the women are but mere toys compared to you."[6]

Libbie was not beyond teasing her husband by sharing her experiences with many of the officers at the posts while he was away. It angered him. Libbie understood the flirtatious nature of both George and Tom and even attributed it to their half-sister who was romantically leading a young man on, stating that she was "gifted with the Custer trait-—an inclination to coquette."[7] In 1870, George apologetically admitted to some infidelity, writing:

> That however errattic [sic] or unseemly my conduct with others may have been you were still to me as you always have been the one great all absorbing object of my love . . . measured by the strict law of propriety or public opinion I was wrong. I knew it then as plainly as I know it now . . . I might extenuate but not defend it.[8]

The marriage would withstand these storms, such as possible indiscretions by George himself and his gambling debts.[9] During the Yellowstone expedition he wrote to Libbie and asked:

> Have you invited some young lady to stay with us next winter? Do not let it be of past generation. Moral worth is superior to beauty, but a young lady visiting a garrison should be possessed of both, else your officers might regard escorting as fatigue duty.[10]

Whether he sought company for his lonely wife at the garrison, was matchmaking for bachelor officers, or had a more personal reason is uncertain. In other words, he wanted any women staying in their house to be moral as well as young and beautiful; it was like a bizarre shopping list. Two months later, he wrote:

> If you want two girls with you why not bring Mollie T? Our social life will depend on the life within the garrison--and how much this is enhanced by girls attractive and intelligent of fine character. It is a shame such should be kept in the dark in nice, pleasant, but dull old Monroe.[11]

In the spring of 1875, Custer stopped in dull Monroe and visited two attractive sisters—Emma and Nellie Wadsworth. Escorting the sisters around town and to dances in Monroe, he invited them to Fort Lincoln. Nellie Wadsworth appears in seven published group photos with the Custers, and in one group photo on horseback, she is right next to Custer, while the other three women, including Libbie, sit nearby. She also appears in two images taken in the parlor of the Custer residence listening to a piano concert. In one, she is seated in their living room with the other listeners, while in the second, she stands next to George as he turns music pages for the piano player. In almost every image showing both of them, one is staring directly at the other. [12]

Fig. 39. This photograph includes Fred Calhoun looking over Emma Wadsworth's shoulder as she holds a puppy. Custer, in a provocative pose, eyes Nellie Wadsworth (far right). This image, combined with a love note from someone named "Nellie," creates some intriguing questions about their relationship.

Courtesy of the George Kush

In one intriguing photo, the two sisters sit with Fred Calhoun and George. While Calhoun is seated a respectable distance from Emma, Custer leans into Nellie's personal space with an expression of longing that is hard to disregard.[13]

An incriminating love letter discovered in Custer's personal papers by Libbie is revealing. Libbie wrote "Nellie" on the back of the note that reads:

> *Dear Gen[eral] Try and come down tonight if possible for I have many things to say to you. Remember I love you for one. Oh do not disappoint me for the love of heaven burn this as soon as you get it and oblige your own loving.*

For some reason the note survived and was not burned.[14] Does it indicate a torrid love affair, or the romantic fantasies of a young admirer? It is impossible to know, even when taken in context with the other accusations, images, and letters.

In 1874, another storm was brewing that would eventually engulf the 7th Regiment and the entire country: an expedition into the Black Hills. The Lakota translation of the area was *Paha Sapa*, and it was the early home of the Kiowa, Crows, and later, the Cheyenne. The more powerful Lakota drove them all out by 1775 in a squeezed migratory trickle-down displacement, created by ever-increasing pressure from white civilization and intertribal warfare. The Lakota adopted the Black Hills as their sacred refuge, and the incursion of the whites or other tribes was met with tremendous resistance.[15] The name of the region is derived from the lush, dark pine forests that encompass the hills, creating an illusion that they are shades of ebony. The higher elevations of the Black Hills were abundant with timber, beautiful creeks, plentiful trout, and wild game. Rumors of large deposits of the yellow metal—gold—that made white men crazy had been circulating for years.

From the U.S. Government's perspective, the Black Hills had 'officially' become Lakota land after some chiefs signed the Fort Laramie Treaty of 1868. This treaty was a concessionary conclusion of recent battles in the Powder River country, known as Red Cloud's War. Two major Lakota personalities would emerge during this period: a distinguished looking native named Red Cloud and a Brule with a deceptively passive appearance named Spotted Tail.

Fig. 40. Oglala Chief Red Cloud would fight a successful war with miners and soldiers along the Bozeman trail. Contrary to the belief of many, he was willing to sell the Black Hills, but a consensus on the price with the U.S. Government could not be reached. This led to the Sioux campaign and Little Bighorn where he was not present.
Red Cloud. National Park Service, Little Bighorn Battlefield National Monument, LIBI_00312_11174, D. F. Barry.

Born around 1822 near the Platte River in Nebraska, *Makhpia-sha,* or Red Cloud, was also known as 'Scarlet Cloud.' Those close to Red Cloud noted his father was Brule and his mother was from the Saone band. The couple likely named their son after a fiery meteorite that streaked across the night sky around his birthdate.[16] He was briefly recognized as a medicine man, but better remembered as a ferocious fighter against enemy tribes that included the Utes, Crows, Shoshonis, and Pawnees. By the 1860s, he was dedicated to stiffening the resistance toward invading whites.[17]

Spotted Tail, known to his people as *Sinte Gleska,* was born sometime around 1823 near the White River in the Dakota Territory. A multi-patterned raccoon fur gifted to the young man by a trapper inspired his name. By age 30 he was honored as a 'shirt wearer' by his Brule tribe for his bravery in battle and wisdom in tribal affairs. Originally friendly to the whites, he eventually joined the bands opposing them. His endurance and strong personality only helped to elevate his status as a Brule chief, when he returned to raid against white settlers after the Sand Creek Massacre.[18]

Gold was discovered in Montana Territory in 1862; soon small mining settlements like Virginia City and Bannock sprouted along the rocky creeks. In 1864, John Bozeman identified an old Indian trail that was a short cut that trimmed nearly four hundred miles off the longer journey to the gold fields. Nearly 3,400 gold-seekers followed the dangerous path through the Lakota's prime buffalo hunting grounds in the Powder River country.

By 1866, the army constructed three remote forts along the dangerous trail: Fort Reno, Fort Phil Kearny, and Fort C. F. Smith. Afraid to attack the fortified soldier posts, Red Cloud's warriors methodically attacked the small hay and wood cutting parties that ventured outside the stockaded walls, hoping to pressure the soldiers into leaving. Three of the most famous western Indian fights can be attributed to his efforts at resistance. In December 1866, William Fetterman and 80 men were drawn into a trap that resulted in their deaths on the snowy ridge outside of Fort Kearny. In early August of 1867, only one day apart, the synchronized Wagon Box and Hayfield fights were less successful attempts by Red Cloud to wipe out small parties near Fort Kearny and Fort Smith, respectively.

The U.S. Government determined the loss of life and the huge financial costs were just too great, so a treaty was initiated at Fort Laramie in April of 1868 to settle the issue. Over 200 Lakota, including Spotted Tail, signed the treaty in July, while a cautious Red Cloud waited until all three forts were abandoned and burned to the ground before signing four months later. The Great Sioux Reservation was created that included and protected the Black Hills from white intrusion. The reservation boundary was about half the current state of South Dakota, west of the Missouri River. A huge area of unceded hunting territory with an ambiguous boundary was established west of the reservation.[19]

While 15,000 Lakota moved inside the reservation boundaries, and many signed the treaty, Congress did not understand the individual nature of the tribal structure or its leadership values. Sitting Bull, Crazy Horse, and many other important leaders felt no obligation to abide by the treaty since they did not 'touch the pen.' The boundaries of the unceded territory, where the non-reservation Lakota called 'roamers' settled, contained perhaps a third of the entire Lakota population. A more important impetus for avoiding the reservation was the determination by the roamers to maintain their independent culture, religion, and nomadic way of life that had existed for decades. It was important that the horse raids, coups, and battles with other enemy tribes continue as it had for most of their lives.

In 1874, the federal government decided to search for a proposed fort site in the Black Hills to monitor the tribes and explore the region. Custer proclaimed his own thoughts about the upcoming excursion:

> The expedition is entirely peaceful in its object, it being the intention to explore the country known as the Black Hills and gain some knowledge as to the nature of the latter. For many years it has been believed from statements made by Indians that the Black Hills are rich in minerals . . . I say that our object is a peaceful one but I have no idea that our trip will be so . . . I will have a well equipped force; however, strong enough to take care of itself

Fig. 41. 7th Cavalry posing with its large wagon train in the Black Hills. The military mission to locate a potential site for an army fort and the discovery of gold there were the sparks that helped ignite the campaign of 1876.
National Park Service, Little Bighorn Battlefield National Monument, Little Bighorn Battlefield National Monument, LIBI_00011_07197

> and of its friends who will honor us by their presence. . . . We expect to discover a rich and valuable country.[20]

Such comments made it clear that Custer, like many other Americans, believed the economic opportunities were more important than treaty obligations. Custer, in a post-expedition interview with a newspaper correspondent from the *New York Herald*, expressed his feelings about opening the Hills, "I think it is fine country, and am anxious to see it opened to the whites, as I am satisfied it will afford a livelihood for thousands."[21] To the Lakota, the land didn't need discovery; they were more than familiar with it.

The 7th Cavalry was given only two months to get into the Hills, explore the region, and get out. On July 2, 1874, an immense wagon train with soldiers from 10 companies of the 7th Cavalry left Fort Lincoln to the chords of *Garryowen* and *The Girl I Left Behind Me*.[22] Custer encountered one small group of Lakota and seized an elder named Stab holding him hostage. Stab was used as a guide and a possible body shield until released with needed provisions when the soldiers left the Black Hills. While no fighting took place, the expedition miners discovered gold. Trustworthy scout Charley Reynolds delivered the news, and soon, miners hoping to strike it rich quickly headed toward the Lakota oasis. The tribesmen believed that the expedition was a violation of their treaty, and Red Cloud stated, "I do not like General Custer and all his soldiers going into the Black Hills, as that is the country of the Ogallala [sic] Sioux. . . . General Custer also kept Stabber, one of my Indians, in his camp for five or six days."[23]

After returning from the Black Hills expedition, Benteen and his company were detailed to deal with unrest in New Orleans. In April 1875, Kate gave birth to a son, Theodore. By midsummer, Benteen was rounding up illegal miners in the Black Hills but was unsuccessful in stopping the deluge. He once again spent a bleak winter at Fort Rice. During the winter of 1875, another tragedy stuck the Benteens when Theodore died. In a repetition of the deaths of three of his other children, Benteen was absent when this child died. Mrs. Benteen would once again have to suffer the death of another child alone and without her husband's comfort.[24]

In March of the same year, Custer invited 5,000 hungry Lakota to Fort Lincoln in an attempt to provide food that had not arrived at the Standing Rock Agency. Custer sympathetically noted, "The question of how these Indians are to be provided for threatens soon to become one not of establishing regulation—but of humanity."[25] Problems between the government entities for such relief and a last minute arrival of cattle to the reservation halted Custer's rescue attempt. Ironically, some of the same Lakota and the peace-seeking commander would soon face each other at Little Bighorn.

The Black Hills seemed to be a solid solution to the thousands of Easterners who were unemployed due to the country's ongoing economic depression. In more modern terms, the Black Hills provided a jumpstart to the nation's lagging economy, and the U.S. Government opened the door for further conflict. Custer, Sherman, Sheridan, and Grant had unwittingly set up the conditions that created the Sioux War of 1876. Even though the land was given to the Lakota by treaty, and an invasion of white miners was illegal, the army could do little to stop the flood of gold seekers.[26] President Grant and Congress decided to appease the tense situation by buying the Black Hills from the Lakota. After several attempts to negotiate with Red Cloud and Spotted Tail, a consensus on the price could not be reached.[27] An 1875 annual report created by a federal commission naively stated:

> Except under extraordinary provocation, or in circumstances not at all to be apprehended, it is not probable that as many as 500 Indian warriors will ever again be mustered at one point for a fight; and with the conflicting interests of the different tribes, and the occupation of the intervening country by advancing settlements, such an event as a general Indian war can never again occur in the United States.[28]

The extraordinary provocation would soon arrive in the form of a military ultimatum, advancing white settlement into the hills, and unity within the tribes rather than conflict. Almost prophetically, when asked if he anticipated any trouble from the Sioux in the fall, Custer stated, "Hardly this fall, as I think they will wait until the grass is plenty next spring. They will probably raise a trouble then and make traveling a little dangerous around the Black Hills country."[29] He would find trouble of his own making, not in the Black Hills, but along the Little Bighorn River, instead.

A frustrated President Grant was politically ill-equipped to deal with what he called his 'Indian problem' and abandoned his peace policy. Any legal treaty to relinquish rights to the Black Hills required three-fourths of the males to sign them over, and such was impossible with many of them roaming off the reservation.[30] In a reactionary move, he now attempted to get Sitting Bull and Crazy Horse to come in with their non-reservation bands. Grant gave all Lakota an ultimatum to return to the reservation from unceded hunting territories by January 31, 1876, or be listed as 'hostile.' The non-reservation families had no intentions of submitting to the government demands; instead, Sitting Bull and Crazy Horse were appealing to their reservation friends to leave. Since many on the reservation were suffering from hunger due to promised food and annuities that did not arrive, they were successful in getting dissatisfied Lakota to join them.[31]

While military hardliners like Sheridan and Sherman freely spouted the word 'extermination' when referring to certain groups of warriors who would not submit to reservation life or had killed soldiers and settlers, Custer's sentiment was different. In an 1876 article, Custer stated:

> 'No person who at all comprehends the necessities of the Indian question' . . . let alone anyone who belonged to 'a Christian and civilized nation' . . . could 'say a word in favor of extermination.'[32]

Campaign Against the Lakota and Northern Cheyenne
Montana, Wyoming, Dakota, Unceded Territories

When Sitting Bull's Lakota failed to surrender, Grant determined that a punitive expedition was the best course of action, and the Sioux War of 1876 commenced. Sherman and Sheridan, who had

Fig. 42. General George Crook led the Wyoming column that battled the Lakota and Cheyenne two times before the Battle of Little Bighorn. His less than stellar performance at the Battle of the Rosebud only increased the perseverance and boldness of the tribes.
Courtesy of the National Archives

successfully planned the winter campaign against the southern tribes in 1868, now planned the same against the Lakota and their allies, the Northern Cheyenne. Just as in 1868, three independent military commands would head toward the winter villagers with the hope that one or more would be able to engage the Lakota. Leaving Fort Laramie, General George Crook, Commander of the Department of the Platte, would head north into current day Wyoming and Montana. Colonel John Gibbon's force would advance east from Fort Ellis in Montana Territory, while General Alfred Terry with Custer, would depart from Fort Abraham Lincoln, Dakota Territory, and head west, along the Yellowstone River.[33]

Ohioan George Crook fought the Indians of Washington and Oregon. He wore two red badges of courage: the first received because he was shot in the hip with a poisoned warrior arrow, and the other, a rebel bullet wound from leading his Ohio infantrymen during the Civil War. He quickly rose to the rank of brigadier general, and after the war, with creatively packed mule train and Apache scouts, he whipped the Arizona Apache in 1873. Crook, while brave, was also eccentric and sometimes wore his distinctive long side-whiskers braided under his chin. He often preferred riding a mule rather than a horse. Sometimes uniformed in canvas coveralls, he painted a distinctive image that stood out like Custer, but in a dissimilar manner.[34]

Crook's subordinate Colonel Joseph J. Reynolds was the first to draw blood when he attacked a surprised Cheyenne camp along the Powder River in Montana Territory on March 17. Cheyenne warriors and their families quickly escaped the uncoordinated attack; however, the bluecoats captured 700 ponies and destroyed the village.[35] Northern Cheyenne, Wooden Leg, who was 6 foot, 4 inches tall, had been born in the Black Hills, and he recalled, "Women screamed. Children cried for their mothers. Old people tottered and hobbled away to get out of reach of the bullets singing among the lodges. Braves seized whatever weapons they may have had and tried to meet the attack."[36]

Reynolds withdrew after receiving pressure from returning warriors, and his troops were angry that their dead comrades were left unburied and mutilated by the Cheyenne. Perhaps most disheartening was the abandonment of Private Lorenzo Ayers, who had been wounded and left behind. As Reynold's soldiers withdrew, they witnessed the wounded Ayers with about 10 to 15 warriors menacingly dancing around him. The Cheyenne later recaptured 500 of their horses and struggled through the freezing cold to find Crazy Horse's Oglala village nearby, who shared their belongings.[37]

Reynolds attempted to deflect the blotched affair onto two of his subordinates by filing charges against them; however, his own poor performance was cause for a court-martial, initiated by Crook, that effectively ended his career.[38] The 1876 campaign had a pitiful start and gave an early warning both to the Lakota and Cheyenne that the bluecoats were truly hunting them down.

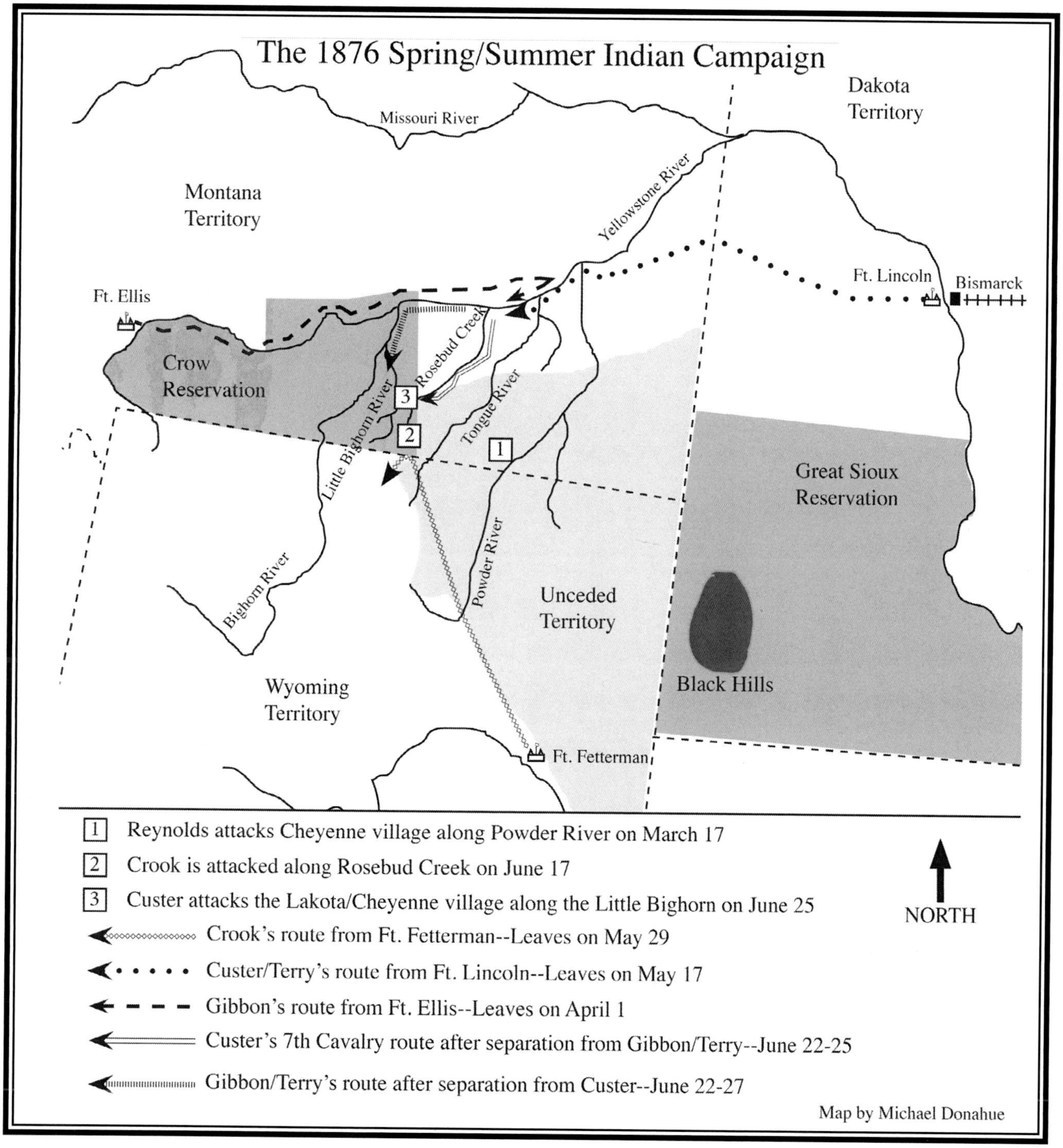

Fig. 43. Map: The 1876 Spring/Summer Indian Campaign.

At about the same time, Custer was in another battle of his own making in a series of events that would literally bring him to his knees. Secretary of War William W. Belknap assigned post traderships at the many western military establishments, and such a patronizing system set the stage for graft, corruption, and abuse. Kickbacks and pilfering were rewards for the politicians and the traders, while both soldiers and Indians suffered lost, inferior, and overpriced goods. Custer was aware of the situation and was called to Washington to testify. Belknap was indicted, and Orvil Grant, brother of the president, was also implicated. The 7th commander's hearsay testimony was critical of both, and his actions enraged the defensive president, who decided to keep Custer from participating in the upcoming campaign. The lieutenant colonel was held in detention in Chicago instead of heading back to Fort Lincoln.[39]

Grant refused to see Custer, who began an appeal process through Sherman, Sheridan, as well as Alfred Terry—chosen to lead one of the three prongs of the campaign. All three wrote letters to Grant on Custer's behalf, and Terry, who was inexperienced in Indian campaigning, wanted him along. Terry recalled that his subordinate pleaded on his knees, and "with tears in his eyes, begged my aid. How could I resist it?"[40] Custer in a letter to Grant, pleaded:

> I appeal to you as a soldier to spare me the humiliation of seeing my regiment march to meet the enemy and I not share its dangers.[41]

Fig. 44. The ever-cautious General Alfred Terry allowed Custer an independent command and had difficulty with field service due to its physical demands. He immediately began to deflect blame for Custer's defeat after his troops discovered the dead.

Courtesy of the National Archives

Grant, who was being roasted by the press for his treatment of the former Civil War hero, received the appeal from Terry that "Custer's services would be very valuable with his regiment."[42] The president finally agreed to let him go with the stipulation that he was to be under the command of the older Terry, who was to lead the forces from Fort Lincoln.

The intellectual and kind-hearted Alfred Terry was born in Connecticut in 1827. He was awarded the rank of brigadier general after his regiment, the 7th Connecticut, took Fort Pulaski, Georgia. The bearded Terry became the commander of the Department of Dakota, which included portions of Dakota and Montana Territories as well as part of Minnesota.[43]

Custer had often kept both Reno and Benteen at arm's length by distant assignments, but now all three officers would have to cooperate and rely on one another. To make matters worse for Custer, he would once again have to serve under another older officer: the ever-cautious former lawyer, Terry. Such were the tensions among the ranking officers as the Sioux campaign commenced.[44]

Fig. 45. Colonel John Gibbon led the Montana column along the Yellowstone River in an effort to keep the tribes from moving northward. Like Custer, he could not get his troops across the river to attack a village located on the southern side of the wide stream. He eventually combined forces with Alfred Terry and helped in the removal of the 7th Cavalry wounded.
Courtesy of the National Archives

While the Terry/Custer command tardily prepared to move west from Fort Lincoln, Pennsylvanian Colonel John Gibbon's Montana column took the field in April with about 450 soldiers. Gibbon graduated from West Point, and during the Civil War, the brigadier general was wounded both at Fredericksburg and Gettysburg. "No Hip Bone" or "One Who Limps," as the Indians knew him, became a colonel in command of troops at Fort Ellis, Montana Territory. Gibbon was in command of five 7th Infantry companies, as well as four 2nd Cavalry companies, under the crusty Major James Brisbin.[45] The overweight and cranky Brisbin was on crutches from rheumatism and unable to ride a horse, thus forcing him to travel by ambulance.[46]

As Gibbon passed the Crow Agency, he secured the services of the famous mixed-race scout Mitch Bouyer, as well as some Crow scouts. Bouyer was the son of a Santee Lakota woman and French father, and was able to speak Crow, Lakota, and English. A photograph of the confident scout captures his dark complexion, wide nose and smiling dark eyes that seem to reveal a pleasant personality. He was dressed in native garb that included a fur cap adorned with two woodpeckers and a single upright feather, as well as a beautifully adorned blanket.[47] Civil War infantry veteran Lieutenant James H. Bradley recorded his experiences with the Crow scouts placed under his direction; a trying job that became a clash of military discipline versus the traditional ways of the warrior. Bradley wrote that he enlisted 23 Crow guides and two 'squaw men' interpreters.[48]

Second Lieutenant Charles "Charley" Austin Booth accompanied the Montana column, and as he departed, his main concern was his wife's health since she was about to deliver their second child after having had difficulty with her first delivery. Booth was born in 1850 and graduated from West Point. He wrote, "I want to see both my wife and new baby so much; but it cannot be; instead of coming towards you I must keep going on and on away from you, for no one knows how long."[49]

Gibbon's column advanced along the Yellowstone River and received disappointing orders from the outranking Terry to halt at the mouth of the Bighorn River. The Crook and Terry commands could not take the field until mid-May, and there was a fear that the warrior numbers could be too numerous for his small command to handle. While Gibbon waited, Captain Edward Ball of the 2nd Cavalry was ordered to explore the Bighorn and Little Bighorn valleys for Indian signs.[50]

In a strange coincidence, a halt was made in the same valley where Sitting Bull would later be camped when attacked by Custer. Scout Jack Rabbit Bull took a discarded hardtack box and drew on it with a piece of charcoal. Green grass was placed in the cracks of the box as a sign to the Lakota that the Crow and the soldiers would 'clean them out.'[51] Booth recalled, "All the Sioux seem to have left this part of the country, for some unknown region to the East of here."[52]

An impatient Gibbon decided to proceed eastward, which was a timely move, since officers and scouts were beginning to question his motives and initiative. The Montana column moved along the Yellowstone until it reached the confluence of Rosebud Creek, at which point Bradley took a risky scouting venture under the cover of darkness toward the Tongue River to search for the Lakota.

As Gibbon neared the Tongue River on May 16, columns of smoke were seen rising in the distance, and it was obvious that a village was nearby. The tepid Gibbon decided it was time to cross the river and make an attack with about 400 men. Boats found near Fort Pease were used to carry saddles, equipment and men across the Yellowstone River.

With the melting snow, the river was an icy, muddy torrent 300 yards wide, which caused difficulty crossing the horses. Booth sarcastically recalled:

> We made preparations for going out to strike the camp, but all fell through because the cavalry could not get their horses across the river. They drowned five horses, and only succeeded in getting part of the horses of one Company across, though they worked about five hours. They have been grumbling for some time because they could not get a chance to go after the Sioux camps, and blaming Gen. Gibbon for their inactivity, but now, after having had a chance, and proving unable to improve it, I am inclined to think they will talk a little less.[53]

In a repeat performance by an equally determined officer named Custer, the Yellowstone River had whipped them both and left warriors on the other side watching the humorous scene. Gibbon had been on the trail for a month and a half, had located a large village, and was ready to give up his plan of attack. The river had stopped him. It was Wednesday, May 17. Hundreds of miles east at Fort Lincoln, Custer was climbing into his saddle to look for these same warriors.

Chapter 7

Two Trails to Destiny

After months of delay, General Terry led the Dakota column that included all 12 of the 7th Cavalry companies. The column also included two companies of the 17th Infantry, one company of the 6th Infantry, a battery of Gatling guns, Arikara scouts, civilian guides, and various interpreters. The departure from Fort Abraham Lincoln on May 17, 1876, began as a tardy spring campaign rather than a winter expedition.[1]

Fort Lincoln consisted of a large rectangular grassy field about 100 yards long that was surrounded by wooden structures on each side. Long plain barracks for the enlisted men stood on one side, while small beautifully crafted Victorian-style houses occupied the other. A real chasm existed between the ranks of the officers and enlisted men, not only psychologically, but physically as well.

Custer's home was situated in the middle of the neat row of officers' quarters. It was a large two story, beige-colored symmetrical frame house with white trim and four red brick chimneys jutting into the blue sky. Inside was a huge piano, wallpaper, drapes, stuffed animal trophies, books, weapons, and ornately framed photos throughout the house.[2] The genteel and soft-spoken Terry entered, and George requested that Libbie join them. She recalled:

> He [Custer] shut the door, and very seriously and impressively said, 'General Terry, a man usually means what he says when he brings his wife to listen to his statements. I want to say that reports are circulating that I do not want to go out to the campaign under you . . .
>
> . . . but want you to know that I do want to go and serve under you, not only that I value you as a soldier, but as a friend and a man.' The exact words were the strongest kind of declaration that he wished him to know he wanted to serve under him.[3]

Custer had much on his mind as he prepared for the campaign. He was anxious to redeem himself after the fiasco in Washington, and his regiment was ill-prepared for the expedition, yet his personal financial status was perhaps even more troubling. During the winter of 1875-76, Custer's poor decision-making in purchasing wild stock options had created a personal debt of $8,500, forcing him to sign a six-month promissory note at 7% interest to a Wall Street firm. Successfully asking for some time away from military service, twice, to placate the situation, Sheridan refused a third extension to Custer; even though he begged, "If I am forced to leave now I will be thrown into bankruptcy with a positive loss of over ten thousand dollars."[4] Custer was associated with entrepreneur Ben Holladay who always seemed to have some strike-it-rich-quick plan that was normally tainted and often failed. Investments for stage lines into the Black Hills, a horseshoe company, and silver mining all turned sour for Custer.[5]

Custer did have a promising financial boon that loomed in his future from a Boston speaker's bureau. The *Redpath Lyceum Bureau* was willing to sign a contract with him to lecture five nights a week at $200 a night for four to five months.[6] Custer had one big problem: he was not a public speaker. With help from Libbie and his good friend, actor Lawrence Barrett, he needed to overcome his stammering speech problems and match his writing talents with speaking skills.[7] The possibility of making $1,000 a week on his return from the campaign would have solved his financial woes, but he had to do something of notice during the summer months.

The evening before departure, both Custer and Varnum had their hair cut in anticipation of the long campaign. The long curly locks of his earlier years had disappeared. The 36-year-old Custer wore a shorter hairstyle, and the balding lieutenant colonel's hair would never become a trophy of a Lakota warrior.[8] Before leaving Fort Lincoln, Terry wrote:

Fig. 46. Reporter Mark Kellogg's desire to be near Custer's actions cost him his life during the battle. Using diaries, maps, photos, and field research, the author was able to discover Kellogg's death site in 2003 outside the current battlefield property line.

Courtesy of Sandy Barnard

> Several independent sources seem to establish the fact that the Sioux are collected in camps on the Little Missouri and between that and the Powder River . . . it is represented that they have 1,500 lodges, are confident, and intend to make a stand.[9]

Given the fact that the officers knew that each lodge normally held more than one warrior, the number they would likely encounter would be at least 1,500-2,000 warriors.[10]

The night before the departure of the troops, officers, families, and friends gathered in the officers' quarters with candles and fireplaces burning as rain pelted the thin, wavy glass of the windows. Single officers hoped for an invitation to join married officers or seized an opportunity to intrude with a farewell to those being left behind. One such youngster was 22-year-old James "Jack" Sturgis, the son of 7th Cavalry commander Colonel Samuel D. Sturgis.[11] The boyish and energetic Jack had graduated from West Point and grabbed a place as a second lieutenant

Fig. 47. At Custer's quarters, Fort Abraham Lincoln, November 1873, by Orlando S. Goff. Top row, from left to right, are: Mrs. Thomas McDougall, Captain John S. Poland, Dr. J. Van T. Middleton, Charles Thompson (son of Captain Thompson), Mrs. George Yates, Captain William Thompson, Captain Thomas Custer.
Bottom row, from left to right, are: 2nd Lieutenant Nelson Bronson, 6th U. S. Infantry; 2nd Lieutenant George D. Wallace; Lieutenant Colonel George A. Custer; 2nd Lieutenant Benjamin Hodgson, Mrs. Elizabeth B. Custer, Miss Agnes Bates (seated), Lieutenant Charles Varnum (seated), Captain George W. Yates (seated lower middle), 1st Lieutenant Thomas McDougall (seated upper middle above Yates), Mrs. Margaret Calhoun, (seated), Lieutenant Colonel William P. Carlin, Mrs. Myles Moylan, Mrs. Donald McIntosh, 1st Lieutenant Myles Moylan, 2nd Lieutenant Donald McIntosh.
National Park Service, Little Bighorn Battlefield National Monument, LIBI_00011_07125.

of the gray horse Company E. He was the youngest and last officer to join the 7th only the night before its departure.[12]

Since the garrison could not hold all the wagons, men, and supplies for the expedition, a tent city was erected south of the fort. A heavy rain turned the campground into a mud lake. While the enlisted men slept in the miserable weather, the officers had the comforts of warm beds in their homes nearby.

Among the camp residents was a 43-year-old Canadian widower left to raise two daughters. Mark Kellogg was a former telegrapher, who had moved to La Crosse, Wisconsin. During the Civil War, he reported for the *La Crosse Democrat*. By 1876, he worked part-time for the *Bismarck Tribune*, owned by Clement Lounsberry. The publisher decided it was not wise to leave his business interests behind and sent the bespectacled Kellogg as an embedded reporter. Kellogg's stories would get national attention by being forwarded to the *New York Herald*.

Knowing Custer's connections with the press and how they had defended his recent actions in Washington, Sherman advised Terry, "If you want General Custer along, [the president] withdraws his objections. Advise Custer to be prudent, not to take along any newspaper men, who always make mischief." Terry allowed Kellogg to go, ignoring Sherman's advice.[13]

On May 17, the brassy tones of reveille came at 3:00 a.m. and the soldiers were not in a good humor, as the early morning was dark, foggy, and dull.

One soldier nicknamed "Dutchy" from Benteen's company stood, eating breakfast. Charles Windolph had been born the son of a shoemaker from Bergen, Germany, and when he turned 18, had fled to avoid being drafted into the Prussian Army. Ironically, he found himself ankle deep in mud, about to go fight Indians rather than the French.[14]

General Terry suggested that Custer parade through the fort so that the worried women and children there could see for themselves what a strong fighting force was departing.[15] Sixty Arikara scouts led the column; both the white man's smallpox, and their traditional enemy, the Lakota, had decimated their tribe. This was an opportunity to get away from the reservation, make some money, and seek revenge against their foes. One private was unimpressed with the dark and medium-sized Arikara, sometimes known as 'Ree' scouts, noting, "They looked much like antiquated negro washerwomen."[16]

The other scouts in the large contingency under Varnum's charge included veterans of the 1873 expedition, Bloody Knife and Charley Reynolds. Also, along with this group, were interpreters Isaiah Dorman and Frederic "Fred" Gerard. Dorman was likely the only black man among the mounted review, and one of the few who knew Sitting Bull personally. He was born a free man and moved west to live with Inkpaduta's Lakota, eventually marrying a Santee Lakota woman. Working as a woodcutter and later a mail carrier from Fort Rice, he was eventually hired as an interpreter for $75 a month.[17] Frederic Gerard had been employed as a trader and clerk for the American Fur Company in Dakota Territory and spoke fluent Lakota and Arikara. Employed as an interpreter for the Ree scouts, Gerard also sold hand-grown vegetables to the Fort Lincoln soldiers.[18]

Custer rode behind the scouts on his white-faced sorrel, 'Vic,' [short for Victory], who had three white fetlocks as well as a white face. The proud commander pranced, with his bobbed hair carefully tucked under a low crowned brownish-gray hat to help endure the ever-changing weather of the plains. He wore one of his three fringed buckskin suits and was armed with at least one (maybe two) revolvers(s). His 16-pound octagonal-barreled Remington rifle swung from a leather scabbard that hung horizontally across the pommel of his saddle, and his Bowie knife hung in a fringed buckskin case from his tightly cinched belt.[19] Mrs. Custer was mounted on her husband's large brown horse, Dandy, and Mrs. "Maggie" Calhoun rode beside her. Both women were allowed to join their husbands for one night as they journeyed west and were to return the next day to Fort Lincoln.

Nearby, rode Custer's soldier-servant wearing a white felt hat that his commander had donated to him from his personal collection, the dark-headed and whiskered John Burkman, nicknamed 'Old Nutriment.' Burkman was responsible for taking care of the general's menagerie of dogs, horses, and the officer himself, but he had few friends in the regiment.[20]

The officer corps was next in the procession, followed by the band on their distinctive gray horses. Thereafter, the proud troops rode in column. Windolph recalled:

> You felt like you were somebody when you were on a good horse, with a carbine dangling from its small leather ring socket on your McClelland [sic] saddle, and a Colt army revolver strapped on your hip; and a hundred rounds of ammunition in your web belt and in your saddle pockets. You were a cavalryman of the Seventh Regiment. You were a part of a proud outfit that had a fighting reputation, and you were ready for a fight or a frolic.[21]

The massive train of over 200 wagons and nearly 100 pack mules strained in a serpentine formation westward through the mud toward the Heart River. The numbers of officers, enlisted men, and civilians were approximately 925 strong. Libbie and Maggie peered back at the troops and noticed an unusual mist rose above the plain that reflected a seemingly symbolic mirage of a 'ghost' column marching in the sky above the real soldiers.[22] Custer had the band play *The Girl I Left Behind Me* and *Garryowen*.

> We are the boys that take delight in
> Smashing the Limerick lights when lighting,
> Through the street like sporters fighting
> And tearing all before us.
>
> Instead of spa we'll drink down ale,
> And pay the reck'ning on the nail
> No man for debt shall go to jail
> From Garryowen in glory.

The Dakota column traveled 14 miles and camped early that afternoon next to the Heart River. Custer's younger brother, Boston, known as "Bos," was hired for an incredible $100 a month as a guide on the upcoming campaign, which was obviously a case of nepotism since he did not have such training or skill. The 28-year-old civilian became the brunt of many jokes played on him by his brothers.[23]

Captain Myles Keogh, another participant in the journey, was born in County Carlow, Ireland and, at age seven had survived the potato famine. The dapper Catholic Irishman with the goatee and moustache served in an Irish battalion of the Papal Army, fighting

Fig. 48. Captain Myles Keogh fought in over 30 battles for the Union Army in the Civil War, but missed the Washita and Yellowstone Indian fights. Evidence found after the Little Bighorn fight, as well as his position on the battlefield, indicate he may have been killed or wounded early in the battle. Photographed by D.F. Barry, circa 1872.

National Park Service, Little Bighorn Battlefield, LIBI_00011_07896 in Italy for the Pope where he distinguished himself before becoming part of the colorfully uniformed Papal Guard in the Vatican. Looking for adventure, he came to America and fought in over 30 battles with the Union Army, where he came to the attention of George McClellan, under whom he served as a staff member with Custer for two months. Keogh had a drinking problem and was known to be somewhat of a brute when intoxicated. At age 36, Keogh was sensitive, suspicious, emotional, moody, and the embodiment of the romantic sensibilities of the era. As Keogh trotted on the campaign, a Catholic medal bounced around the officer's neck. A photo of fellow officer Captain Thomas McDougall's sister was tucked in his pocket.[24]

At the first camp, a young immigrant with an unlucky past blew his F trumpet command for 'Assembly.' Born in Sala Consilina, Italy, Giovanni Crisostimo Martino was abandoned by his parents near a church as an unwanted child. The clergymen quickly baptized the baby who later became known by the names 'Giovanni Martini' and the anglicized John Martin. In March 1873, he immigrated to America landing in New York. Finding employment difficult and convinced that the army was the answer to survival in America, he found himself a trumpeter in Benteen's company.[25]

As the men camped their first night away from Fort Lincoln, the enlisted men received two months' pay from the paymaster. Custer had intentionally delayed the payments until they were away from the red-light district gambling and drinking establishments in nearby Bismarck. While he had avoided rounding up drunken soldiers and deserters, the men had no place to spend their pay, and according to Windolph, "There was a lot of grumbling by the men."[26] Paymaster William Smith disbursed a remarkable $50,000.[27]

The next morning Libbie and Maggie returned with the paymaster and his escort to Fort Lincoln. The time had unknowingly come for Libbie and George's final farewell. Burkman recounted the separation in his poor English:

> It brings a lump into my throat, rememberin' how she clung to Custer at the last, her arms tight 'round his neck, and how she cried. She want [wasn't] one to take on usually but seemed like she jist couldn't go back and leave him that day. Thar was tears in his eyes too and he kept tellin' her she was a soldier's wife, she must be a brave little woman, soon he'd be back and then we'd all have good times at Fort Lincoln agin.[28]

Finally, Libbie leaned down and placed her hand on Burkman's shoulder encouraging the orderly to take care of her husband. It was a promise that he would be unable to keep, and one that would haunt him for the remainder of his life. He remembered:

> Then we stood watchin' her ride away across the prarie [sic], the General and me. She looked so little and so young and she was leanin' way over with her head bent and we knowed she was cryin.' We watched till she was just a speck way off on the plains. Custer's face went white and he was awful sober. 'A good soldier,' he said, low and quiet, 'has to serve two mistresses. Whilst he's loyal to one the other must suffer.'[29]

Custer broke the 12 companies into a two-wing command structure with his two senior officers taking six companies each. Reno commanded six of Custer's favorite companies—B, C, E, F, I, and L in the right wing. Benteen, whose excommunication from Fort Rice had now ended, led the left wing of Companies A, D, H, G, K, and M.[30]

The early days of the trip turned into misery, as rains turned the prairie into mush and wagons sunk to their wheel hubs, and in early June, a freakish 36-hour snowstorm halted the command for two days. West Point Engineer Lieutenant Edward Maguire, with his creative ability to cross the terrain, impressed the men as they fought the natural obstacles that presented themselves daily. As Custer scouted ahead with two companies, the men followed the trail they had created with Stanley in 1873.[31]

During the journey, John Martin was ordered by Chief Trumpeter Henry Voss to act as orderly trumpeter for Custer's headquarters as part of a rotation. Voss had a reputation of being somewhat of a bully, and Martin refused the duty knowing it was not his turn. The angry chief trumpeter punished the private by tying him for two hours to the picket line for horses. Benteen intervened for his trooper and Custer promised to look into the matter upon their return to the fort. Weeks later, the same type of conflict repeated itself with life-saving consequences for Martin.[32]

By late May, Crook was once again moving northward in Wyoming Territory to resume his search for Sitting Bull. For the first time, all three columns were in the field simultaneously, but it was far from their planned winter campaign. Each unit was moving independently through an ocean of land, looking for a moving needle in a proverbial haystack. Location and communication between the acting columns was almost nonexistent, and each was on its own if attacked, but from the army's outlook, their main fear was allowing the Lakota *to escape*.[33]

Crook's strength consisted of 1,002 enlisted men and 47 officers before the arrival of a large contingency of allied Crow and Shoshone scouts. Interestingly, one of the rough and tumble teamsters was Calamity Jane, who, disguised in men's clothing was eventually sent back to Fort Fetterman, when she was discovered.[34]

Looking something akin to a grizzly bear, Crook established a camp along Goose Creek in the current city of Sheridan, Wyoming, and pushed on to the Tongue River. On the evening of June 9, his troops had a brief skirmish with Lakota warriors, with no serious injuries. Afterward, the camp took on the appearance of a picnic, with men playing the popular games of poker, whist, and checkers on homemade boards, while others hunted and fished for trout at the base of the snowcapped Bighorn Mountains. On June 14, the Wyoming column was excited by the arrival of 176 volunteers from the Crow nation, under chief Old Crow. Soon afterward, Texan and former Confederate cavalry officer Thomas Cosgrove led 86 Snake or Shoshone fighters and scouts into the camp.[35]

Meanwhile, in late May, Gibbon, who was still on the north side of the Yellowstone River, ordered Lieutenant James Bradley to scout the southern tributaries where he found evidence of the village along

Rosebud Creek. Bradley and other officers were disappointed in Gibbon's passiveness when he did not attack the camp. Indian numbers in the village were later guessed to be 400 lodges, which could be over 800-1,000 warriors, "pretty big odds," as Bradley noted. The timid Gibbon instead concentrated on sending a dispatch to Terry, who he assumed was somewhere to his east.[36] Soldier messengers were loaded into boats and directed to contact Terry or locate one of the steamboats in the area. Both the *Josephine* and the *Far West* were owned by the Coulson Line and were contracted to carry supplies for the campaign. Pilot Grant Marsh commanded the *Far West*, and had delivered supplies for the 1873 Yellowstone expedition. In 1875, Marsh commandeered the *Josephine* on a record journey up the Yellowstone, surpassing the previous record by 250 miles. Starting as a cabin boy at age 12, Marsh was legendary in his lifetime for his ability to 'navigate a steamer on a light dew.'[37]

Gibbon's messengers discovered Marsh who told them that Terry was near the Powder River, where the steamboat had unloaded supplies. After further communication Terry sent experienced civilian scout George Herendeen to tell Gibbon to remain where he was. Born in Ohio and orphaned at 13, Herendeen fought in the Civil War before trying the occupations of prospector, cowboy, and scout. The scout delivered the message to a relieved Gibbon on June 9th; the two commanders finally knew the location of the other.[38]

The intelligence about the approximate location of the tribes set the base camp at the mouth of the Powder River into a hub of activity. Fort Lincoln post trader, John Smith, who had come up the river on the *Far West*, set up a bar for enlisted men and officers at the new camp, where whiskey was sold for one dollar a pint. The heavy and cumbersome sabers carried by the troops were packed up in crates to be shipped back home. The soldiers purchased light and comfortable straw hats. Soldiers and wagons were left along the Powder River to protect the temporary depot. Custer wanted to take the band beyond the Powder River, but Terry would not consent to it; the musicians were dismounted and left behind, with their mounts given to soldiers who needed horses.[39]

Terry decided on a *reconnaissance in force*, and with the new intelligence from Gibbon that the village that had originally stood along the Tongue had moved to the Rosebud, he created a new strategy. Custer was to take six companies of the 7th as well as some infantry, along with pack mules, and scout the north side of the Yellowstone River, while Terry, with his staff and some infantry, paralleled the mounted troops, enjoying the comfort of the *Far West*.[40]

The ever-cautious Terry wanted a scouting party to search all the waterways south of the Yellowstone River and east of the Rosebud: the Powder River, Little Powder, Mizpah Creek, Pumpkin Creek, and the Tongue. Reno was chosen for the job and was to link up with Custer near the mouth of the Tongue. The major was ordered not to go in the direction of the Rosebud, as it might "flush the covey." The impatient Custer wanted to go straight toward the village along the Rosebud but Terry over-ruled him.[41] Godfrey and other officers noted, "It has been the subject of conversation among officers why General Custer was not in command but no solution yet has been arrived at."[42]

Reno took the right wing, as well as one Gatling gun pulled by two formerly condemned cavalry horses. Food for horses and men consisted of 12 days' rations. Before the Reno scout departed, it was decided that he should take a mule train to carry supplies. Unlike George Crook, who was the master of the pack train, the 7th had no idea what they were doing. Mitch Bouyer, along with some Arikara and Dakota scouts familiar with the country, accompanied the expedition.[43]

Two surgeons accompanied the Reno scout—Dr. Henry Porter and Dr. James DeWolf. Twenty-eight-year-old New Yorker Henry Porter had gotten his medical credentials from Georgetown University of Medicine and received a commendation from Crook in the Apache campaigns of 1873. He served several army contracts and was settling down in a business venture as a druggist in Bismarck when he decided to sign a three-month contract for service in 1876.[44] DeWolf was born in Pennsylvania and during the Civil War was an artillerist in the Peninsula campaigns. He later graduated from the Harvard University School of Medicine, and the doctor with the boyish looks was able to secure service as a contract surgeon.[45]

Reno and his men left on his ten-day mission on June 10, proceeding up the muddy east bank of the Powder River before cutting across to Mizpah Creek. Finding only older and abandoned Indian camps, the six companies headed toward the Tongue, where Indians were known to have camped only weeks earlier.

Riding a sorrel horse near Reno was a 5-foot, 8-inch, brown-haired Scotsman named Peter Thompson, who had moved with his family to Pennsylvania at the end of the Civil War. He listed his former occupation as miner, but it was likely that he had only done some farming, which he despised. He signed up as a private in Tom Custer's company one year earlier. Thompson quickly observed something about Reno's special relationship with his company: "Whether it was a dislike to our company or to [Tom] Custer that made him cranky with us I do not know, but the conclusion that the members of the company came to was that for the Custers, Major Reno, had no love."[46]

Tom remained with his brother, acting as aide-de-camp, rather than join Reno. Commanding the company in Tom's absence was a young West Pointer orginally from Albion, New York, named Henry Moore Harrington. A photo of the droopy-eyed and youthful 27-year-old shows him attempting to grow a thin moustache. The second lieutenant was the lowest ranking officer in the 7th that commanded an entire company, and his inability to stand up for his company would cause his men to question his 'grit.'[47] While Harrington was inexperienced, the rest of the Company C non-commissioned officers were an intriguing group of experienced veterans. Ohioan First Sergeant Edwin Bobo, a former clerk, was a veteran of the Washita campaign. The hazel-eyed, brown-headed 31-year-old was not the healthiest of officers and suffered various ailments shortly before the campaign.[48]

Thirty-two-year-old Sergeant August Finckle was also a former clerk whose national origins were betrayed through his almost incomprehensible German attempt at English. Standing a little over six feet tall, with a dark complexion, he was one of the tallest men in the regiment. Born in Berlin, he had been a captain in the Prussian Army until his enlistment in the 7th, in 1872.[49] Irish born Jeremiah Finley arrived in the U. S. in 1860 and had fought for four years in the Civil War. A Washita veteran, the former laborer had another talent that Custer discovered, he tailored the buckskin coat that his commander wore on the 1876 campaign.[50]

As Reno ventured toward the Tongue River, the Gatling gun flipped in a ravine, injuring three of the crew and continued to be a constant headache. He discovered an abandoned village site and a lodgepole trail heading west toward Rosebud Creek. Reno decided to take the initiative and go in the direction of where Bradley had spotted the village on May 27. It was this clear disobedience of orders that would change the course of history.[51]

◆

On June 17, Bouyer and the Arikara scouts found an abandoned Lakota village site along the Rosebud that contained evidence of four hundred lodges. Bouyer incorrectly told Reno that the village could be overtaken in a day's ride, but the major decided that, with low rations and large numbers of warriors ahead, it was time to return to Terry.[52] As Reno journeyed toward the mouth of the Rosebud, fresher adjoining trails indicated that Lakota were leaving the agencies and the village population was growing.[53]

While Reno did not know the exact location of the warriors, Crook did. As the major was searching the Rosebud, a battle was taking place with the elusive warriors 100 miles away on the same creek. Leaving his base camp on June 16, the optimistic Crook carried only four days of rations, since he expected to live off the captured goods of the village. Some Cheyenne hunting buffalo stumbled upon his bluecoats and galloped back to warn their village along Sundance Creek near the Little Bighorn River.

News that bluecoats were coming was unsettling, and the frightened women packed emergency supplies as others took down their lodges. The chiefs counseled, "Young men, leave the soldiers alone unless they attack us."[54] While some of the elder chiefs, including Sitting Bull, cautioned restraint, the young warriors decided to take the offense and keep the invaders away from the village.

On June 17, the meandering blue column sent the Crow and Shoshone scouts ahead to search for signs, and around 8:00 a.m., a halt was made to allow the strung-out column to close up along Rosebud Creek. While Crook and his men were playing cards and having morning coffee along the banks, a force of approximately 1,000-1,200 Lakota and Cheyenne warriors under Crazy Horse hid on the high pine covered ridges overlooking the soldier camp.[55] Oral history from Crazy Horse and his family revealed the plan of attack:

> A small decoy party would be sent out to lure the soldiers into the trap, just as they had done successfully against William Fetterman in 1866. The warriors had not expected the soldier's allied scouts to reconnoiter the area before the decoy force could be mustered, and spying the huge numbers of warriors in the trees, the scouts quickly rode for their lives.[56]

Crook looked up to see his scouts coming back, "as fast as their ponies could carry them, followed closely by the hostiles, both yelling at the top of their lungs, giving the war whoop that caused the hair to raise on end."[57]

Crook had never witnessed such large numbers of warriors attack troops with as much extreme boldness. At some point early in the battle, an act of bravery occurred that etched itself into Cheyenne oral history. Cheyenne warrior Comes-In-Sight's horse was shot from under him, leaving him in danger of being killed by the bluecoats and their allies. A rider quickly galloped through the hail of lead and the young warrior was able to jump on the back of the pony behind the rider and trot to safety. The daring rescue was performed by none other than his sister Buffalo-Calf-Road-Woman, and to this day, the Cheyenne refer to the affair as the battle 'Where the Girl Saved Her Brother.'[58]

Crazy Horse believed his medicine made him invincible, and he galloped around a little hill, only 300 yards from the soldier firing line. One of Crook's interpreters and guides, Batiste Pourier, knew the great warrior and, "fired at him four times but missed him and I was a good shot [in] them days too."[59]

This was an action that Crazy Horse would reenact only a week later against Custer. A seesaw fight by warriors and soldiers continued with soldiers driving the warriors before them, only to have the warriors take the initiative and attack the soldiers as they withdrew.[60] Sitting Bull's nephew White Bull decided to dress in his best clothing in case he was killed. Attired in matching leggings and moccasins of striped blue and white beads, he also wore a borrowed long red flannel headdress with alternating red and white feathers that touched the ground. Such fine war clothes "Made a man more courageous," he thought. White Bull honored his people by wounding one Shoshone scout with his repeating rifle, as well as rescuing three warriors in distress.[61]

About 225 men and officers, in an uncanny semblance to the numbers that the 7th Cavalry would have a week later, charged the warriors. The warriors continued to retreat and the aggressive bluecoats continued to stretch their blue line like a rubber band about to break. In a foreshadowing of Little Bighorn, the Indians, seeing the separated troops, closed in on the small groups of soldiers, like a bear trap. The soldiers withdrew with great difficulty and sustained casualties in the almost hand-to-hand combat. The Southern Cheyenne Lame White Man encouraged the men to be brave, and it was noted by Wooden Leg that, "Our young men had high regard of him. The Cheyennes came out from hiding and went flocking to him . . . it then became the turn of the soldiers and their Indians to get out of our way."[62]

Fig. 49. White Bull. Sitting Bull's nephew fought bravely in the Rosebud and Little Bighorn fights. He counted numerous coups during the Little Bighorn fight over the entire battlefield. During the 1873 Yellowstone campaign, he was in Sitting Bull's camp and in the proximity of both August fights with Custer.
National Park Service, Little Bighorn Battlefield National Monument, LIBI_00153_06700. G.J. Murray, 1926

The Battle of the Rosebud was over by 2:30 p.m. Crook had suffered 10 killed and 21 wounded. While there are no specific numbers of warrior casualties recorded, it is likely they experienced losses that were similar.[63] One Sans Arc casualty was taken back to the village on Sundance Creek and buried in a beautiful burial lodge.[64] Crook, locked in his

Civil War mentality, believed he had won a victory, since he held the battlefield. The unemotional, cool general had a horse shot from under him in the melee and limped back to his base camp along Goose Creek, where he contented himself with hunting and fishing. The warriors, having halted the soldier advance, felt they had won a great victory and were emboldened by their success. More importantly, the battle convinced the warriors they were invincible and gave them valuable experience about how the cavalry fought on broken and hilly terrain. They would use that knowledge a week later along the Little Bighorn.

Meanwhile, Reno left Rosebud Creek and rode toward the mouth of the Tongue for a rendezvous with Custer and Terry. Custer hovered miles away, near the area where he had almost been killed on August 4, 1873. Scaffold burials, as well as remains of a soldier, were discovered in an abandoned village site, and Custer stood silently staring at the bones of the cavalry soldier. He, along with his relatives, plundered the dead warrior graves.[65] One Arikara scout recorded that Custer ordered Isaiah Dorman to take a warrior body from a scaffold to look at, and Dorman later threw it into the river, using it as bait as he fished in the area.[66] Boston explained, "Armstrong, Tom Custer, and I pulled down an Indian grave the other day. Autie Reed got the bow with six arrows and a nice pair of moccasins which he intends taking home."[67]

Eighteen-year-old Harry Armstrong "Autie" Reed was proudly named after his famous uncle. Custer's nephew had been hired to herd cattle, but by the time they arrived at the Little Bighorn, he was no longer needed in that capacity and enjoyed a freedom of sticking close to his famous relative. The frail youth was sometimes sickly, but was liked by the officers.[68]

Godfrey was disgusted by the desecration of the graves, and later wrote:

> A number of their [Indian] dead, placed on scaffolds, or tied to the branches of trees, were disturbed and robbed of their trinkets. Several persons rode about exhibiting trinkets with as much gusto as if they were trophies of their valor, and showed no more concern for their desecration than if they had won them at a raffle. Ten days later I saw the [soldier] bodies of these same persons dead, naked, and mutilated.[69]

As Reno neared the Tongue after his ten-day scout, he sent a messenger to Terry stating:

> I am in camp about eight miles above you. I started this a.m. to reach your camp, but the country from the Rosebud here is simply awful . . . I enclose you a note from Gibbon, whom I saw yesterday. I can tell you where the Indians are not, and much more information when I see you in the morning. I take it the Tongue River is not fordable at the mouth, and I will necessarily have to camp on this side. . . . My command is well. I will be at the Tongue River opposite your camp about 8:00 a.m. [June 20]. Many animals are leg-weary and need shoeing. We have marched near to 250 miles.[70]

Reno would soon wish the river was wider, after the reception he received from his infuriated superiors. He believed that his intelligence would outweigh his disobedience, but such was not the case. Hoping for surprise, both Custer and Terry were afraid that Reno's unauthorized scout might have warned the Lakota that they were closing in. Terry's anger was recorded in a letter to his sisters:

> To my great surprise I received a note from Colonel Reno which informed me that he had flagrantly disobeyed my orders, and he had been on the Rosebud, in the belief that there were Indians on that stream and that he could make a successful attack on them which would cover his disobedience. . . . He had not the supplies to go far and he returned without justification for his conduct unless weary horses and broken down mules would be that justification. Of course, this performance made a change in my plans necessary.[71]

According to Thompson:

> . . . confirmed by those in position to know, that when Reno made his report . . . Custer upbraided him very bitterly for not finding out the exact number and the direction the Indians were taking instead of supposing and guessing. There were some sharp questions and short answers; but Terry interposed and smoothed the matter over.[72]

Custer thought that Reno should have attacked while on a hot trail, and Reno was damned for what he did as well as for what he didn't.[73]

In a letter from Cooke to his family, he stated the situation was serious and that formal charges were pending. Both Terry and Custer severely censured Reno for jeopardizing the entire campaign with his disobedience, and Terry wanted to punish Reno by leaving him behind. Reno went to Custer, begging to accompany the command. Ironically, Custer, who had just gone through a similar humiliation with Grant, appealed to Terry on the major's behalf. With reservation, Terry finally agreed to let Reno accompany the command in an autonomous capacity. Keogh and several other officers protested the decision, but Custer wanted to let nothing mar the campaign's success.[74] Custer stated, "I feel hopeful of accomplishing great results. I will move directly up the valley of the Rosebud."[75] The Indian trail headed toward the headwaters of that stream or possibly toward the Little Bighorn.

Gibbon's column finally united with Terry's command at the mouth of the Rosebud on the afternoon of June 21, as the *Far West* moored alongside Custer's camp and they unloaded forage and rations for a 15-day march. A conference of the officers was held on the steamer, and peering over the 1859-60 Raynolds-Maynadier map, Custer, Terry, Gibbon, and Brisbin pierced it with tacks, indicating their respective marches to the Little Bighorn Valley.[76] Gibbon and Terry decided to approach the Little Bighorn Valley from the north, while Custer approached from the south. The plan of action was to make 30 miles each day, but the slower infantry would have to pace themselves, however, as no one knew for certain the location of the village. The 7th was to march southward to the headwaters of the Tongue, in a delaying move, in order to place the two converging units in the vicinity of the Little Bighorn Valley. A perceptive Terry warned Custer that he, "should keep constantly feeling to his left, and even should the trail turn toward the Little Big Horn [that] he should continue southward along the headwaters of the Tongue." Terry feared that Custer's zeal would "carry him forward too rapidly."[77]

Another favorite camping site for the Lakota was a small waterway called Tullock Creek. Knowing this, Terry ordered Custer to scout the area and send George Herendeen to the combined Gibbon/Terry column with intelligence as they passed the area. The adventurous scout was anxious for this task, since he was offered a $200 bonus to accomplish this dangerous mission.[78]

Lieutenant Bradley stated, "It is understood that if Custer arrives first, he is at liberty to attack at once if he deems prudent. . . . We [infantry] have little hope of being in at the death, as Custer will undoubtedly exert himself to the utmost to get there first and win all the laurels for himself and his regiment."[79] Terry in one of his dispatches hoped, "That one of the two columns will find the Indians. I personally go with Gibbon.[80] Kellogg seemed to reinforce such a move in his last letter:

> Tomorrow, June 22nd, General Custer with 12 cavalry companies will scout from its mouth up the valley of the Rosebud until he reaches the fresh trail discovered by Major Reno, and move on that trail with all the rapidity possible in order to overhaul the Indians whom it has been ascertained are hunting buffalo and making daily and leisurely short marches. Gibbon's part of the command will march up the Big Horn valley in order to intercept the Indians if they should attempt to escape from General Custer down that avenue.[81]

Terry gave Custer written instructions that indicated he wanted him to explore the area south of the Little Bighorn to keep the warriors from escaping in that direction; on the other hand, he also stated that he did not "Wish to impose precise orders" that could hamper Custer when near the enemy. The phrase, 'Conform to them unless you shall see sufficient reasons for departing from them,' was so subjective in nature that Custer must have smirked with the inherent freedom of the line.[82]

Mary Adams, Custer's cook, supposedly overheard Terry say to Custer, "Custer, I don't know what to say for the last." Custer replied, "Say whatever you want to say." Terry then said, "Use your judgment and do what you think best if you strike the trail. And whatever you do, Custer, hold on to your wounded."[83]

Terry knew that Custer was still stinging from the recent controversy with Grant and told Brisbin after the conference that Custer was, "smarting under the rebuke of the President [Grant] . . . wants an independent command, and I wish to give him a chance to do something."[84] He also stated honestly, "I have had but little experience in Indian fighting, and Custer has had much, and is sure he can whip anything he meets."[85]

Calhoun also knew that the Grant chastisement was on the mind of his brother-in-law and stated, "In my next letter you may expect to hear of a great success. All are in good spirits, and the leader [Custer] is not feeling so very much disgraced as it was thought he would."[86]

Major Brisbin suggested to Terry that he lead the Rosebud advance and believed the 7th Cavalry should be supplemented with the four companies of his 2nd Cavalry. After a short debate, Terry, the ultimate gentleman and politician, suggested that Brisbin could offer his companies to Custer. The lieutenant colonel's rejection was curt and to the point: The 7th can handle anything it meets.[87] After Brisbin departed, Lieutenant John Carland, who overheard the offer, recalled Custer proclaiming, "That was very clever of Brisbin to offer to go with me but you know, Carland, this is to be a 7th Cavalry battle, and I want all the glory for the 7th Cavalry there is in it."[88]

Custer originally accepted the Gatling gun detachment, but after overhearing of Reno's problems and getting advice from Gerard to leave them behind, he decided against bringing them.[89] Officers' call was sounded and there was a large assembly at their commanding officer's A tent situated near the banks of the Yellowstone River. Godfrey recalled:

> [Custer] gave us our orders. We were to transport, on our pack-mules, 15 days' rations of hard bread, coffee, and sugar; 12 days' rations of bacon, and 50 rounds of carbine ammunition per man. Each man was to be supplied with 100 rounds of carbine and 24 rounds of pistol ammunition, to be carried on his person and in his saddle bags.[90]

Each man was to carry on his horse 12 pounds of oats and extra forage for the mules. Godfrey and Moylan grumbled about carrying so much weight.[91] An angry and excited Custer sarcastically warned them:

> Well, gentlemen, you may carry what supplies you please; you will be held responsible for your companies. The extra forage was only a suggestion, but this fact bear in mind, we will follow the trail for 15 days unless we catch them before that time expires, no matter how far it may take us from our base of supplies; we may not see the supply steamer again.

He also told them that he would follow a hot trail, "If it took us to the Indian agencies on the Missouri River or in Nebraska."[92] As he turned to enter his tent, Custer added, "You had better carry along an extra supply of salt; we may have to live on horse meat before we get through."[93] The officers took it literally and packed extra salt.

Second Lieutenant Winfield Scott Edgerly, a tall West Pointer and veteran of both the Yellowstone and Black Hills expeditions, stretched out in his tent. Custer came by to converse with him and stated, "I expect a great deal from you young officers."[94] Edgerly assured him that he would not be disappointed. The two-battalion wing formation was abolished and each company commander was made responsible for the condition of his men and the grazing and watering of horses. Benteen was simply in charge of his own company, and Reno was banished into command limbo.[95]

Approximately 178 mules were selected to carry food, forage, ammunition, and other supplies to speed up the pursuit. Each company was required to pack their own mules, under the direction of a few civilian packers.[96] Soldiers quickly wrote letters home, realizing that it might be the last chance to send mail for some time via the *Far West*. Some of the men seemed to have premonitions of death and disaster; some drew up wills, while others made arrangements with friends for their belongings, should the worst occur.

The commander of the 7th requested some Crow scouts from Gibbon, knowing how important their extensive knowledge of the unknown terrain would become. The limping colonel begrudgingly gave up six of his best guides, as well as their interpreter Mitch Bouyer. The older Half Yellow Face led the group, which included White Man Runs Him, Hairy Moccasin, White Swan, Goes Ahead, and the youngest, Curly. The Arikara, as well as some Dakota scouts, Bloody Knife, and their interpreters saddled up for the journey.[97]

As men prepared for the upcoming expedition, a critical encounter took place on the evening of June 21st that opened old sores. One officer recalled:

> Custer and Benteen engaged in some personalities and recriminations. Benteen said that if they were to get into a fight he hoped he would be better supported than he was at the Battle of the Washita. Custer then twitted Benteen of shooting an Indian boy in the battle, and Benteen went on to explain why he had to do so to defend his own life. [Lieutenant Richard] Thompson says the discussion of matters between Custer and Benteen . . . waxed rather warm at this time, and it was plain to be seen that Benteen hated Custer.[98]

Here, only days before a major battle and eight years after the event, memories of the Washita fight were in the minds of both officers. The friction between Custer and Benteen festered for the next four days, and the white-headed captain's hatred of Custer would continue until June 22, 1898, the day of his own death.

While Custer penned a letter to Libbie, some of the officers decided to have a poker party aboard the *Far West*. After the poker game, Keogh had former lawyer and infantry officer John Carland create his last will.[99] He was not the only officer who decided that it might be important to get life insurance from New York Life. George Custer, Yates, Calhoun, and Lieutenant James Porter all purchased policies worth $5,000 each, while Keogh extended the value of his policy to $10,000.[100]

That Montana night was dark and cloudy, with a few raindrops pelting the dirty canvas of the tents. Strong breezes rustled the cottonwood leaves along the riverbank where they were camped. Burkman described finding Custer the next morning, "He must o' jist dropped off. He was hunched over on the cot, jist his coat and boots off, and the pen still in his hand."[101]

Before his departure at 11:00 a.m. on June 22, Custer wrote his final note to his wife:

> *My Darling—I have but a few moments to write as we start at twelve, and I have my hands full of preparations for the scout. Do not be anxious about me. You would be surprised how closely*

I obey your instructions about keeping with the column. I hope to have a good report to send you by the next mail. A success will start us all toward Lincoln. I send you an extract from Genl. Terry's official order, knowing how keenly you appreciate words of commendation and confidence in regard to this movement, and, were it not impossible to do so, the Department Commander places too much confidence in your zeal, energy and ability to impose on you precise orders which might hamper your action when nearly in contact with the enemy.

Your devoted boy, Autie.[102]

Captain Marsh ran into Charley Reynolds who was in tremendous pain because he had developed an infection on his right thumb, which was bandaged up by Dr. Porter. Marsh encouraged him to stay on the boat since his hand was useless; however, the handsome scout touted, "Captain, I've been waiting and getting ready for this expedition for two years and I would sooner be dead than miss it."[103]

Near the confluence of Rosebud Creek & Yellowstone River, Montana Territory
Thursday, June 22, 1876, noon

Custer, always a champion of spectacle, paraded his troops on the damp ground. Gibbon, Terry and Brisbin were mounted for this review and a trumpeter sounded *Boots and Saddles.* There was the squeak of leather as the men swung into their hard McClellan saddles, pocketing their boots into the hoodless stirrups and their carbines into small leather sockets. The parade was an impressive assault on all the perceptual senses, as the red, white, and blue guidons from each company popped in the breeze, and the Arikara scouts sang the strange dirge of their death songs suggested by Custer. Such a moment was always accompanied by the band offering up a familiar tune; however, this time the verse had to come from the soldiers' own voices and collective memories. The last verse seemed the most fitting for this departure.

Our hearts so stout have got us fame,
For soon 'tis known from which we came;
Where'er we go they dread the name
Of Garryowen in glory.

Instead of spa we'll drink down ale,
And pay the reck'ning on the nail;
No man for debt shall go to jail
From Garryowen in glory.

The regiment advanced with Custer leading, when suddenly, he wheeled around on his horse, Dandy, and joined the officers watching the procession. Prancing behind their commander were two sergeant color bearers, one carrying the blue regimental standard with an eagle in the center, and the other his personal colors: a swallow-tailed banner with white cross sabers overlapping two bands of color; the top of the which was blood red and the bottom cobalt blue.[104]

Custer sat with a crimson tie dangling from a large-collared shirt, wearing a light buckskin coat with matching trousers. Marsh recalled he was, "A figure to attract attention anywhere. His form was straight and slender, his speech quick, his movements restless."[105] Sergeant Samuel Alcott recalled the lieutenant colonel's manner of dress on June 22, "He was dressed in a suit of fringe buckskin and one of his favorite hunting shirts. He wore a pair of high boots with red tops and a black slouch hat that had seen better days. It was a form of dress that was familiar to all of us who knew him."[106]

Each officer rode past Terry and saluted. Windolph remembered, "I don't believe many of the soldiers were very worried. We knew there would be some hard fighting, but a soldier always feels that it's the other fellow who's going to get it. Never himself."[107] The melancholy and anger of the previous evening had disappeared as Custer smiled and laughed. Gibbon recalled, "He appeared to be in good spirits, chatted freely with us, and was evidently proud of the appearance of his command."[108] Custer shook hands with each officer, and Gibbon, in a pleasant tone, warned him against being greedy, and with a gay wave of his hand he [Custer] called back, "No, I will not," and rode off after his command.[109]

The 7th commander's main concern was the possibility of having his command usurped by new developments or by bumping into Crook. There was no Hancock, no Sully, no Sheridan, no Stanley, no Sturgis, nor Terry to siphon command control from him. Terry remarked shortly after Custer's departure, "Custer is happy now, off with a roving command of 15 days. I told him if he found the Indians not to do as Reno did, but if he thought he could whip them to do so."[110]

Gibbon and Terry soon departed together, west along the Yellowstone to the Bighorn River and southward through horrendous badlands. Custer would follow the fragrant buds of the wild rosebushes along Rosebud Creek. Both trails would head toward the plush valley known as the Little Bighorn, with one trail leading to a deserted village camp of refuse and the other to immortality.

Chapter 8

Into the Valley of the Shadow of Death

On June 22, 1876, an excited Custer galloped to the head of the column with Bloody Knife and Mitch Bouyer. Windolph observed, "He had plenty of dash. And he was mighty proud of his regiment."[1] The cautious Arikara divided into two groups that flanked the troops in case of a peripheral attack, as the Crows took the lead. They were entering a region where the rivers run north and all dump their muddy contents into the Yellowstone. The tributary called Rosebud Creek was scented with the buds of wild rosebushes and covered with brush, cottonwood, and willow trees, indicating a source of life in the otherwise barren landscape.[2] Burkman remembered, "A queer sort o' glumness hung over us all. . . . The steady, quiet ridin' through them hills kinda got on a man's nerves."[3]

Custer's pride quickly turned to anger when pack mules began to lose their loads before they got out of camp; however, four carefully packed straw-covered bottles carrying amber colored whiskey for the officers survived.[4] Godfrey recorded, "Our pack trains proved troublesome at the start . . . and during all that day the mules struggled badly."[5] The problem was so severe that Custer made camp after traveling only 12 miles, and his plan to travel quickly by using a mule train as well as leaving the Gatling guns behind, was soon derailed.

He had relied on Lieutenant Mathey to take care of the train during the Washita campaign and called on him again.[6] A frustrated Custer decided to punish anyone who slowed his progress. The half dozen civilian packers had the impossible task of quickly training soldiers in the art of packing some of the most stubborn animals on the plains.

The regiment camped below some tan, chalky bluffs covered with pine as well as cedar. Interpreter Gerard recalled that Bloody Knife was intoxicated that evening. The scout told Gerard, "Tell Gen. Custer that I don't believe he is hunting for Indians; more likely on a pleasure trip. If he should find them, he would not dare to attack them." Knowing the verbal jabs that both men were capable of and the mutual respect they held for one another, it is clear that the firewater was talking. Custer replied, "Tell him that I am on an independent command and shall fight the Indians wherever I find them."[7]

Officers' call sounded near sunset and a serious mood confronted the men that squatted in front of Custer's cot. Mathey reported the companies that had packed their mules poorly. Escorting the pack train would no longer rotate daily with each company taking turns—the poorest performers would escort the mules as punishment the following day. The idea of eating dust all day long for consecutive days, as well as pushing the straggling and unwilling beasts, was a detested task.[8]

Custer informed his officers that he declined the offer of Gatling guns and the 2nd Cavalry. He also explained that to bring them along would have created jealousy, and he preferred harmony with the 7th getting the credit for any victory. He explained that he believed the guns would impede their march, and that they could handle the 1,000 to 1,500 warriors they might discover.

As if on cue, the harmony that Custer tried to instill disappeared with Benteen's confrontational, second-guessing remark, "I, for one, am very sorry you didn't take them. I think we will regret not having them."[9] Benteen was not afraid of his superior. He said, "I never fought Custer in any way but the most open handed manner; always going face to face for it."[10] According to Benteen:

> After arrival of the last officer, General Custer commenced his talk, which was to the effect, that it had come to his knowledge that his official actions had been criticized by some of the officers of the regiment at headquarters of the Department, and that while he was willing to accept recommendations from the junior second lieutenant of the regiment, he wished the same to come in a proper manner, calling our attention to the paragraph of Army regulations referring to the criticism of actions of commanding officers; and said he would take the necessary steps to punish, should there be a reoccurrence of the offense.[11]

Edgerly summarized, "I shall not allow grumbling, and shall exact the strictest compliance with orders from everybody."[12] Benteen again challenged Custer, as he had done in 1869, saying, "It seems to me you are lashing at the shoulders of all, to get at some; now, as we are all present, would it not do to specify the officers whom you accuse?"[13] Perhaps Custer, who had learned a lesson from Benteen's unexpected admittance in the Washita episode, did not wish to repeat the embarrassment of the whipping incident. According to Benteen's recollections, Custer responded, "Colonel Benteen, I am not here to be catechized [sic] by you, but for your own information, will state that none of my remarks have been directed towards you."[14] Edgerly's recollection was different, quoting Custer as stating, "I want the saddle to go where it fits."[15] It was clear that the tension of the evening before had not subsided.
Custer issued orders:

> . . . no more trumpet calls unless in an emergency or when in contact with the enemy; the command would rise each morning at 3:00 a.m. and march 25 and 30 miles a day beginning at 5:00 a.m. [and Custer would determine when the column would halt each evening]. In the event of an attack, the stable guards, packers, and cooks were to go out at once to the horses and mules to guard them and keep them safe. All units were under orders to stay within supporting distance of one another and not get ahead of the scouts. No firing was to take place until ordered by an officer, and the troopers were ordered to sleep with their weapons.[16]

The 7th commander warned against the units straggling, as it was dangerous to do so in this situation. The officers synchronized their pocket watches.

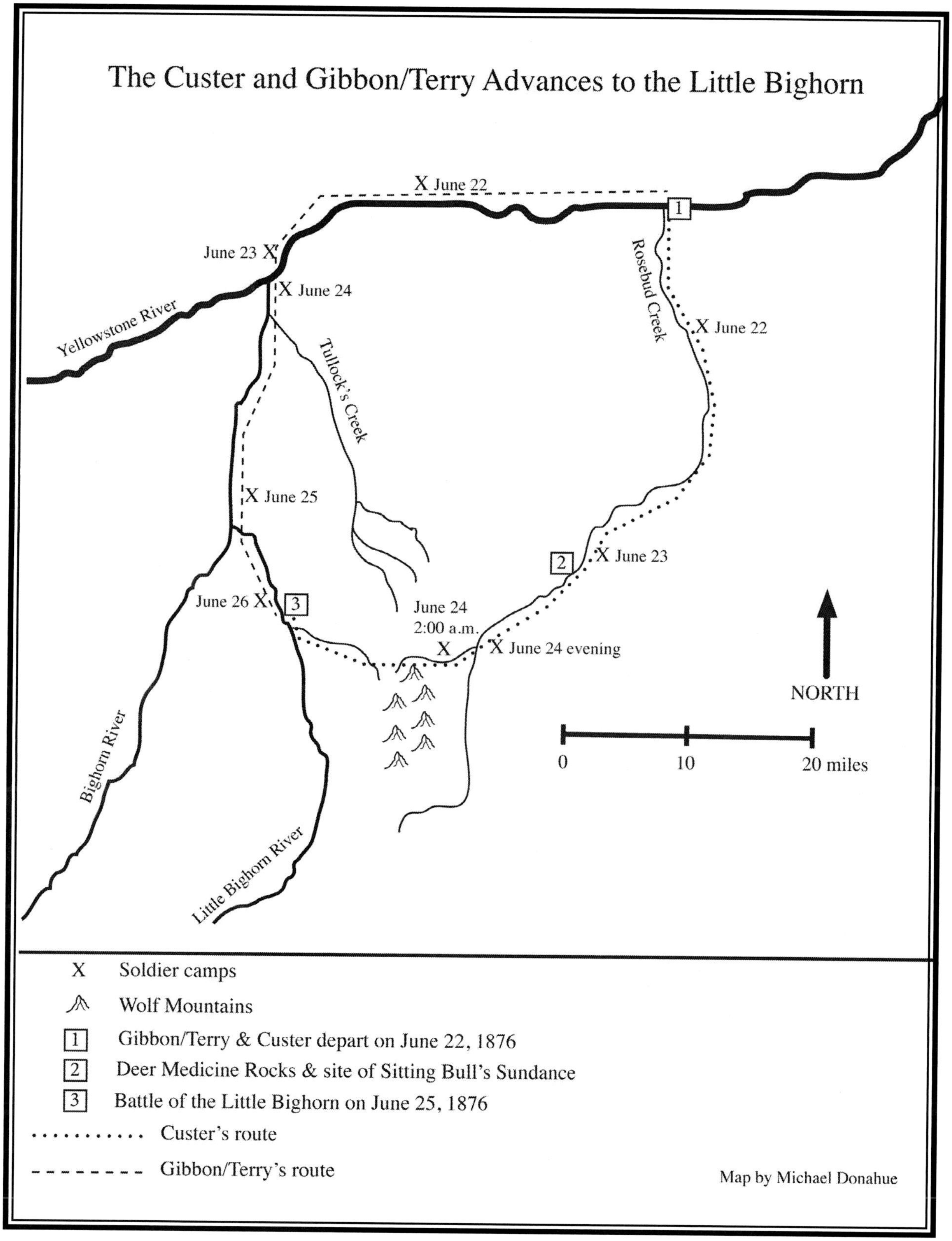

Fig. 50. Map: Custer and Gibbon/Terry Advances to the Little Bighorn.

Custer wanted everyone to feel they had an important role to play and needed everyone's cooperation. There would be no favoritism, and everything would be done to the letter of military regulations and rank. Even Keogh, who had been on a recent drinking spree and had clashed with Reno on his scouting expedition, was given a clean slate. Keogh had promised his superior that he would do better if given an opportunity.[17]

At the conclusion of this meeting, Crook then became the subject of the conversation, with Custer making a divisive comment. Godfrey remembered:

> Without any seeming occasion, [Custer] remarked to the assemblage of officers, that he understood Gen. Crook had been known to say that on campaigns he got better efficiency all around out of the companies commanded by lieutenants than from those commanded by captains. In other words, the greater enthusiasm of the junior officers tended to better efficiency.[18]

He did not dwell on the subject further, but when the officers were dismissed, Capt. Benteen spoke up, "Well, how about this captain business?"[19] Custer replied by a wave of his hand saying, "Oh nothing personal was intended, Capt. Benteen, and you need not take my remarks as applying to this occasion at all."[20]

Godfrey said the officers commented on the incident a good deal; they wondered what Custer's motive may have been in bringing the subject up at that time.[21] Edgerly remembered the episode a little differently, with Custer stating that another regimental commander had commented that it could be "a good one if only he could get rid of the old captains and let the young lieutenants run the companies."[22]

"It was not his [Custer's] habit to unbosom himself to his officers . . . in it he showed concessions and a reliance on others; there was indefinable something that was not Custer," wrote Godfrey. "His manner and tone, usually brusque and aggressive, or somewhat curt, was, on this occasion, conciliatory and subdued. There was something akin to an appeal, as if depressed, that made a deep impression on all present."[23] Lieutenant Wallace projected, "I believe General Custer is going to be killed. I never heard him talk that way before."[24]

Edgerly recounted:

> He said he had perfect confidence in the officers and men of the 7th and didn't believe it was of any use for anybody or any regiment to try to do what we could not, that his intention was to treat everybody as fairly as he could and make the trip as pleasant and successful as possible, and I want to say here that during the entire march from Lincoln I didn't hear one word of fault found with any of Custer's actions and gave Keogh command strictly in accordance with his rank on the morning of the fight.[25]

The 7th Cavalry did have some energetic young lieutenants, which included Winfield Edgerly, George "Nick" Wallace, and John Crittenden. Curly-headed Edgerly missed the Civil War but was a

West Point graduate. Assigned to Company D, he was Weir's right hand man and often led the company in lieu of his sentimental captain.[26] The homely George Wallace was a West Pointer who participated in the Yellowstone and Black Hills expeditions. Wallace was the official recorder of this trip—recording times, distances, and other important information as chief engineering officer.[27] Crittenden came from a prestigious family legacy, but the delicate West Point cadet flunked out in his junior year. He was able to get a commission in the regular army, probably due to the influence of his father, Lieutenant Colonel Thomas L. Crittenden. While firing a shotgun, a cartridge exploded, blinding him in the left eye. In order to give his son some actual military experience, his father convinced Custer to allow him to replace a missing officer on his staff. The 22-year-old officer was detached to the 7th Cavalry and stood out with his large drooping beard, glass eye, as well as being the only infantry officer mounted with the regiment.[28]

Fig. 51. Second Lieutenant Winfield Scott Edgerly. Edgerly, along with Company D, rode farther than any other unit in an attempt to join Custer on June 25th. His trek into the upper reaches of Medicine Tail Coulee allowed him to see some of Custer's dead until he was driven back by large numbers of victorious warriors.

National Park Service, Little Bighorn Battlefield National Monument LIBI_00015_664

Strolling back to his company, Godfrey was asked by the scouts how many warriors they expected to encounter. The lieutenant stated that they might meet 1,000 to 1,500 and that they could be whipped. Bouyer asked if Godfrey had fought the Lakota before, and the scout proclaimed, "Well, I can tell you that we are going to have a damned big fight."[29]

Officers gathered around Edgerly's shelter tent and their mood lightened as they sang songs for about an hour, serenading the enlisted men.[30]

On Friday, June 23, at 3:00 a.m., Mathey reported the three companies whose mules were unmanageable the day before, and Benteen's mules had been the biggest problems. Custer detailed the captain to the rear of the column.[31] The twelve companies advanced at 5:00 a.m. and were forced to ford Rosebud Creek, not once but five times, due to the nature of the rough terrain and winding creek. It became a daytime nightmare for Benteen. The pack train was soon out of sight of the main command.[32] "We rode hard . . . there was no foolishness," recalled Windolph.[33]

Soon the soldiers came upon a large abandoned village campsite that both Reno and Bradley had discovered earlier. Everybody was busy studying the signs and trying to figure out the age of the trail and the Indian camp.[34] A second abandoned village site was located around 10:30 in the morning. By 1:30 p.m., the column found a third vacated camp not far from a large towering butte only 400 yards from the creek.

At the bottom of the butte were a series of flat-faced totem sandstone boulders covered with pictographs of Indian figures and shamans. The area had been a campsite for native peoples for hundreds of years. The regiment kept moving and went into camp, at 4:30 p.m., after covering 33 rugged miles.[35]

Benteen's service as mule train escort had not gone well, and it was soon stretched for over two miles; his main concern was an Indian attack. He halted the mule train and, "One troop was put in advance of it, one troop on the right flank of the center of train, nearest the hills, and the remaining troops in the rear of the train."[36] Such an arrangement showed a disregard for Custer's orders, but then again, his commander certainly had not stayed within supporting distance as promised.

Arriving at the camp shortly before sundown, the frustrated Benteen immediately tracked down Cooke. He told the adjutant to relay to Custer how he had solved his problems with the mules. Benteen wanted to be cautious not be perceived as 'grumbling' but admitted that he was pushing the limits. Cooke, however, suggested he take up the issue directly with Custer the next morning.[37]

Well-educated, assistant surgeon Dr. George E. Lord, originally with the 6th Infantry and ferried from Fort Buford on the steamer *Josephine*, joined Custer's command on the Yellowstone River.[38] He was not used to the rigors of such an expedition and had attempted to ride with Benteen, but he suffered from dysentery and found himself in a lonely and dangerous position behind the slow mules. Benteen remembered that Lord arrived late in camp and, "Declined tea, and wanted nothing to eat or drink. I state this to show what must have been the physical condition of the doctor on June 25th, on going into the fight, after an almost continuous march of 84 miles."[39]

Few accounts deal with the common soldier's activities on this day, but it was another long hard day of routine activities: watering and feeding horses, gathering wood for fires, performing duties for officers, acting as guards and pickets, and cooking the same meals over again. Some of the men suffered in their worn-out Civil War trousers that were amended with wool blouses, unsuitable for the 100-degree days. Hats rapidly wore out on such a long campaign of rain, wind, and sun, and the men were anything but uniform in appearance. The men from several companies wore kepis, straw hats, white civilian coverings, and droopy French-styled black hats atop their uncombed hair. The flour-sack patching of trouser seats, the mingled odor of human sweat combined with that of their horses, poor dental hygiene, and sore backs caused by the McClellan saddles did not lend itself to the romantic visions of today's cinematographers.[40]

The unique blend of men came from all walks of life: laborers, teachers, candymakers, farm boys, businessmen, adventurers, criminals, and unemployed immigrants. Most joined the army not from a sense of patriotism, but as a means of survival during the continuing hardships caused by the Panic of 1873. The pay of $13 dollars a month was welcomed, but few realized the hardships they would encounter, which led to abnormally high desertion and suicide rates.

For those who did not speak English well, the army was an opportunity to learn the new language of America. Forty-three percent of the 7th Cavalry was foreign born and a mingling of accents was to be heard near the campfires: Italian, French, Northeastern, Southern, but remnants of the German and Irish language predominated. Many Germans had left the strife of the Franco-Prussian War,

Fig. 52. Lt. Varnum, Lt. Bronson, 6th Inf., Capt. Benteen, and Capt. French, studio portrait. Unknown photographer, circa 1873.
National Park Service, Little Bighorn Battlefield National Monument. LIBI_00015_00612

and many of the Irish had ended up in America fleeing the continued results of the Great Famine of the mid 1840s.[41]

A large number of immigrants in the regiment were from Ireland and had come across the Atlantic with their own language and customs. One private guiltily confessed in a letter home that he had not been a participant of a Catholic mass in months. Many wished to avoid the memory of thousands of their friends and family being buried in unmarked mass graves: an ironic twist since many would end up in one in Montana territory.

Just how many of the Irish knew or cared that it was June 23 is not known, but it is likely that under the big sky of Montana some of the Irish must have remembered it was St. John's Eve or Bonfire Night. Private Jean B. Gallenne a former clerk not from Ireland, but France, remembered his patron saint's feast day.[42] It began as a pagan midsummer ritual centuries ago and was catholicized after the arrival of St. Patrick. As was the custom on this evening each year, a toast was made around a fire, and the men sacrificed some of the precious whiskey they had purchased only two days earlier or grabbed their tin cup of black coffee, raising it, as well as reciting the customary words: "*Go mbeirmid beo ar an am seo aris*" which, translated in English, meant "May we all have life this time next year." Many of those gathered around the fires would make this last toast in vain.[43]

The Crow scouts probed the dusk four miles away and discovered a huge circle where the earth had been trampled down by a large number of Lakota. Curly recalled, "We found the scalps and the beards of white men" and decided such evidence should be communicated immediately.[44] Near the deserted Sundance site were symbols etched in the sandstone bluffs by warriors and their ancestors. The residue of a lighting bolt had struck the rock and a cragged blue mark from electricity ran for 40 feet; until it strangely ended in the torso of a six-inch-long etching of a deer, complete with antlers. They were standing below the Deer Medicine Rocks. This sacred site had scores of drawings made over 3,000 years by native peoples.[45]

The scouts returned to show the gruesome relics to Custer. He asked them whether the trail separated—they answered in the negative. Through interpreter Bouyer, Custer explained:

> These Sioux have been killing white people, and I have been sent by the Great Father to conquer them and bring them back to the reservation. I am a great chief, but I do not know whether I will get through this summer alive or dead. There will be nothing more good for the Sioux—if they massacre me, they will still suffer, and if they do not kill me, they will still suffer because they have disobeyed orders. I do not know whether I will pass through this battle or not, but if I live, I will recommend you boys and you will be the leaders of the Crows.[46]

These comments reveal Custer was contemplating his own mortality. A perceptive Wallace had picked up on it the previous evening. Custer had cut his hair short to appease Libbie's fears of his scalping, and he had signed a large life insurance policy. The size of the deeply rutted travois trail, the large campsites, and the numbers of warriors projected may have begun to weigh on his mind.

On the morning of June 24, a worn-out Benteen approached Custer to explain how his previous instructions to protect the packs were faulty and that the regiment was out of sight all day. Custer, surprisingly, complimented him on his initiative and stated, "I am much obliged to you, Colonel, and I will turn over the same order of march for the rear guard to the officer who relieves you."[47] This response likely surprised Benteen and indicated that his superior was living up to his word of taking suggestions. Mathey reported which company's mules had been problematic the previous day, and the penalized units belonged to Yates, Keogh, and French.[48]

Custer warned Varnum with his scouts "not to let any trail get away from us."[49] Around 6:30 a.m., the command arrived at Sitting Bull's Sundance camp where they saw on the ground "figures drawn indicating by hoof prints Custer's men on one side and the Dakota [or Lakota] on the other," recalled Arikara scout, Red Star. Between them were drawings of dead men lying with their heads toward the Dakotas. "The Arikara scouts understood this to mean that the Dakota medicine was too strong for them and that they would be defeated by the Dakotas."[50] While at the Sundance site, a strong southerly wind gust suddenly toppled Custer's personal flag. Godfrey picked it up and wedged it into the ground next to sagebrush to help support the flagstaff, but once again, it fell to the rear. A superstitious Wallace speculated it was an omen of defeat.[51] The Crow had continued forward and returned to the troops at the Sundance site with a scalp they had found along the trail.[52] Supposedly, it belonged to one of Gibbon's soldiers, and Sergeant Finley placed the macabre souvenir in his saddlebags.[53]

At mid-day, the troops came upon another deserted village site that had been a campground only 12 days earlier. A four-hour halt was ordered at this location, and the men were allowed to eat some lunch while the scouts searched ahead for signs. Varnum was ordered to backtrack to look at a divergent trail that Godfrey claimed was missed. It turned out to be only a side trail.[54] According to Herendeen, "Custer did not want to lose any of the lodges, and if any of them left the main trail he wanted to know."[55] Second Lieutenant Luther Rector Hare was ordered to help Varnum with the scouts. Hare was a reliable, young officer who had been raised in Texas and graduated from the Military Academy.[56]

Custer chatted with his Arikara scouts and confirmed that their role was to find the village, and the soldiers would do the fighting. They were told that they could take all the Lakota ponies they could capture, and that he was proud to have them along. He also told them, "If we are victorious," and, "If we are successful," then the scouts might get special honors and a trip to Washington. Similar to his talk with the Crow, usage of the word 'if' by a normally confident commander indicated some self-doubt. It was certainly a different tone than the rhetoric he had expressed only days before when he left the *Far West*.[57] The troops continued their march, while Crow scouts Curly, White Man Runs Him, Goes Ahead, and Hairy Moccasin rode ahead.[58] The command marched a distance of 28 miles when Custer called a halt at 7:45 p.m. along Rosebud Creek.[59]

The soldier camp was hidden in a beautiful part of the Rosebud Valley near some white sandstone bluffs, and as darkness descended, the soldiers busied themselves with cooking supper. Burkman remembered:

> I believe we had . . . a kind of presentment of what we were headin' toward. Maybe 'cause Custer was unusual quiet and stern, not cuttin' up tricks with Captain Tom or Boss Custer or Autie Reed the way he'd been doin' ever since we left Fort Lincoln.[60]

All during the day, the troops feared being observed by warriors in the proximity and such was the case. Two Cheyenne from Little Wolf's band on their way to Sitting Bull saw the soldier camp, but in their leisure did not ride ahead to warn the village.[61]

Curly returned and told Custer that the Indian trail left the Rosebud and turned toward the Little Bighorn Valley. Being the youngest of the scouts, he had been busy as a courier several times that day and for an unexplained reason, the Crow scouts had a fight and sent him back alone.[62] Custer had to ponder his next move, since his orders were to proceed down Rosebud Creek even if the trail turned toward the Little Bighorn Valley.

Several men seemed aware of Custer's serious mood as the trail got fresher. Once, when Burkman and his commander were far ahead of the column, the nervous orderly expressed his concern about the warrior numbers, but his commander said:

> 'We're close on their heels and looks like they's a lot of 'em, more'n we figgered on.' I says, 'Not too many to lick, though.' He turned then, quick, and flashed a smile at me over his shoulder. 'What the Seventh can't lick,' he said, proud-like, 'the hull U. S. Army couldn't lick.' We dismounted and let our horses rest and graze whilst we waited fur the troops to catch up. We lay stretched under a tree. Custer didn't talk none, he jist lay starin' up into the sky. That's the last time we was alone together.[63]

Private William O. Taylor barely reached five feet in height when he enlisted at age 17, having lied about his age; he also noted Custer's serious demeanor later that evening, recalling:

> My troop was quite near to Custer's headquarters, a single 'A' tent, before which he sat for a long time alone and apparently in deep thought. I was lying on my side, facing him, and was it my fancy or the gathering twilight that made his face take on an expression of sadness that was new to me. Was it because his thoughts were far away, back to Fort Lincoln where he had left a most beloved wife, and was his heart filled with a premonition of what was to happen on the morrow?[64]

Benteen wanted to bed down early since mosquitoes had kept him awake the previous night, and he had a premonition that Custer would approve a night march.[65] Reporter Kellogg was in a debate with other officers about the Democratic convention occurring in St. Louis and the effects of rumored cutbacks in military spending.[66] First Lieutenant Charles Camillus De Rudio, known as "Count No Account," was entertaining some officers with tales of his past exploits.[67] De Rudio was descended from an ancient Italian noble family and had graduated from the Austrian Military Academy in Milan. As a revolutionary activist, he had escaped both the guillotine blade and a penal colony in French Guiana, only to end up in the American Civil War and now the 7th Cavalry. De Rudio was in line to take command of Company E in 1875, but Custer placed his friend Algernon Smith in the position instead. This soured the relationship that, up to that time, had been fairly positive.[68] Officers sang some somber songs but ended with an upbeat, *For He's a Jolly Good Fellow*.[69]

Varnum remembered, "After we camped, the Crow Indians came in and had a long talk with the General in a clump of trees near my camp. The General then sent for me."[70] The other scouts reinforced what Curly had said earlier, that the trail led toward the Little Bighorn Valley, and they informed Custer of a high vantage point called the Crow's Nest that presented a panoramic view of the entire region. Custer ordered them to the promontory and he, along with the command, would follow them under cover of darkness and hide behind the high hill. The signs were fresh and the village was believed to be within 30 miles.

Even though he had traversed over 50 miles that day, Varnum was asked to join the Crow scouts, and the lieutenant asked for Reynolds to accompany him. Bouyer, as well as six Arikara scouts, joined the party that departed at 9:00 p.m. and arrived at about 2:00 a.m. at the Crow's Nest.[71]

Around 9:30 p.m., shortly after the scouts had departed, Custer initiated an *Officers call*. They were commanded to have their companies ready to march at 11:00 p.m. Mathey was asked which company's mules had been problematic on that day. He answered that he did not like to make comparisons, seeing that all were doing the best they could, but if required to do so he would have to name the packs of Companies G and H. McIntosh took the criticism good naturedly, but it made Benteen angry.[72]

A tardy Benteen, who had argued with Custer on the evening of June 21st as well as on the 22nd, and confronted him unsuccessfully on the 23rd, missed the Officers' call. He had the nasty task of escorting the mules the day before and now would have to deal with the unpleasant task on the following day, June 25th. Benteen chaffed knowing he would again be with the mules twice in three days of marching.

At this point, Custer decided not to follow Terry's instructions of proceeding toward the Tongue River headwaters. It will always remain conjecture as to why he decided to alter the plan. His temperament as the hunter who could not give up the fresh trail may have played a role. Fresh pony tracks near the soldier advance created the fear that the troops had possibly been discovered, and he may have feared the village being warned as well as it escaping. Custer may also have feared losing his command seniority by running into Crook.

At around 11:00 p.m., the men, horses, and mules began a slow gait in the pitch-black darkness. Half Yellow Face, Gerard, and Bloody Knife were told to lead the column and followed the Indian trail. Custer emphasized once again that he did not want any divergent trails overlooked, especially any that led to the south. Bloody Knife sarcastically replied that Custer should not be so particular about the smaller camps, "We'll get enough when we strike the big camp."[73] Custer and Cooke struck matches and lit candles at the front of the column in order to see the Indian trail.[74] "It was dark as pitch . . . You couldn't see 20 feet in front of you," remembered Windolph.[75] Soldiers attempted to follow the clanking metal hitting the saddles of the men in front of them. Burkman recalled, "We couldn't see one another. Some of us got lost. Some fell, horse and all, into ravines and had to get out alone the best way they could. The moon had gone under a cloud."[76]

Since it was still June 24th, Keogh was again stuck with the duty of escorting the mule train that became a nightmare in the darkness. A hungry Keogh swore and cussed as it took him almost two hours to get the mules across a creek.[77] A new strategy was initiated according to Thompson, "Each company had to lead its own pack mules; it was too dark to see to drive them, no moon, simply the faint starlight to guide us."[78] Some of his company's packs came loose and were cut rather than delay the command. During the night, a bundle of clothing from Company F unknowingly dropped from a mule as well as a box of hardtack.[79]

Mathey warned Custer that the mules were tired and strung out. His superior told him to do the

Fig. 53. Tom Custer. The competitive younger brother of George Custer earned two Medals of Honor in the Civil War and perished near him on Custer Hill. Tom Custer was mutilated beyond recognition and recognized by a tattoo on his arm.

National Park Service, Little Bighorn Battlefield National Monument, LIBI_00015_00197 best he could and that they would not advance much farther.[80] Using sign language, the 7th commander asked Half Yellow Face if they were going to make the divide before daylight. The Crow leader signed they would not and could not go over the divide in the daylight without being discovered. A disappointed Custer learned that there were places to hide the troops near the Crow's Nest, which was the best he could attempt under the circumstances.[81]

Tom Custer, in an entirely different mood from his brother, took advantage of the darkness to play a joke on De Rudio, who had, earlier that evening, bragged of his heroics. De Rudio was told to go back and check on his company mules, and when he returned was surrounded by whooping Indians and drew his pistol. Tom arranged for the friendly Arikara scouts to 'surround' the older lieutenant and give him the scare of his life. Tom and others laughed in the darkness.[82] The night was a strange mixture of Keogh's Irish curses, Tom's pranks, and the tired grumblings of soldiers half asleep in their saddles.

The regiment halted around 2:00 a.m. However, due to the darkness and exhausted state of the troops, the distance seemed greater than six miles. Ryan remembered, "We rode down into a ravine and went temporarily into camp without unsaddling the horses. We simply dropped down, threw the bridle reins over our arm and went to sleep, which was not an uncommon thing for a cavalryman to do on a campaign."[83] Custer also quickly dropped from his horse and fell asleep.[84]

As Custer and his men awoke, approximately 20 miles away some other young men were going to sleep after a long night of social dancing and partying. About 20 young Cheyenne men, along with a few Lakota, had done a special dance that was a serious ceremonial event called the 'dying dance.'[85] They had vowed to give their lives in the next fight with any bluecoats they might encounter.

Chapter 9

Reunion along the Greasy Grass

Legend has it that the Bighorn Mountains, along with its drainages, were named after the Bighorn sheep that roam the region, while others note that the dramatic bends of the nearby canyon are shaped like their horns. In these beautiful snowcapped mountains, that the Lakota called the White Mountains, a trickle of melting ice drains down the hills around the Wind River until it becomes a small watery tributary surrounded by the cottonwood trees that are sentinels to the water's existence. This small waterway, known today as the Little Bighorn River, grows and empties into the larger, trout-filled Bighorn River, which in turn merges into the widening Yellowstone. This raging river turns into the powerful Missouri, which becomes the mighty Mississippi and transfers its power into the warm waters of the Gulf of Mexico.

These puddles, creeks, and rivers supported human survival, and the people of the Plains—native and non-native—respected this. When the army searched for the elusive tribes, they knew they had to be near one of the waterways. The bluecoats could not stray far from this life-sustaining source for more than a day or two. All four of Custer's Indian fights were next to rivers, and his name will always be linked to the oddly named waterway on which he died, the Little Bighorn River. With its contradictory name, it seems to be the perfect reflection of the Ohioan with an equally paradoxical personality. To the Lakota, it was known as *piji sla wakpa*—the Greasy Grass River. Origins of the designation of 'Greasy Grass' are debatable, but the likely explanation is that the wonderful luxurious grasses in the valley were 'greasy or fattening' for their horses.

The village that had migrated along Rosebud Creek to the Little Bighorn may have sustained 10,000-20,000 horses, not to mention its 6,000-8,000 inhabitants, all in need of water. Indian ponies were needed for transportation, hunting, warring, and as a symbol of individual wealth. Horse raiding was an integral part of this social network. Contrary to common belief, the horse culture known so well to Hollywood celluloid today lasted only about 150 years. A wealthy warrior had a multitude of horses, and each horse drank 20 gallons of water a day—more if used heavily. The horse was more than just a possession used for utilitarian purposes, and young boys learned to ride almost as soon as they learned to walk. A Crow chief summed up the symbiotic relationship:

> My horse fights with me and fasts with me, because if he is to carry me in battle he must know my heart and I must know his or we shall never become brothers. I have been told

Fig. 54. Cheyenne Wooden Leg holding a soldier's Springfield carbine. The young warrior fought both in the Rosebud and Little Bighorn fights and left detailed accounts to Dr. Thomas Marquis who published his stories.

Courtesy of John Doerner

that the white man who is almost a god, but a great fool, does not believe that his horse had a soul. This cannot be true. I have many times seen my horse's soul in his eyes.[1]

Many of the ancestors of the natives in the camp had, at one time, lived on the large freshwater lakes in Minnesota. Driven westward by whites and native enemies from that region in the 1860s, the village that gathered in Montana territory included members of every tribe commonly called, 'the Sioux.' The Lakota, also known as the 'Teton Sioux,' were the majority in the village and were linguistically close to the Dakota and Nakota. The word Sioux was supposedly derived from an Ottawa or Ojibwa term, "*na-towe-ssi*," adapted as *Nadouessioux* by the French. The name supposedly meant enemies or 'Little Snakes,' since the Ojibwa were in constant warfare with these people. The Lakota referred to themselves as the 'allies.' The Sioux, at different periods in their history, had fought the Pawnee, Utes, Ponca, Omaha, Blackfoot (different than the Blackfeet Lakota), Cree, Gross Ventre, Assiniboine, Hidatsa, Mandan, Crow, Shoshone, and the Arikara. The fact was, they had many more enemies than allies.

The Lakota were divided into seven sub-tribes that lived in parts of what today encompasses Wyoming, Montana, North Dakota, South Dakota, Nebraska, Minnesota, and Iowa. The far western band, known as the Tetons (Dwellers on the Prairies) were situated in Montana and Wyoming, as well as North and South Dakota. The Teton subgroups consisted of the Oglala, Sans Arc, Miniconjou, Brule, Two Kettle, Blackfeet, and Hunkpapa.[2]

The Indians were frustrated over the invasion of the Black Hills, poor or non-existent rations at the agencies, and the urge to take part in religious Sundances. They wanted to live the old life again. The tribesmen slowly trickled away from their reservations, where they joined factions under the great warrior Crazy Horse and Sitting Bull. Most of the warriors did not believe Grant's ultimatum. Wooden Leg said:

> We thought somebody had told him [the messenger] what was not true . . . I had not seen a white man for many months. We were not looking for them, we were trying to stay away from white people, and wanted them to stay away from us.[3]

The Cheyenne fled to their Oglala friends after the attack on the Powder River, and both bands joined the consolidated Hunkpapa and Miniconjou camp led by Sitting Bull. Soon the Sans Arc bands joined the group, making five major bands. Other natives were joining the group daily, and few were leaving to go back to the agencies. As they neared the Powder River, the Blackfeet Lakota joined the group. A few Santee tribal members attached themselves to the wandering camp. Some Assiniboines also inflated the size of the Hunkpapa camp, while the Brule attached themselves to the Oglala and Blackfeet bands. A welcomed new arrival was Southern Cheyenne, Lame White Man, and his family. He was not a chief but was respected by the other Cheyenne as 'a wise and good man.'[4] As the village followed Rosebud Creek, they had six growing circles that would eventually extend about a mile in length.

Fig. 55. Hunkpapa Lakota spiritual leader Sitting Bull was the charismatic force that prophesied a victory over soldiers after his Sundance in June of 1876. His plea to avoid the belongings of the white men went largely unheeded.

Courtesy of Christopher Kortlander

In early June of 1876, Sitting Bull participated in a Sundance ceremony below the Deer Medicine Rocks, and his prayer was offered to *Wakan Tanka*, the Everywhere Spirit, to bring bounty and unity to the Lakota. As Sitting Bull offered a sacrifice of an entire buffalo on behalf of his people, Lakota Jumping Bull took a small awl and began to pierce the arms of the prophet. Blood trickled down as 50 small bits of flesh were removed from each arm in the half-hour ceremony, then Sitting Bull danced under the pole holding a buffalo skin. A 'crimson blanket' ran down the spiritual leader's arms into Mother Earth.

After dancing for an extended period, he stopped and, as he was being helped to the ground, he told of a vision he had received of "bluecoats, like grasshoppers, falling upside down into his camp." The soldiers had no ears and refused to hear, but death to them was certain by the hands of the villagers. While there was joy in the victory announcement, Sitting Bull warned, "Do not touch the spoils. If you set your hearts upon the goods of the white man, it will prove a curse to this nation."[5]

After the Rosebud battle, the village moved in the direction of the Greasy Grass River.[6] Every day, more dissatisfied and starving agency natives joined the village, inflating its numbers rapidly. The atmosphere of the camp was much like a family reunion, with relatives from the reservations having opportunities to see their extended families. A rich carpet of green grass stood under the cottonwood

Fig. 56. Two Moons was a minor Cheyenne chief. He bravely fought soldiers at the Powder River fight as well as the Rosebud and Little Bighorn battles.
National Park Service, Little Bighorn Battlefield National Monument. LIBI_00015_00712

and ash trees that lined the Greasy Grass where the village stood. Two large bends of the gravel-bottomed river forced the tipis to form circles in an irregular formation.

To the far north, the *Tsistsistas*, better known as the Northern Cheyenne, were one of the few allies that joined the Lakota in their fight for their traditional way of life. Their name likely came from the word, *Shahiela*, which was the Lakota name for 'red talkers' or 'people of alien speech.' The majority of the Cheyenne in the village were Northern Cheyenne; however, a few had southern origins.[7]

Two Moons, sometimes called 'Two Moon,' was only 29 years old and was a respected Kit Fox society leader. The distinctive looking warrior with the down-turned mouth was born around 1847, and his family name connoted a birth during a month of having two full moons. Such competitive fraternal organizations within the tribe offered honorary status to its members, and sometimes they were easily recognized by different dress, paint, or methodology of warfare. His nephew Young Two Moon was also in the camp.[8]

Twenty-six-year-old Cheyenne, White Shield, wore a stuffed kingfisher and knew the bird would protect him as well as his pony when he painted its image on the horse's rump. A photo of White Shield shows a handsome and physically fit man with a bone choker around his neck and hair that draped his face on both sides down to his chest. He had proven his bravery along the Rosebud by killing a Shoshone scout, as well as a soldier in the fight with Crook's bluecoats.[9]

Antelope, also known as Kate Bighead, was a Southern Cheyenne woman visiting friends in the camp. She had been born along the North Platte River and was part of Black Kettle's camp along the Washita. Running barefooted through the snow, she had been one of the lucky ones to escape during the fight. She had also been in the Cheyenne camp that Custer chased into the Texas panhandle, where he had smoked the pipe as well as taken hostages. Antelope was one of the few women in the camp who knew Custer, since she had seen him on several occasions and thought he was handsome. She had been in the Powder River camp when attacked by the bluecoats under Reynolds and had escaped once again.[10]

Also, in the Cheyenne circle was Yellow Nose, who was about 22-years-old at the time of the battle. This Ute was captured along with his mother who escaped and left her son behind. He became a renowned warrior among the Cheyenne and survived a gunshot wound to the chest when he was only eleven.[11]

The adjacent camp of the Sans Arc, or *Itazipcho*, had approximately 300 lodges. Early French traders named them the 'No Bows,' due to the fact that they had once been ordered by a holy man to leave their weapons behind during a fight. They were soundly thrashed and the name stuck.[12]

The next camp circle was that of the Miniconjou whose name meant 'those who plant by the stream.' Lame Deer was one of their main chiefs, and this band was more populated than the Sans Arc camp.[13] In the camp, Miniconjou, White Bull, the nephew of Sitting Bull, was a skilled horseman who killed his first buffalo cow at age 12.[14] At age 16, during a fight with enemy scouts, he counted coup three times by bravely 'striking' or killing his enemies.[15] He participated in the demise of Fetterman's troops in 1866 and rescued a friend in the famous Wagon Box fight the following year.[16] White Bull was evidently a ladies' man and had 15 wives during his lifetime.[17]

One large camp circle was set back from the river, the Oglala. The name translates as 'those who scatter on oneself,' due to a ritual of scattering dust on their bodies.[18] Their famous warrior *Tasunke Witko*—Crazy Horse—was approximately 35 years old, and the teenage warriors called him 'The Old Man Warrior.' Born around 1841, close to Bear Butte near the Black Hills, the light-skinned boy named 'Curly Hair' was the son of an Oglala holy man, Crazy Horse, and Miniconjou, 'Rattle Blanket Woman,' who died during his childhood. He eventually was given his father's name, as he showed prowess as a warrior at an early age, and grew into one of the most courageous men among the tribe. A vision had told him how to dress for battle in order to remain bullet proof: his hair was to remain unbraided and dots were to be painted on his body, representing hailstones. A lightning bolt was to be emblazoned on his face, while dust from a gopher hole was to be scattered on his body and on his war pony.[19]

His first coup was at age 16, and his first kill, supposedly dictated by a holy man, was not to be a male. While on a horse raid to capture Pawnee horses, he killed a Winnebago woman. A great deal of mystery remains about this episode, but it evidently caused him to leave the Brule, with whom he had been living at the time, either by his own choice or theirs.[20] The slimly built warrior stood about 5-foot, 8-inch tall, with a thin face and a narrow, almost sharp nose. He never wore a warbonnet or showy dress and went into combat nude, except for moccasins and breechcloth.[21] He became a fearless warrior against traditional Indian enemies, as well as whites traveling through his domain.[22]

Like Sitting Bull, he was adamant about stopping the invasion of the non-Indians into the lands on which he had roamed throughout his life. Crazy Horse was known for his solitary manner, which some of his own people called 'queer.' He eventually became a shirt-wearer, one of the greatest honors in the tribe. Such a designation placed the tribe's welfare above his personal desires and that was soon tested.

Crazy Horse courted and eloped with another man's wife, 'Black Buffalo Woman,' who was the niece of Red Cloud. The great warrior was chased down and shot in the face by her jealous husband. Crazy Horse survived with a permanent scar on his face, as well as a fall from grace, since his actions

Fig. 57. Rain-in-the-Face was a Lakota from the Hunkpapa band and was in the Little Bighorn fight. He was likely one of the warriors that killed civilians Honsinger and Baliran along the Yellowstone River on August 4, 1873.

Courtesy of the author

were not in the best interest of the tribe. In a strange similarity to the career of Custer, he was looking for redemption in the eyes of his people.[23] Both men had lost their prestige and rank over their love of a woman. Crazy Horse eventually married a young Lakota named Black Shawl and had a baby daughter who died from disease. Her death devastated the mighty warrior.[24]

Also in the Oglala circle was 13-year-old Black Elk, born near the Little Powder River. The youth had ironically been in the Black Hills when Custer's 7th Cavalry rode through the region in 1874. Little did he know that he would once again get a chance to see some of the same bluecoats up close in late June.[25]

While the exact location of the small Blackfeet band's camp on the Little Bighorn is disputed, they were probably adjacent to and associated with the Hunkpapa camp circle. The historical nomenclature for the *Sihasapa* or 'Blackfeet' has been lost to history. The southernmost and largest circle was Sitting Bull's Hunkpapa camp, located next to another large river bend that jutted into the valley floor. The word *Hunkpapa* was appropriately translated as, 'Those who camp by the entrance,' even though they were bringing up the rear when moving in the summer of 1876.[26]

By the time of the American Civil War, Sitting Bull had replaced some of his traditional enemies for a new one: the ever-encroaching pioneers from the East. He was just 33 when he confronted troops under Alfred Sully near Killdeer Mountain in June 1867. The results were devastating, A large number of troops killed about 150 of Sitting Bull's people, but only two soldiers died.[27]

In August 1872, Sitting Bull, with approximately 1,000 Brule and Hunkpapa, confronted Major Eugene Baker, who was helping the Northern Pacific Railroad engineers. During the fight, the charismatic Indian leader showed his bravery and leadership by confidently seating himself amid a hail of lead bullets and calmly smoking his pipe. Inviting others to join him, only White Bull, Gets-The-Best-Of, and two Cheyennes timidly joined him. They quickly fled after a few quick puffs. Sitting Bull cleaned his pipe before slowly walking away from the scene, impressing all that witnessed this bold act of courage. His friend Crazy Horse had his horse killed from under him during the same encounter.[28]

The Lakota spiritual leader was busy. Sitting Bull's wife had twin sons within a few weeks of moving into the valley of the Greasy Grass, and the proud father was also physically recovering from the recent Sundance ceremony.[29] Wooden Leg commented:

> I am not ashamed to tell that I was a follower of Sitting Bull. I have no ears for hearing anybody say he was not a brave man. He had a big brain and a good one, a strong heart and a generous one. In the old times I never heard of any Indian having spoken otherwise of him.[30]

The Hunkpapa circle seemed to contain a disproportionate number of great warriors. One warrior of renown had received his name as a child when he ate the raw gallbladder of an animal. *Pizi*, also known as Gall, was raised by his mother after his father died, and he was like a brother to Sitting Bull. In 1865, Gall had been arrested by soldiers for a crime he likely did not commit and, during an escape attempt, he was bayoneted and left for dead. The stout warrior miraculously survived the serious wound and was a participant in Red Cloud's War.[31]

Another well-known Hunkpapa warrior in the camp had an intriguing name. Blood streamed down his face during a boyhood scuffle with some Cheyenne in a sham fight; thus, the name "Rain-in-the Face." His reputation had been made in Red Cloud's War in the killing of Fetterman's troops. Rain-in-the Face had fought the 7th Cavalry before, along the banks of the Yellowstone, when he helped kill Baliran and Honsinger. Tom Custer arrested him for these murders, and evidence indicates he was allowed to escape by the hands of a sympathetic soldier from Fort Lincoln.[32]

Twenty-year-old Hunkpapa Moving Robe Woman's tipi sat among Sitting Bull's circle, next to a big river loop. Later known as Mary Crawler, she was sometimes referred to as a 'warrior woman,' which meant she fought beside the men.[33]

One group that was interspersed among the camp circles was the *Oohenonpas*, or Two Boilings, also known as the Two Kettles, because they had a tradition of sometimes cooking meals twice. Others joining the camp were the *Sincangu*, also called the Burnt Thighs or Brules. Enemies had once burned their village, and as the tribal members escaped, their legs were burned. French visitors who witnessed the results called them *Brule*, translated into 'burned.' Members of the Yankton, Yanktonai, Wahpekute, Sisseton, Mdewakanton, and Wahpeton were also present in differing but smaller numbers.[34]

Recent scholarship indicates the actual length of the village based on warrior maps extended for a little over a mile and a half.[35] A more realistic number, based on the number of lodges that could be situated within the area, is probably between 1,500 to 2,000 warriors.[36] Regardless of the actual numbers in the

Fig. 58. Low Dog. This previously unpublished photo of Oglala warrior Low Dog shows him as a younger man. He was convinced that soldiers would not attack his large village along the Greasy Grass River.
Courtesy of Putt Thompson

village, it was memorably massive. The Arapaho, Waterman, who was a prisoner in the camp, remarked, "It was the biggest camp I ever saw."[37]

Another Oglala, Low Dog, admitted the numbers gave the village a false sense of security. "I did not think it possible that any white man would attack us, so strong as we were."[38] The village was ironically located within the boundaries of the Crow reservation, according to the 1868 Fort Laramie treaty; however, to the more powerful and numerous Lakota, that meant little.[39]

The Greasy Grass River was a favorite summer camping place for these bands. As the snow melted during the 'moon of the ripening chokecherry' (June); the water flowed cool and fast up to the horse's bellies in most places. The warriors were well armed with as many as 200 repeating weapons (Spencers, Winchesters, and Henry rifles) acquired by trading hides with white traders or given to warriors as a treaty obligation. Obsolete flintlock rifles left over from the American Revolution, along with many percussion revolvers and rifles that were surplus from the Civil War were in the tipis.[40]

The materials within the camp, made by the white man, showed the rapid transition into a different mode of life that was changing their culture. The natives traded hides for items such as beads, axes, shovels, knives, and firearms. Agency woolen blankets were stacked next to hide blankets; china dishes and tea kettles rested next to buffalo horn spoons and elk horn scrapers. Traditionally, buffalo stomachs were stretched across four upright sticks, creating a stew kettle, but brass kettles, skillets, pots, and pans from the agencies were quickly replacing them. Women's dresses and men's shirts were often made from calico and cotton, replacing many of the animal skins they had utilized for generations. While the tribesmen were willing participants in using the white man's weapons and tools, the very idea irked a traditionalist like Sitting Bull. Such a concept permeated his Sundance vision only weeks before.[41]

The Lakota were a spiritual people with a complex hierarchy of gods, with *Wakan Tanka*, the 'Great Mystery,' predominating. The land and beings were often manifestations of various gods, therefore, the Lakota had reverence for all of their surroundings. Spirituality and order were evident in all aspects of everyday life, from the order of setting up lodge poles to killing animals for food. The Lakota and Cheyenne realized they were part of a larger universe, in which everything was alive and connected, functioning as a cyclical system.[42]

The bison they hunted were used for over 60 everyday items, and that dependency on a single, vanishing resource would be devastating to their way of life. The bison had always existed, and the tribes could not have imagined the comprehensive invasion of the northern plains by white settlers, fortune hunters, and thrill seekers, who would augment the depletion of this crucial treasure. There were an estimated 30 million bison on the North American continent in 1500; by 1890, there were less than 500 in existence.[43]

The immense, multi-colored pony herd, used in hunting the bison, grazed on the western benchlands overlooking the valley. The small and tough descendants of wild mustangs were normally watered in the morning, mid-day, and in the evening. One white observer estimated 10,000 in the herd, while another believed there were 20,000.[44]

A month earlier, Sitting Bull had a dream in which a great dust storm from the east headed toward a white cloud that resembled an Indian village situated at the foot of snowcapped mountains. Behind the dark gale, soldiers slammed into the village, causing lightning, rain, and a tremendous thunderstorm. The storm disappeared, leaving the cloud intact. To Sitting Bull, this was a message of a soldier attack from the east that would be unsuccessful, ending with a Lakota victory.[45] After the Rosebud battle, Wooden Leg remembered, "We had beaten the white men soldiers . . . we did not know of any other soldiers hunting for us. If there were any, they now would be afraid to come. There were feasts and dances in all the camps."[46]

While there was a general mood of happiness in the camp, there was a mistrust of incoming strangers arriving in the village. Strict orders were given by the war chiefs not to fire weapons in the camp. The *akicita*, or police force, made up of Cheyenne as well as Lakota, patrolled and punished those who did not abide by the rules.[47] Cheyenne historian, John Stands in Timber, recalled:

> It was agreed that both the Sioux and Cheyenne military societies would stand guard . . . to keep individual warriors from riding out to meet soldiers. It was a great thing to do that—to go out and meet the enemy ahead of the rest—and they did not want this to happen.[48]

Kill Eagle, a Blackfoot Lakota, arrived with his band before the Rosebud fight and were made prisoners because he was hesitant about joining the fight against the whites. As punishment, some of the party's horses were killed, lodges burned, and the Blackfoot men were abused and beaten by the Lakota. Kill Eagle and his band were captives during the battle of the Little Bighorn and were unable to participate.[49] Five Arapaho, including Waterman and Left Hand, encountered similar treatment when they arrived two days before the Little Bighorn fight. The Lakota believed they were scouts for some soldier party and took their horses, weapons, and clothing until the Cheyenne intervened on their behalf.[50]

As villagers came out of their lodges, they could see the high bluffs above the buffalo berry trees, willows, dogwood, and buckbrush growing near the water. "It seemed that peace and happiness was

Fig. 59. One Bull, nephew of Sitting Bull and younger brother of White Bull. He was present at the prayer offerings left by the Hunkpapa leader on the ridges above the Little Bighorn camp shortly before the battle. Oral history states that he later betrayed Sitting Bull after their return from Canada.

Courtesy of the author

prevailing all over the world, that nowhere was any man planning to lift his hand against his fellow man," surmised Wooden Leg.[51]

Not knowing strife would soon engulf the peaceful valley, Sitting Bull and his nephew, One Bull, went to the high ridges to the east of the camp. The handsome One Bull had been adopted by Sitting Bull at the age of four and now camped next to him. As a member of the Fox Warrior Society, he was one of Sitting Bull's favored bodyguards.[52] The stern black eyes of the spiritual leader took in the peaceful and impressive view. It was his leadership, promises, and prophetic visions that had galvanized these people together in the valley below. The duo sat near a soft buffalo robe, holding a dark red pipe and tobacco offerings wrapped in small buckskin bundles tied to chokecherry sticks and placed on the ground.

Wind blew through the needle-and-thread grass, causing it to move like waves across the rounded hills, setting the landscape in motion. Below, yellow-brown buffalo hide tipis and white canvas lodges wheezed smoke from their tops into the calm evening air, while ponies shifted in a speckled mass on the opposite bench lands. The landscape literally appeared as 'Mother Earth' with the hilly terrain appearing as a human incarnation: knees, breasts, and hips in the dramatic, caressing, long golden shadows of evening.

Once again, Sitting Bull would beseech the Everywhere Spirit on behalf of his followers below:

> *Wakan Tanka*, pity me. In the name of the tribe I offer you this peace-pipe. Wherever the sun, the moon, the earth, the four points of the wind, there you always are. Father, save the tribe, I beg you. Pity me. We want to live. Guard us against all misfortune or calamities. Pity me.[53]

While there was a contemplative mood upon the ridge, east of the river, the village below was active. Some warriors had come into camp after raiding against the Crow.[54] One warrior recalled, "War societies all through the villages were having scalp dances all night. I was about in [the] center of the big village, dancing until daylight when I went to my father's lodge to take sleep as I had been up all night dancing."[55] In the Hunkpapa camp, the older women were packing and bundling food in preparation for a planned move the next day, while the young people were preparing for a dance for young people to sing, dance, eat, and mingle.[56]

In a clearing, in one of the Lakota camps, a ceremonial dance took place known as the 'Dying Dance.' A Cheyenne historian mentioned:

> The Sioux boys had just announced that they were taking the suicide vow, and others were putting on a dance for them at that end of the camp. This meant they were throwing their lives away—they would fight till they were killed in the next battle . . . a few Cheyenne boys had announced their decision to take the vow at the same time, so a lot of Cheyennes were up in the crowd watching . . . none of them knew for sure that night that the soldiers were coming the next day. They were just suspicious.[57]

Cheyennes Wolf Tooth and Big Foot were determined to prove their manhood by breaking the rules set forth by the tribal elders. The close friends liked mischief, and back in 1867, they successfully derailed a freight train in southern Wyoming. Killing the engineer and brakemen, they seized axes and broke open the wooden cars in which they discovered huge bolts of calico that they tied to the tails of their horses. Soon a long string of cloth was flapping across the plains, accompanied by Indian laughter.[58] Under the cover of darkness, they grabbed their horses and rode north about a mile from the Cheyenne circle, hiding along the river in the brush to wait for daylight. They were going to look for soldiers as soon as the sun rose.[59]

During the night, a Sans Arc Lakota named Brown Back and a youngster named Deeds were searching for lost horses along Sundance Creek.[60] Seven Oglala that included Black Bear had left the Red Cloud Agency looking for stolen horses that they tracked to the village. They were returning to the agency and had left the huge party along the Greasy Grass behind. As the sun came up, they continued their journey toward the Wolf Mountains near the Crow's Nest. Other young men were in the same area hunting for buffalo.[61]

The sun rose around 4:00 a.m. on June 25th. Along with his friend and a brother, the exhausted and sweaty Wooden Leg walked wearily back to his lodge with a full stomach and pleasant memories. Mrs. Spotted Horn Bull recalled:

> The country was good; there was rich grass for the ponies, and sweet water; the fields glowed with prairie flowers of yellow and red and blue; there were buffaloes in the valleys and Indian turnips on the hills for digging . . . we had no thought of a fight with the white man.[62]

Black Elk was awakened by his father and told to take the ponies out to graze. He and his young friends left their cousin to watch the horses and, as the day warmed, they decided to go swimming. He could see the women on the western hills digging for turnips. He had a strange premonition "That something terrible was going to happen."[63] Red Feather was a 21-year-old who had stayed up late dancing with and chasing the girls. He was sleeping in and remembered, "The sun was quite high, but I was sleepy."[64] Women hollering for the young men to take their horses to pasture had awakened him.[65] One of the women digging wild turnips in the early morning light was Moving Robe Woman. She and some of her young friends were prying the ground with 12-inch-long pointed ash sticks. This allowed the underground, strange-leafed vegetable to be probed beneath the ground and pried upward to the surface.[66]

As the day got hotter, many villagers headed toward the river. Crazy Horse and other warriors bathed in the river and Yellow Nose joined them.[67] Antelope found her friends bathing in the river and decided to swim as well. There were other groups of men, women, and children playing in the water, while some boys were trying to catch fish for breakfast. She stated, "All of us were having a good time . . . nobody was thinking of a battle coming."[68]

There was a special parade on this clear morning in the Cheyenne circle. The Lakota and Cheyenne young men, who had vowed to give their lives for their people the night before in the 'Dying

Dance,' were honored. Cheyenne Twin Woman said, "The boys went in front, with an old man on either side announcing to the public to look at these boys well; they would never come back after the next battle. They paraded down through the Cheyenne camp on the inside and back on the outside, then returned to their own village."[69]

Wooden Leg had slumbered outside his lodge in order that he would not wake his family. After eating breakfast, the fragrant teenager, who had danced all night, was told by his mother to go take a bath. Wooden Leg and his brother went to bathe among the crowds already in the river. Afterward, the brothers crawled under the shade of a cottonwood to talk about the fun of the previous night until they fell fast asleep.[70]

As the sun rose, Cheyenne warriors Wolf Tooth and Big Foot came out of their hiding place to find almost 50 young Cheyenne ready to search for soldiers. The group carefully avoided the posted guards near the village and headed down a creek.[71] At the very moment the gallant band of young Cheyenne were searching for the bluecoats, the soldiers and their Indian scouts were straining their eyes toward the village 15 miles away. In a strange twist, the collision of these two groups would begin the early stages of one of America's most famous battles.

Chapter 10

Red Dawn at the Little Bighorn

Totally exhausted and hatless, Varnum rubbed his blood-shot eyes as the distinctive aroma of pines filled his nostrils. As he awoke, he could scarcely make out his dark surroundings, but the dawn's glow allowed him to view his bed of rock and pine needles. He had been startled and roused by one of his Indian scouts who had tapped his languid body. A simple hand gesture from the scout communicated the desire to have him climb to the top of the Crow's Nest. The trek to the top was very steep, and Varnum found the ascension difficult due to his leather-soled cavalry boots that slipped on the rocks. At the top, he joined Bouyer, and Reynolds and 11 Indian scouts (five Crow and six Arikara). They peered through a cheap spyglass into the distance.

The group was perched on a horseshoe-shaped limestone ridge that resembled a giant bird's nest from above. The 100-foot high point they stood on was the peripheral edge of a range known as the *Cheetish* or 'Wolf Mountains.' Through the distant bluish haze, the scouts scanned the benchlands on the far side of the Little Bighorn Valley, with the morning sun behind them. Looking back in the direction that they had ridden during the night, they could see smoke along Davis Creek where the soldiers were camped.[1]

The keen-eyed Crow guides told Varnum they could see portions of a pony herd 15 miles ahead and projected that there must be a village nearby; however, intervening ridges hid the huge encampment of tipis.[2] Varnum, with his fatigued eyes, could not see anything, even with the cheap telescope. All he could see were two tipis in the distance along the creek that ran below them—one lodge was broken down, and the other still standing. They urged him in broken English to "Look for worms in the grass," indicating that the brown and white mass of ponies was slowly shifting, like slow-moving maggots on a corpse. Convinced by their strong opinion, Varnum decided to send back Arikara scouts Bull and Red Star with a note to Custer.[3]

As the scouts discussed their discovery at 4:13 a.m., the sun rose. It would be the last sunrise that many of the distant warriors and these soldiers would witness. It was a red dawn on this Sunday morning, June 25th. Red Star described, "The prettiest red colors that we ever saw and as we sat and gazed at these changeable colors of the horizon, it seemed that blood color was stronger than any other color."[4] Red Star's mother and five-year-old sister had been killed by the Dakotas over four decades earlier, and he had no hesitation to scout against them.[5] Taking out his pipe and filling it with tobacco, he puffed it and offered it to the rising sun: "Great Father take this smoke with us and that you will guide us today. We are in fear because you have showed us the blood sign this morning."[6]

From the Crow's Nest, the angry Crow scouts felt the soldiers were begging to be discovered as they witnessed smoke rising from their breakfast campfires.[7] Varnum and the remaining party of scouts spotted two different groups of enemy warriors below their high perspective, oblivious of their presence.[8] One party was the young Lakota boy Deeds and elderly Brown Back searching for lost horses.[9] White Swan suggested going down to kill the two Indians nearest them, but Varnum, along with Reynolds, lost them in the rough terrain below. After returning to the rocky promontory, Varnum and the allied scouts once again saw the warrior parties duck behind a hill, indicating the discovery of the bluecoat camp. The old man and boy turned, riding quickly back in the direction of the supposed village encampment.[10]

Several miles away and unaware that they were being watched, some of the exhausted 7th soldiers snored by their unsaddled mounts, while others choked down a breakfast of hardtack and nasty, alkaline coffee.[11] Burkman woke the snoozing Custer to give him a cup of coffee and hardtack. After the brief repast, Custer went back to sleep. An excited Isaiah Dorman walked over and again woke Custer to let him know that two Arikara messengers had just arrived.[12] The inquisitive leader quickly interrogated the Rees, who handed over Varnum's note.[13] Custer went over to Captain French and nudged him on the side. Making a bowling analogy, Custer said, "Get up quick; we'll make a ten strike this day. There are thousands of ponies on the hills."[14] An invigorated Custer shared the intelligence with other officers, ordering them to be prepared to march at 8:00 a.m. His plan was to hide his troops for the day and surround this village under the cover of darkness, as he did at the Washita, and attack it on the following morning.[15]

Two of the pessimistic Arikara scouts, Bull in the Water and Red Bear, had a premonition of a disaster and decided to eat a good breakfast. "Let us get breakfast for if we are to go to the happy hunting grounds we should go with a full belly."[16]

Bloody Knife and several of the other Arikara scouts seemed equally uneasy. Adorned with a gift from Custer (a black bandana with blue stars on it), Bloody Knife warned George and Tom that based on the trails they had been following, there were likely "several hundred tepees and about 5,000 Sioux," and that there were enough warriors to "keep us fighting for two or three days." Custer verbally jousted, "I guess we'll get through them in one day."[17]

Perhaps sensing paranoia in Tom and Bloody Knife, Custer attempted to lighten the mood. In a rather bizarre exchange, George jokingly challenged the bravery of his brother, Tom, who was listening to the discussions. Custer told Bloody Knife, "Your brother there [Tom] is frightened; his heart flutters with fear; his eyes are rolling from fright at the news of the Sioux. When we have beaten the Sioux, he will then be a man."[18] Here was another opportunity to kid him about something he knew was not true and downplay the scout's concern.

Soon, a thinly-mustachioed private with an Italian accent interrupted them. Trumpeter John Martin appeared, even though it was not his turn to act as the orderly trumpeter for the headquarters staff. "He [Custer] just looked at me and nodded," recorded Martin.[19] Peter Thompson recalled that they pulled out of camp at 8:30 a.m. and, "Having had a few hours sleep, the men began to be talkative and speculation ran high on how soon the campaign would end." The soldiers began to talk about what they would do with Sitting Bull when he was captured, and one old trooper laughed, "Custer will take us with him to the [Philadelphia] Centennial . . . we will take Sitting Bull."[20]

Thomas McDougall had the advance and believed his company would surely be involved if there was action in store.[21] The handsome captain was the son of a Civil War medical doctor, Brigadier General Charles McDougall. Thomas served in the infantry and was seriously wounded at Vicksburg. He had some experience fighting Apaches in Arizona before joining the 7th Cavalry.[22] George and Tom Custer, Bloody Knife, John Martin, Fred Gerard, Red Star, Bob-tailed Bull, and Little Brave took the lead some distance in front of the troops in order to look for a better place to hide the troops and to see the lay of the land on their way to the Crow's Nest. According to Thompson, "Custer rode some distance ahead of us."[23] After five miles, the troops were ordered into a large ravine to the left of the Indian trail. The men were allowed to make coffee, once again, below the base of what is known today as the Wolf Mountains.[24]

From the high hill, an impatient Varnum saw the command approaching and rode down to meet the resting column. "Well, you've had a night of it," Custer observed. "I said yes, but I was still able to sit up and notice things," responded the exhausted scout.[25] As they spoke, Tom and Calhoun, perhaps thinking they had privileges as part of the Custer clan, rode up without permission, but Custer angrily ordered them back to their companies.[26]

As Custer and Varnum rode to the high point, the chief of scouts relayed the bad news that they had seen two parties of warriors, and the 7th had likely been detected. The report angered a doubting Custer who had attempted for four days to carefully conceal his troops along the Rosebud, and he insisted that they were undiscovered.[27] He could see nothing with his own eyes even with his field glasses, and rejected the report. He was in a state of denial. There was no village ahead and his troops had remained hidden.

Bouyer emphatically declared, "If you can't find more Indians in that valley than you ever saw together, you may hang me."

Custer, who was not known to curse, showed his repressed anger, "It would do me a damn sight of good to hang you, wouldn't it?"[28]

Arikara scout Strikes Two mentioned something that no one else recorded, "Mitch Bouyer told us he said to Custer, 'Why not wait a day until Gibbon cooperates and there are too many Sioux.' Custer said, 'There are no more than I handled in another battle.' "[29]

Charley Reynolds pointed with his finger to the area beyond the bluffs, and Custer nodded a confirmation. To cement his observations,

Fig. 60. Charley Reynolds was the reliable scout who brought the news of the discovery of gold from Custer's Black Hills expedition of 1874. He had a premonition of his demise at Little Bighorn and was killed when Reno's troops fled from the valley timber.

National Park Service, Little Bighorn Battlefield National Monument. LIBI_00015_07111

Fig. 61. Red Star was one of the Arikara scouts who allied himself with the army to fight against their traditional enemies, the Lakota. He brought back the first message to Custer that the village pony herd had been spotted from the Crow's Nest.

Gerhard Sisters, photographers. ca. 1905. Retrieved from the Library of Congress, Item/90710100. https://www.loc.gov/item/90710100/. (Accessed March 22, 2018.)

Reynolds then offered up his personal field glasses and, as Custer peered through them, nodded again. Custer was finally convinced there was something in the valley.[30]

Extending toward the supposed camp was a small, slow-running waterway known by various names: Great Medicine, Ash, Sundance, or Reno Creek. The soldiers scanned the adjacent wide-open countryside consisting of tall ridges, rugged hills, rocky bluffs, and brush filled ravines that might hide small parties of warriors, but it was clear from this vantage point that no massive village was hidden to their immediate front. Gerard, Bull, Red Star, and two other unnamed Arikara peered into the distance as well.[31] Gerard pulled out his field glasses and saw the pony herd, including the dust they were raising.

Looking to the north, Tullock Creek became the subject of the conversation since Custer was to explore this area for possible warriors. There was no evidence of camps along the creek, and the troops had seen no divergent trails.[32]

Red Star recorded an event that does not appear in later Crow testimony. It was an angry discussion between Custer and Crow scout White Man Runs Him when the subject of troop discovery was brought up again.

The Crow scouts via Bouyer asked Custer what he thought the Lakota had seen.

Custer replied, "This camp has not seen our army; none of their scouts have seen us."

White Man Runs Him argued, "You say we have not been seen. These Sioux we have seen at the foot of the hill, two going one way, and the other four the other, are good scouts, they have seen the smoke of our camp."

Custer responded, "I say again we have not been seen. That camp has not seen us." He then told White Man Runs Him what he planned, "I am going to carry out what I think. I want to wait until it is dark and then we will march; we will place our army around the Sioux camp."

White Man Runs Him argued, "That plan is bad and should not be carried out."

Red Star recalled:

> [I] first thought that Custer's plan was good. The Crow scouts insisted that the Dakota [Lakota] scouts had already seen the army and would report its coming, and that they

would attack Custer's army. They wanted him to attack at once, that day, and capture the horses of the Dakotas and leave them unable to move rapidly.

His strongly-voiced opinion must have convinced Custer, who finally conceded, "Yes, it shall be as you say."[33] Custer's remark to the Arikara scouts a short time later revealed his thoughts, "There are other soldiers coming to attack the village, and I want to be there first so that we will get the honor of fighting the Sioux. I want you to charge the camp first and capture all the horses you can."[34]

Later, Pikuni Blackfoot scout William Jackson also heard an attempt by Bouyer to dissuade an attack. He heard Bouyer proclaim, "General, we have discovered the camp down there on the Little [Big] Horn. It is big one! Too big for you to tackle! Why, there are thousands and thousands of Sioux and Cheyennes down there." Custer gave him an angry stare and challenged his bravery, proclaiming, "I shall attack them! If you are afraid, Bouyer . . . " Before he could continue the sentence, Bouyer replied, "I guess I can go wherever you do."[35]

While the great debate was going on at the Crow's Nest, the soldiers were a half-mile away relaxing with tobacco and coffee in a wide ravine. After their reprimand, Calhoun and Tom joined Godfrey, Cooke, Moylan, and Edgerly to smoke cigarettes and pass the time. Cooke laughingly chided young Edgerly that he would "have a chance to bathe his maiden sabre that day."[36] While the soldiers rested, a sergeant backtracked, searching for some clothing dropped during the night. An entire box of hardtack had unknowingly fallen off a mule as well. A band of Cheyenne under Little Wolf, while following the troops, stopped to see what was in the wooden box. The surprised soldiers came upon the equally startled Cheyenne, attempting to hack open the crate. The soldiers fired on the Cheyenne who escaped and the soldiers returned to share the unsettling development.[37]

As Custer and the scouts returned, Tom delivered the bad news about the scampering warriors. Such a proclamation only confirmed the likelihood of their discovery.[38] Some of the scouts looked at one another solemnly.

Finally, Bloody Knife looked up at the sun and with his hands signed, "I shall not see you go down behind the mountains tonight."[39]

Custer received another bit of bad news. Herendeen had spotted Indians watching the 7th from a nearby ravine, which was verified by the evidence of unshod hoof prints in the sandy soil.[40] A shocking sound shattered the silence of the brushy ravine: the sound of a trumpet call from Martin commanding the officers to gather.[41] Custer reclined on the ground in a relaxed manner with his officers gathered around him. Edgerly remembered that their commander told them of his plans to hide his troops and attack the next morning, but, "All hopes of a surprise were gone, and the only way of catching the Indians was by marching at once."[42]

Six enlisted men and one non-commissioned officer from each company were ordered to help Mathey prod the mules along. Approximately 150 men would advance alongside the mules on this Sunday morning, which seemed a sufficient ratio. Charles Windolph overheard Reynolds talking to the officers, saying, "It was the biggest bunch of Indians he'd ever seen over there."

Benteen impugned, "Hadn't we better keep the regiment together, General? If this is as big a camp as they say, we'll need every man we have."

Custer curtly responded, "You have your orders."[43]

Since no division of troops had been ordered at this point, Benteen believed that the transferring of additional troops to the mules and outdistancing the train was a mistake. Standing nearby was Canadian Donald "Tosh" McIntosh, who was a former army clerk, who was in command of Company G. He looked like an Indian in a blue uniform as his features displayed that his father was white while his debated native ancestry from his mother was possibly tied to the Seneca or Cherokee tribes.[44] Benteen and McIntosh were dreading this day, as it was their job to escort the mules.

Custer, who always seemed to have a sense of fair play, gave them a way out of this predicament. According to Benteen, the company commanders were ordered to check their companies to make sure they had 100 rounds of Springfield carbine ammunition and 24 rounds of Colt revolver ammunition on their persons. They were also to make the selection of the six privates and one non-commissioned officer to escort their company mules. The first company commander to accomplish this and report back to Custer had the advance. The slowest would have the sordid assignment of mule train escort.

In a sly game of daily one-upmanship, Benteen, without checking his company's readiness, took a few steps and about-faced with the proclamation, "H Company is ready."[45] Custer was caught off guard by this action, but had no choice but to honor his promise. Benteen was to lead the regiment and be in a prominent position in any fight, or so he thought. McDougall, who had the advance earlier, received the distinction of being least prepared.

Custer gave the Arikara an encouraging talk about bravery, and that they were to stop at nothing to stampede the Lakota horses as an important part of his strategy. Arikara scout Stabbed encouraged the younger men, "Young men, keep up your courage, don't feel that you are children: today will be a hard battle. We have been told there is a big Sioux camp ahead."[46] He had each man come by and lift his shirt. Stabbed brought some special earth on the journey and spitting on it, made clay that he rubbed on the chest of each scout. Such protection would appear to work on all but three of the scouts in the upcoming struggle.[47]

Wallace, the official timekeeper of the expedition, noted it was noon when the command proceeded across the divide toward the Little Bighorn Valley.[48] Benteen proudly accelerated the advance on his fast-walking horse, 'Dick.' Custer rode up to Benteen and told him to slow down; not to be outdone by the captain, Custer took the lead.[49]

Right over the ridges to their right was the Tulloch Creek watershed, and Herendeen rode up to notify Custer that he was ready to deliver a message to Terry. According to Herendeen, Custer responded rather impatiently:

> There was no occasion for sending me there as the Indians were known to be in his front, and that his command had been discovered by them . . . it was also my opinion that Custer wished to fight the

Indians with the 7th Cavalry alone, and he was clearly making every effort to do this. It appeared to me at the time that Custer was right and that there was really no use in scouting Tulloch's fork.[50]

This was one of several examples of Custer's unwillingness to communicate with other commanders, until it was too late. A partially-skinned buffalo was seen near the trail and indicated that the troops had surprised the warrior who was in the process of butchering the animal. Such evidence further convinced the officers that the soldiers had been spotted, and the village was about to be warned of their presence.[51]

As they reached the almost imperceptible divide between the Rosebud and the Little Bighorn drainages, the higher ground offered a good view of the Reno Creek valley ahead. De Rudio owned a pair of Austrian-made field glasses that were superior to any in the command. Cooke asked if he could borrow them for the commander's use, and the Italian-born lieutenant reluctantly agreed. De Rudio never saw them again. Custer peered in the direction of the village ponies and came back reporting he had seen "some cloud-like objects which the scouts said were the pony herds."[52] It is clear from Custer's remark and actions that he still had some doubts and wanted more accurate intelligence.

Lieutenant Francis Gibson was second in command of Company H and an important participant in this new excursion in search of the Indian camp. Born in Philadelphia, the 29-year-old Francis Gibson's experience included the Washita, as well as the Yellowstone and Black Hills expeditions.[53] He wrote in his obscure report:

> We proceeded but a few miles, when dense clouds of dust were seen about 12 miles in our front. At this juncture the Column was again halted; here Battalions of three Troops each were formed, commanded respectively by Cols. Reno, Benteen, Keogh, and Capt. Yates. Benteen's Battalion composed of Troops H, D, and K was ordered to proceed across the big hills to the south, some five or six miles, to ascertain if the Indians were trying to escape up the valley of the Little Bighorn River, and if so, the orders were to intercept them and drive them down towards the location of the village. In case no Indians were seen up the Valley, we were to hurry back and reform the Command, as soon as possible.[54]

Gibson was the only officer that gave the reason for the troop division. As Custer and Cooke dismounted, the adjutant got out his notebook to put down on paper a plan the lieutenant colonel had used before. Custer decided to scour the country before him as the soldiers advanced—a *reconnaissance in force*. He divided up the command and announced the arrangement of the battalions.[55] In 1868, he had used such a strategy when attempting to locate Indian camps and he noted this scheme:

> The country was divided into four sections, and to each detachment was assigned one of these sections, with orders to thoroughly scout the streams running through it. It was hoped that one of these parties might, if in no other way, stumble upon a campfire or other indication of the rendezvous of the Indians.[56]

A battalion consisted of at least two companies of men with about 40 men in a company, but could consist of more. Weir had one of the numerically strongest companies in the regiment. The 7th commander planned to have it accompany him as the regiment divided. The cantankerous Benteen challenged his superior by claiming Company D. Custer blew up this time at the insubordinate spirit, yet he acquiesced. Custer said rather impatiently, "Well, damn it to hell, take D Co[mpany]."[57]

Custer wanted the confrontational captain away from him, and Benteen was ordered with his three companies, about 125 men, to the high ridge to his left. Benteen had held the advance for only 10 minutes. Companies D, H, and K were in this group with company officers Weir, Benteen and Godfrey commanding. According to Benteen, his orders were to send, "An officer and a few men as advance guard, 'to pitch in' to anything I came across, and notify him [Custer] at once."[58] Since Custer was not sending any scouts along with Benteen, the small detachment would perform that duty.[59] As they departed, Benteen disregarded Custer's order to send the small contingency forward.

Reno watched the regimental split up with interest, since he had no companies to command after leaving the Rosebud. As second in command, he was given three companies—A, G, and M—about 140 enlisted men and officers.[60] His company commanders were Moylan, McIntosh, and French. Keogh was third in command due to seniority and accepted the battalion of companies C, I, and L. Since Tom Custer was on detached service with this brother, Harrington and Calhoun were the company commanders under his direction. Keogh was hoping to lead an attack during this fight, and Custer wanted to give him a chance. Keogh had missed the Washita and Yellowstone fights; his entire career was lean in terms of any real Indian fighting. Yates was given command of companies E and F, again due to seniority; the company officers were Algernon Smith and himself.[61] Custer's headquarters group could have ridden with any of the three battalions, but he chose the five companies that were run by people he felt would support him without reservations. According to Burkman, Custer had a special affinity for the gray horse company.[62]

As the command fragmented, civilians, and some soldiers without unit designations, made the decision to remain with the pack train, or decide which of the three battalions they would accompany. Little did any of them realize that the decision would become a matter of life or death. Custer recommended to the ill Dr. Lord that he remain at the rear with the pack train, but he refused and stayed with the headquarters group. Burkman was ordered to saddle up Custer's horse, Vic, and care for his other horse, Dandy. His commander placed his hand on Burkman's shoulder and said, "You've been doin' guard duty three nights in succession. You're tired out. Your place is with McDougall and the pack train. But if we should send fur [sic] more ammunition you can come in on the home stretch."[63]

Custer then approached chief trumpeter Voss and according to the orderly, "I heerd [sic] him say laughin,' Martin's a good bugler, Voss, but today I'm goin' to need one that can understand the English language, since I can't talk Italian."[64]

Nephew "Autie" Reed was excited. Like a concerned father, Burkman urged him to stay behind but noticed, "The youngster threw back his head and laughed. He pulled his foot from the stirrup and give me a friendly kick. 'Me stay! . . . When they's goin' to be a fight! Why, John, you're crazy! That's what I come out West fur, to see an Indian fight . . . You're mad . . . cause you can't go along."

To rub salt in the wound, Martin turned in his saddle and shouted to Burkman, "Shall we bring you a scalp, John, or do you want a live squaw?"[65] Boston Custer, who had almost stayed behind on the *Far West*, could not resist riding along beside his brothers and nephew and took off with the five companies. Boston had two Indian ponies with the pack train and appeared to have backtracked at some early point to get a fresher horse.[66]

Kellogg, riding a mule, borrowed spurs from interpreter Gerard in order to get his stubborn animal to keep up with the headquarters staff. He also was warned to stay behind with the pack train; however, he apparently expected "Interesting developments and he wished to keep up with the scouts, and report everything he could see ahead." He wrote, " I go with Custer and will be at the death," prophesying the death of the warriors by the army, but instead, his choice would end his career as an embedded journalist.[67]

Burkman recalled that Custer's face was "White and sober-like. He want [wasn't] laughin' none. He knowed he was leadin' his men into a helluva big fight."[68] He watched Custer's red tie blow over his shoulder as he rode off over the ridge on Vic.

Reno rode on the left side of the creek, which consisted of rough hills, while the headquarters contingency of approximately 20 staff members, along with Keogh and Yates, rode on the easiest terrain on the right bank of the creek following the huge lodgepole trail. Varnum and Hare advanced with the scouts in front of all the battalions.[69] Hare was to report to Custer "As soon as he should discover any Indians."[70] Varnum recalled reporting back to Custer several times as the command trotted along the main fork of Reno Creek. "From every hill where I could see the [Reno Creek] valley, I saw Indians mounted."[71] Varnum saw ponies being driven by warriors toward the village as he proceeded toward the Little Bighorn River.

Brown Back, Deeds, and others scampered toward the village in front of the quickly advancing soldiers. They soon passed a single, beautifully decorated burial tipi, from their abandoned camp of June 17, where a Sans Arc warrior killed in the Rosebud fight remained. Warriors camped near it but quickly departed after being warned of the soldier advance, leaving behind smoldering fires and other equipment.[72]

White Man Runs Him said, "The soldiers moved fast down Reno Creek, their horses trotting most of the time." [73] As they rapidly advanced, Custer placed greater distance between himself and Benteen, as well as the pack train. Mathey later testified that the train had immediate problems with mules losing their loads as they proceeded toward the Little Bighorn. Still hoping for surprise, Custer sent Cooke back to the pack train to order them away from the main trail to cut down on the dust cloud they were creating.[74] Burkman said the dust was so bad that a man could not see beyond an arm's length.[75]

Meanwhile, Benteen's battalion was riding over terrible country that was wearing out his horses. After riding several miles and ignoring Custer's orders to send a detachment forward, he finally told Gibson to scout ahead.[76] As soon as Benteen departed, he received two other messengers from Custer who repeated the same orders: "Pitch into any enemy," he found and to keep going all the way to the valley. It is clear that he was afraid that Benteen might take some personal initiative and disobey his orders.[77]

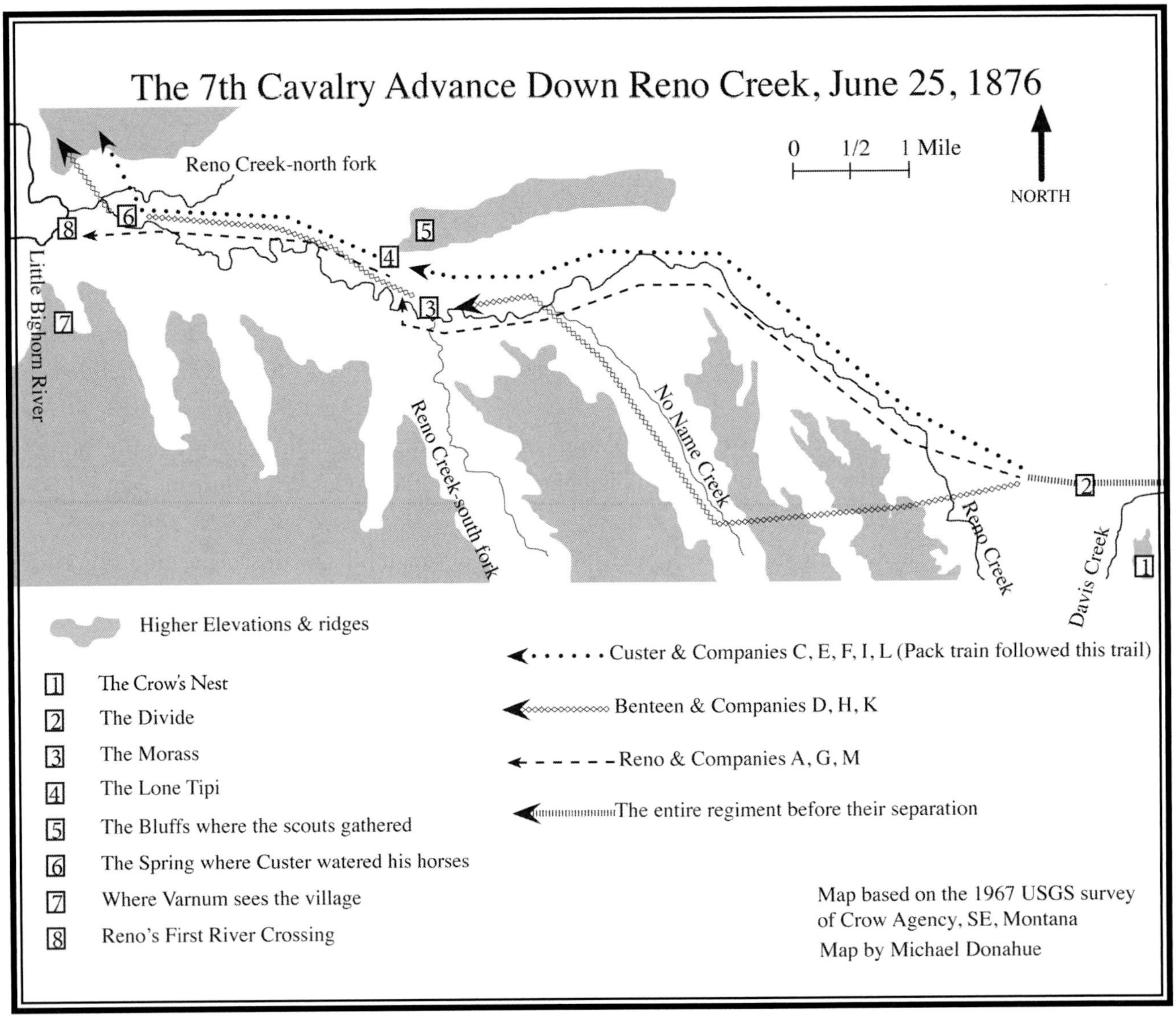

Fig. 62. Map: 7th Cavalry Advance Down Reno Creek, June 25, 1876.

An exhausted Gibson eventually saw cottonwoods along the middle fork of Reno Creek and only more ridges. He stated:

> Up the valley [of Reno Creek] I could see a long distance, but in the direction of the village only a short one, owing to the turn in the valley and the broken character of the country. I hurried back to Benteen, and told him there was no use in going any further in that direction. Therefore in compliance with Gen. Custer's orders, in case no Indians were seen in the valley, we were to return by the shortest route to the trail of the main command, and follow it up.[78]

In reality, they were over five miles away from the river and could not see into the Little Bighorn Valley from that vantage point. Gibson and Benteen's comments make it clear that they could not see into the valley, and they were not in compliance with their orders. Later, in the 1879 Reno Court of Inquiry, when asked if Benteen had complied with Custer's orders, he bragged, "It was scarcely a compliance." Then the questioner queried, "Did you consider it a violation of his [Custer's] instructions?" Benteen said, "I must say I did."[79]

Disobeying his orders, Benteen made an oblique turn to the right in an effort to find the rest of the 7th Cavalry.[80] Soon, his three companies discovered the soldier and Indian trail, as well as a marshy area known as 'the morass.' The men and horses readily sucked down the stagnant water in the cattail-filled pool. Chip Creighton remembered, "It was alkali and would take the hide off your tongue, but it was wet."[81] There was no urgency in the air, so the men took a respite of 15 to 20 minutes to water their horses. Soon, a lone rider trotted past them. It was Boston Custer. He had returned to the pack train and was now attempting to catch up with his brothers.[82] The *New York Herald* reported, "Benteen watered his horses at a pool in the road [trail]. While the battalion was halted, Boston Custer rode up, spoke with several officers and then rode on to the front."[83] He waved and stopped at the morass long enough to converse with Edgerly and others before continuing onward.[84]

As Benteen was climbing and descending the difficult terrain, Custer and Reno had been miles ahead riding on opposite sides of what is today called Reno Creek. Martin recalled, "All the time, as we rode, scouts were riding in and out, and the General would listen to them and sometimes gallop away a short distance to look around."[85] Hare observed, "Custer seemed to be very impatient."[86]

Earlier, the Crow scouts were ahead of the five companies and sneaking through some pine trees on top of some white chalky bluffs overlooking the beautifully decorated lodge. It became known as *the Lone Tipi*.[87] The Arikara joined the Crow scouts; two warriors from the village spotted them and rode away circling to indicate enemies were near.[88] All the scouts and interpreters, with the exception of Varnum, stopped at this point, until the arrival of the five companies, to receive further orders. Varnum continued forward three miles, with only his orderly, to a hilltop overlooking the Little Bighorn Valley. He scanned the valley and observed a few tipis behind timber, about two miles downstream west of his position, but he could not see the extent of the whole village. He was the first soldier to see the village, and he raced back to tell Custer.

The scouts overlooking the Lone Tipi could see the flushed warriors racing toward the river, and Hare wrote a note to Custer that he had finally seen the enemy.[89] The Arikara scouts rode down and slashed open the solitary lodge.[90]. They were disappointed to find no living Lakota—only a dead warrior. One of the Arikara scouts exclaimed, "There is plenty of grub here."[91] Much like an ancient pharaoh's tomb, food had been left for the dead warrior's journey to the other side. Arikara scout One Feather crawled into the fragrant tipi and drank the soup left for the dead warrior and ate some of the meat. While the scouts may have been superstitious about certain things, eating a dead warrior's meal was not one of them.[92]

As Custer and the five companies approached the Lone Tipi, he witnessed his scouts milling around. A detail of about five Company F soldiers had been formed along the way as an exploratory unit and they were asked to look into the tipi. Sergeant Daniel Knipe heard Custer tell his orderly to burn the tipi. As Custer had told Sheridan in the Civil War, "Look for smoke"—if Terry was in the proximity, he would surely see it. Custer waved at Reno across the creek to join him.[93] The Arikara scout assignment of immobilizing the enemy by stampeding their pony herd was not being carried out. According to one scout, the angry lieutenant colonel rode up and hollered, "You were supposed to go right on into the Sioux village!"[94] Their interpreter Gerard rode to a hill about 40 yards from the tipi, and from the higher elevation, he could see the spooked warriors racing over the hills toward the river. Attempting to get Custer's attention, Gerard waved his hat and shouted, "Here are your Indians, running like devils!"[95]

Custer chastised the Arikara, threatening:

> I told you to dash on and stop for nothing. You have disobeyed me. Move to one side and let the soldiers pass you in the charge. If any man of you is not brave, I will take away his weapons and make a woman out of him.

One Arikara challenged him, "Tell him [Custer] if he does the same to all his white soldiers who are not as brave as we are, it will take him a very long time indeed." The scouts all laughed at this rebuttal and signed to Custer that the Arikara were "hungry for battle."[96] The still hesitant Arikara grouped on the north side of the smoking tipi to allow the soldiers to pass. A frustrated Custer told Reno to take the Arikara with him.[97] Any plan of the Arikara stampeding enemy pony herds to dismount warriors began to evaporate due to his scout's fear.

The eight companies were approximately three miles from the Little Bighorn River and could see dust rising over the small ridges to their front. Custer and the troops still were not aware of the huge village in the valley ahead. Shortly after leaving the lone tipi, the lieutenant colonel gave Reno orders that were similar to those presented to Benteen. In typical Custer fashion, he planned to react to the situation as it developed: there was no master plan. While the exact location and wording of the order has been debated, the essence was to attack any Indians in front of him, and he would be supported by the other troops. Some witnesses remembered that Reno received his orders at the Lone Tipi.[98] Reno recounted the order as "General Custer directs you to take as rapid a gait as you think prudent and charge the village afterwards, and you will be supported by the whole outfit."[99]

An uncertain Reno pleaded for assurance that the eight companies would stay together. He asked Cooke if Custer was going to support him, and he answered, "Yes, the General would support him."[100] Needing additional confirmation, he questioned his superior personally and Reno's last words to Custer were, "Let us keep together," to which his noncommittal commander in his jaunty way "lifted his broad brimmed hat as much as to say I hear you."[101]

The rather vague commitment was in keeping with Custer's flexible but uncertain tactical plan. He was flying by the seat of his pants as he had done so many times in the past and planned to react with boldness to any situation. He had asked Benteen to attack anything in front of him and let him know about it. Now he was asking the same thing of Reno.

Custer's greatest weakness as a regimental commander was poor communication with subordinates, as well as superiors. He always kept his cards and plans close to his chest. In his defense, he could not communicate what he did not know. His eagerness to chase the warriors, as well as the timidity of his scouts, did not allow him the luxury of a plan based on sound reconnaissance.

Reno, with his battalion, departed at a rapid gait, and the Arikara scouts had difficulty keeping up.[102] Varnum returned to the column as Reno was leaving and asked where he was going and then shared with Custer his concrete information about the actual location of the village. There is no evidence that Custer relayed this crucial intelligence to Reno after he departed. Varnum and Wallace were given permission to go along with Reno. Fear that had gripped some of the Arikara now spread to some of the soldiers. Two miles past the Lone Tipi, one "set of fours" from Company C began to disintegrate as Thompson's group of four lost two men who returned to the pack train.[103]

With his three companies, Reno trotted toward a small knoll that overlooked the banks of the Little Bighorn River. The swift river was approximately 25 to 30 feet wide and was fairly deep at the crossing.[104] As the soldiers rode around the hill, they could no longer observe the Keogh/Yates battalions behind them. What they did not know was that the other five companies had halted to water their horses approximately a mile behind them.

Cooke and Keogh personally accompanied Reno to the river's edge, and Cooke shouted to Reno, "We are all going with the advance, and Miles [sic] Keogh is coming too."[105] At the stream, some of Reno's men were running their horses and Cooke screamed, "For God's sake men, don't run those horses like that; you will need them in a few minutes."[106] Hare clearly stated that they watered the horses in the river for 10 or 15 minutes.[107]

Brown Back and the young boy named Deeds had just crossed the river ahead of Reno's troops. They were racing as fast as their tired ponies could carry them to warn the camp, but the young boy was unable to outrun the Arikara scouts who had retaken the lead. Thirteen-year-old Deeds was killed near the river, while Brown Back continued toward the camp, screaming a warning that bluecoats were behind him.[108]

As Reno peered across the plain, a small party of warriors zigzagged back and forth in front of him creating a cloud of dust. Edward Davern, Reno's orderly, was near the river and saw "A few Indians [20 to 40] riding around in circles."[109] Another soldier recalled seeing a very large cloud of dust, but no warriors.[110] Reno was the lead force in his first major Indian fight and seemed to have reacted with trepidation as he imagined massive numbers of warriors hiding in the dust about to envelop his battalion.

Gerard was on the other side of the river and could see dust and confusion in the valley beyond. He erroneously warned Reno that warriors were coming to attack him and decided to share the information with Custer. Gerard yelled to Cooke that it looked to him like the Indians were coming to meet them.[111] Cooke promised to ride back and report the new developments, which, unbeknownst to the adjutant and Gerard, were inaccurate. A fearful Reno sent two more messengers within five minutes to Custer with the same message. While Custer was hesitant to communicate his strategy, Reno was more than willing to give updates.[112] Reno's battalion took at least 20 minutes to regroup on the other side of the river.

The five companies with Custer had stopped to water their horses at a freshwater spring ¾ of a mile from the river. He directed the men not to let the horses drink too much, since they had much traveling to do that day.[113] Custer told the Crow scouts, "I am through with the [Crow] Scouts, you have brought me to the Sioux. . . . He called Half Yellow Face and White Swan and told them to go over to the ridge and see if the Sioux camp was there. They started, but instead of climbing the ridge, they followed Reno."[114]

Once again, at a critical juncture, Custer had lost his 'eyes and ears' when he needed them most.

At the watering place, a lodgepole trail headed toward the hills to his right, and Custer had been preoccupied with the problem of side trails along the entire journey.[115] Knipe looked to his right and saw about 60 to 75 Indians on the hill about a mile away watching the command.[116] Lakotas Little Soldier and Standing Bear were among the warriors watching and wondering what the soldiers would do next.[117]

Knipe reported the sighting, and Custer had to make a quick decision—to follow Reno into the valley or address these warriors on his flank. He did not know if this was a decoy, or if there were thousands of warriors hidden over the ridge in a Fetterman-like ambush. If there were large numbers of warriors behind the ridge, they would be at his rear between his attack force and the pack train if he advanced. He could not risk the near catastrophic mistake of the Washita. He had little choice but to ride forward in that direction in order to gather more information.[118]

Hairy Moccasin remembered, "Mitch Bouyer was ahead with the four scouts right behind. Custer was ahead of his command a short distance behind us. Custer yelled to us to stop, then told us to go to the high hill ahead."[119] The five companies rode fast and hard uphill toward the warriors, following the lodgepole trail. Cooke, with Reno's first messenger slightly behind, probably caught up with the column as it was in motion and presented the erroneous report of advancing warriors from the disheartened Reno.

Custer, who rarely divulged his intensions, would pay the price on this day for his negligence to communicate. Though ordered by Terry to send a messenger with a report of his findings along Tullock Creek, he had failed to do so. He had sent no further orders to Benteen or the pack train for over an hour. Three messengers from the Reno command relayed information to Custer, but he did not respond. Not unlike Elliott at the Washita, he would lead a high-spirited confident chase toward a small group of warriors. This decision, in similar manner, would prove to be a fatal one.

Chapter 11

Boys, Hold Your Horses!

A nervous Reno fretted about his crossing of the Little Bighorn River, while Custer and his companies raced toward the bluffs overlooking the Little Bighorn Valley. Sergeant Stanislas Roy, who was lining up his men with Reno, glanced toward the hills to his right as someone shouted, "There goes Custer."[1] Custer's pace was a fast one. John Martin, riding adjacent to George Custer recalled, "The General seemed to be in a big hurry."[2] Like ghostly apparitions, the warriors who were watching them vanished, concealing themselves in nearby ravines, or racing toward the camp below. The lodge pole trail the soldiers followed led steeply uphill to the top of a bluff overlooking the valley, and it turned in front of a long, tall ridge.[3]

Bouyer, Curly, Hairy Moccasin, Goes Ahead, White Man Runs Him, and the Arikara scout Soldier drove their ponies hard to stay slightly ahead, and they were the first to reach the hundred-foot bluffs overlooking the valley.[4] Even though they had warned of a large village ahead, what they witnessed was a shocking surprise—the camp was over a mile in length. Yet as big as it was, they could not see all of it since a taller hill to the northwest blocked their view of a portion of the village. Most astonishing was that the village was intact, and no one was fleeing.

The soldiers arrived next to the tall ridge, known today as Sharpshooter Ridge. It was likely the site where Sitting Bull and One Bull had prayed and left tobacco offerings to the Everywhere Spirit in an effort to protect his Lakota people. In an incredible coincidence, Custer first saw the village from the very area that was covered with Sitting Bull's offerings the evening before. Dog-tired, but adrenaline-filled bluecoats chewed and spat tobacco within sight of the same substance sacredly bagged for *Wakan Tanka*.[5]

Custer took Martin, and both rode to the top of Sharpshooter Ridge to get a better look at the valley, while the soldiers halted in a swale out of sight.[6] Martin recalled that he saw, "children and dogs playing among the tepees but no warriors or horses [in the village] except [a] few loose ponies grazing around."[7] Curly, also described the quiet peacefulness of the scene, and remembered, "When we looked down to the camp we noticed there were not many [warriors] around and Mitch Boyer said he thought the Indians were out campaigning somewhere and suggested we hurry down and fight them."[8] The five companies' rapid pace had afforded a glimpse of the camp before Reno could turn the scene into predictable chaos.

Martin stated, "The General seemed both surprised and glad, and said the Indians must be in their tents asleep." The trumpeter, with his limited range of English slang, likely misunderstood one of

Custer's previous Washita colloquialisms—"We've caught them napping."[9] Most of the warriors were indeed sleeping in their lodges as they themselves testified. Everything he had been told since leaving the divide, as well as information from Reno's messengers, was untrue. "Custer's luck" was amazingly intact. Some of these warriors had embarrassed him along the Yellowstone River; however, this time he had them in his sights, and his comment to Martin reveals that he was reflecting on their previous escape—We have "got them this time."[10]

Martin heard Custer's plan to take hostages of the women and children, disclosing:

> There was then a discussion among the officers as to where the warriors might be and someone suggested that they might be buffalo hunting, recalling that they had seen skinned buffalo along the trail on June 24. Custer now made a speech to his men saying, **'We will go down and make a crossing and capture the village.' And the consensus of opinion seemed to be among the officers that if this could be done the Indians would have to surrender when they would return in order not to fire upon their women and children.**[11] [Author's emphasis]

Daniel Knipe later recounted, "I have always thought that General Custer's plans were to attack the Indian camp at three different points at the same time, to stampede the warriors, to capture the squaws, ponies, and equipments."[12] It is clear that Custer planned to use his successful experience at the Washita again.

As Custer observed the quiet village below, Reno's last messenger, an Irish private named John Mitchell, arrived to tell Custer that warriors were coming to challenge the major.[13] Custer must have wondered where these phantom warriors were, as he could see no warriors racing to confront his fearful subordinate. If he did see something, it was only some ponies and a few herders attempting to get back to the village. Still forming near the river for his attack, Reno was hidden by the bluffs and timber to Custer's left. Martin recalled, "We did not see anything of Reno's column when we were up on the hill. I am sure the General did not see them at all, because he looked all around with his glasses."[14] Martin stated, "He turned in his saddle and took off his hat and waved it so the men of the command, who were halted at the base of the hill, could see him, and he shouted to them, 'Hurrah, boys, we've got them! We'll finish them up and then go home to our station.'"[15]

Custer came down off the hill and told Keogh and Cooke what he had seen.[16] The lieutenant colonel took off his sweaty buckskin coat and tied it to the cantle of his saddle as the temperature neared 91 degrees. "Custer rode up and down the column talking to the soldiers," recalled Ree scout Soldier, and the "soldiers cheered and some tied handkerchiefs around their heads and threw hats away."[17] The Company F detail went forward to see what lay ahead.[18] Officers shouted, "Attention!" "Fours Right!" "Column Right!" then the blue column trotted quickly forward over a small rise in front of Sharpshooter Ridge that allowed for the enlisted men's first glimpse of the huge village.[19] The excited soldiers hollered and gave three cheers as they galloped past Tom and George Custer as well as Martin, above them.

Little Bighorn Battle Map 1

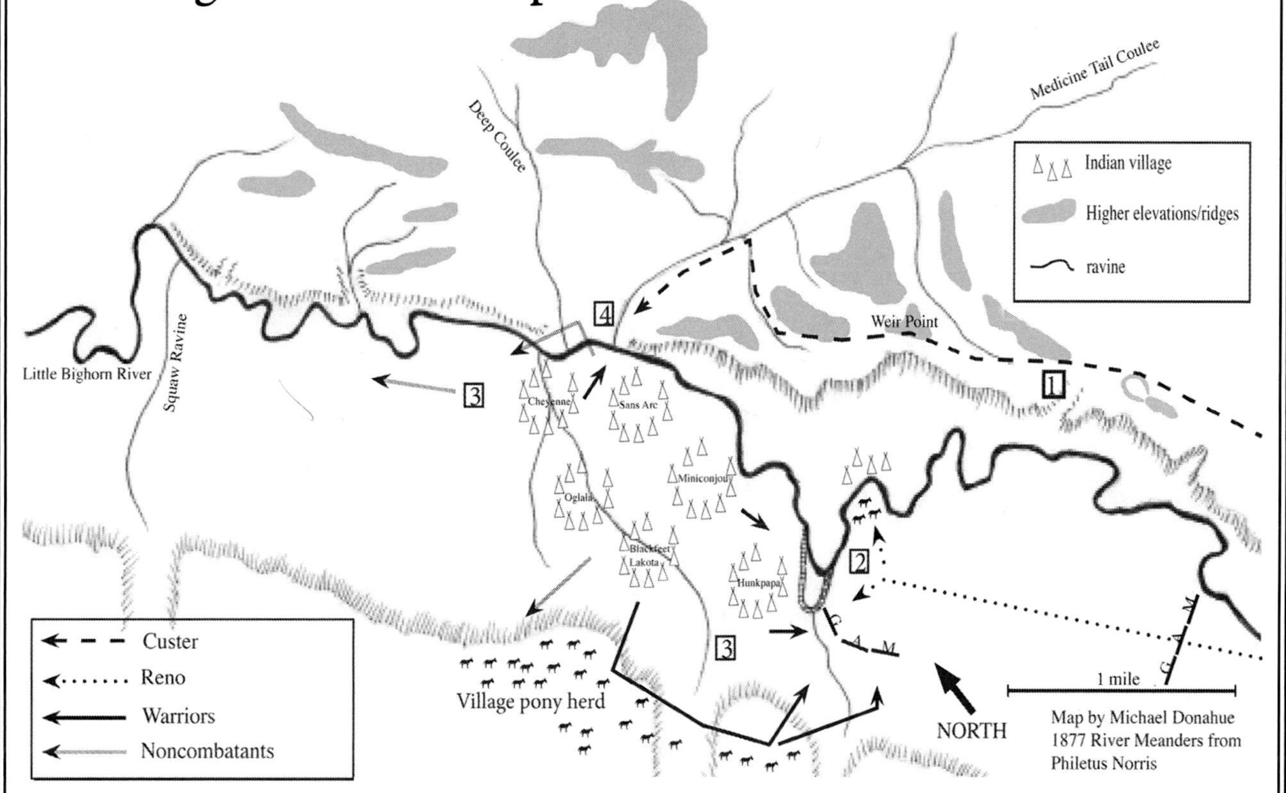

1 Custer with companies C,E,IF,L stop at Sharpshooter Ridge and sees the unaware village. He advances along the bluffs as Reno charges across the valley. Custer continues over the Weir Point into Medicine Tail Coullee.

2 Reno charges with companies G,A,M and dismounts to form skirmish line with M on the left bank, A in the middle and G next to the river. The line advances while the horses are hidden in the nearby timber.

3 Warriors respond to Reno's attack as noncombatants flee across the river and downstream, while a few run to the western bluffs.

4 Ten minutes after Reno begins his skirmish, Custer's five companies engage warriors near the river for approximately 20 minutes.

Fig. 63. Map: Little Bighorn Battle Map 1.

Custer shouted, "Boys, hold your horses, there are plenty of them down there for us all."[20]

The hard climb to the bluff top had taken its toll, and tall sergeant "Gus" Finckle, in his strong German accent, hollered to Knipe that he couldn't make it as his horse was giving out. Knipe shouted, "Come on Finkle [sic], if you can."[21] Thompson as well as another private named James Watson were also unable to keep up and fell out of the column.[22] Ree scouts Soldier, White Eagle, and Bull attempted to keep up with Custer but could not, and for the first time, could see the entire village. Minutes later, they saw fellow scouts driving ponies up to the bluffs nearby and scout, Soldier, joined them as they retreated toward Reno Creek.[23]

Tom Custer turned and gave Knipe a verbal order, "Tell McDougall to bring the pack train straight across to high ground—if the packs get loose don't stop to fix them, cut them off. Come quick. Big Indian camp." To clarify the order, Tom amended the order that if the loose packs were ammunition packs, not to cut those, "and if you see Benteen tell him to come on quick—a big Indian camp."[24]

At that moment, Reno commenced his charge across the valley floor and the panicked villagers began to react to the soldier advance.[25] Martin stated that, at the head of the column, he could not see Reno, or hear gunfire as they went down the ravine, as the intervening landscape made it impossible.[26] Bouyer, with at least three of the Crows, continued along the bluff's edge and went up to the highest-peaked hill on the bluffs, today called Weir Point.[27] It seems likely that young Curly, who had a quarrel with the same men the previous evening, left the soldiers and other scouts near Sharpshooter Ridge, back-tracking down Reno Creek.[28]

The mood of the two groups of bluecoats about to assail the Indian camp could not have been more different. On the higher elevation, the officers and enlisted men cheered as they looked down on the tipi town. Custer, with his speech, got the adrenaline pumping and the soldiers had to be warned to slow down and get behind their leader. These men anticipated a victory only minutes away.

In the valley, only minutes earlier, Reno, shaded by a broad-brimmed straw hat, watched his troops ford the Little Bighorn River. De Rudio spurred his stubborn mount to make the jump into the river near Reno, and water splashed all over the agitated major, who sarcastically shouted, "What are you trying to do? Drown me before I am killed?"[29]

Reno's men felt apprehension, uncertainness, and fear. Some soldiers overheard Gerard's fearful proclamation that the Indians were not running away, but coming to attack them. A few soldiers saw Reno getting his courage from a small pocket flask, which did little to encourage confidence. A few of Reno's troops began to cheer, and the deflated major shouted out, "Stop that noise!"[30] To their front, Reno's men could see nothing except dust, and a few warriors riding in the brown dirty haze. Arikara scout Young Hawk was equally pessimistic. He "had a bunch of loose eagle feathers, [then] he unbraided his hair and brought it forward on his head and tied it with the eagle feathers. He expected to be killed and scalped by the Dakotas."[31]

From the village, One Bull and White Bull could see this activity to the east of the Hunkpapa camp and noticed a small herd of Lakota ponies.[32]

Actually, no warriors were rushing in large numbers toward Reno's men, and a majority of the village was unaware that soldiers were about to gallop across the valley floor.[33] Reno ordered his three dripping companies forward into columns of two men abreast, and their gait increased as they proceeded with Company G trailing Companies A and M. Varnum screamed out, "Thirty days furlough to the man who gets the first scalp." Ryan humorously recalled, "We were anxious for the furlough, but not so particular about the scalp."[34] Varnum was ahead of the charging soldiers and stated, "There was no absolute contact between his [Reno's] command and the Indians."[35] Warriors were riding back and forth creating a huge veil of dust but did nothing to stop the soldiers.

According to Wallace, with Company G, they were eating dust from the other two advancing companies. As the valley widened, they were ordered to move up to the left into line next to the other two companies. "'Left into line, gallop!' 'Forward guide right!' and away they rode as fast as they had ever ridden on horseback," according to one private.[36] Reno had a fiery mount and had to do everything in his power to control his horse and carry his carbine.[37] One private noted that the officers took off their coats, since this designated them as chiefs, and warriors would make them specific targets.[38]

Brown Back, who had witnessed young teenager Deeds being killed by the scouts behind him, arrived in the Hunkpapa camp minutes ahead of Reno to warn the Hunkpapa circle. Almost every Lakota and Cheyenne account affirmed that the village was surprised by the soldier attack. Their victory along the Rosebud, their ever-increasing population, and Sitting Bull's vision of success helped lull the villagers into a feeling of complacency and false security. While there were rumors of nearby bluecoats, it was not enough to be a major concern. Oglala warrior Foolish Elk summed it up, "There was not much concern about the soldiers, as the Indians thought they had enough men to whip any force that could come against them, seeing that the Indians were all together for once—all the different tribes."[39] Lakota Eagle Bear remembered:

> None of the Indians expected a fight. The children were swimming in the river and playing on the bank. The women were cooking and packing their stores as we planned to break camp . . . We were not prepared to fight and even when we saw the soldiers coming, we did not make ready for battle.[40]

Other warriors grabbed ponies and mounted quickly, riding forward to shield the panic-stricken noncombatants, who ran toward the benchlands to the west, while a larger, panicked multitude fled toward the supposed security of other tribal circles.

From the valley floor, both Varnum and warrior Gall got a glimpse of the gray horse company riding along the high bluffs across the river at the same time.[41] Reno's allied scouts broke into two groups and charged after warrior horses grazing near the western hills and the river timber.[42] One group of Arikara, with their captured horses, rode up a steep ravine on the opposite side of the river, ascending to the area near Sharpshooter Ridge soon after Custer's departure.[43] Simultaneously, Bobtailed Bull and a few Ree scouts showed extreme bravery by riding near the western bluffs, beyond soldier support, and skirmished with the warriors near the larger pony herds on the benchlands.[44] These Ree scouts captured a few prized ponies, but step one of Custer's plan to immobilize the Lakota and Cheyenne had failed.

Some of the Arikara may have inflicted the first casualties in the battle. After killing 13-year-old Deeds, they also killed some women and children in the area, near the timber. Gall lost two wives and three children and, from accounts, they may have been killed near the Hunkpapa camp.[45] Herendeen saw the bodies of six women in the ravine, where Gerard had spotted the Arikara scouts. Herendeen declared, "Our men did not kill any squaws, but the Indian scouts did."[46] Similar to the Washita, allied scouts probably accounted for most of the noncombatant casualties.

Three men of Company M could not control their horses as the soldiers got near the village. They yanked viciously on their reins, but they could not stop the momentum of their steeds, and privates George E. Smith and James Turley rode to their deaths.[47] Reno screamed as loud as he could—"Dismount," and "Fight on foot!" The soldiers in the valley could only get glimpses of the tipis and warriors through the brush and gritty dust. The soldier-led horses were placed in some nearby woods for protection.[48]

A shallow ravine, in front of the soldier line, Shoulder Blade Creek, offered protection to the warriors on foot who jumped into the recess. Hunkpapa Iron Hawk described the counterattack:

> By now warriors were running toward the soldiers, and getting on the ponies, and many of the Hunkpapas were gathering in the brush and timber near the place where the soldiers had stopped and got off their horses. I rode past a very old man who was shouting, 'Boys, take courage! Would you see these little children taken away from me like dogs?'[49]

Company A remained in the middle of the three companies as the two outer companies flipped their respective positions, according to the tactical manual of the time. Company M, which had ridden on the right side of the charge, now raced to the exposed left end of the skirmish line. It was not an enviable position, since it extended perilously alone on the open plain. Company G ran over to the river's edge, and far right of the line, and was offered the protection of a swampy morass, to keep warriors away.[50] Reno controlled a scant 90 men on his firing line, against a force of gathering warriors that quickly numbered in excess of 300, and it was growing larger by the moment. The thin blue line advanced on foot about 75 to 100 yards.[51] As men advanced, they were supposed to keep intervals of several yards between each man. One sergeant recalled, "It [Company M] was the only company that had any skirmish line. The other two companies were on its right and huddled into a bunch."[52] Another soldier observed, "The men were in good spirits, talking and laughing and not apprehensive of being defeated, and the Sioux, toward the village were riding around kicking up a big dust but keeping pretty well out of range."[53]

In the village, 13-year-old Black Elk quickly mounted his buckskin pony upon hearing the warning. He stated, "Then out of the dust came the soldiers on their big horses. They looked big and strong and tall and they were all shooting."[54]

Low Dog proclaimed:

> We retreated until our men got together, and then we charged upon them. I called to my men, 'This is a good day to die; follow me.' We massed our men, and that no man should fall back, every man whipped another man's horse and we rushed right upon them.[55]

A Hunkpapa warrior named Hawk Man lead a fatal charge that no one followed, and became an early casualty.[56]

John Ryan commented, "There were probably 500 of them [warriors] coming from the direction of their village. They were well-mounted and well-armed. They tried to cut through our skirmish line. We fired volleys into them and repulsing their charge . . . emptying a number of saddles."[57] Black Moon, Swift Bear, Dog with Horns, and a warrior named White Bull [not Sitting Bull's nephew] were killed in the fight with Reno's soldiers.[58]

Crazy Horse, who was bathing in the river, was definitely unprepared for battle. He raced to his lodge to paint dots representing hailstones on his body, and a distinctive lightning bolt across his face. This lengthy preparation took about 10 minutes, causing concern for the few impatient warriors waiting to follow him into combat. He finally appeared and shouted:

> Here are some of the soldiers after us again. Do your best, and let us kill them all off today, that they may not trouble us anymore. All ready! Charge![59]

During the chaos, Sitting Bull warned the women and children in the village to flee to the other camps and hills to the west. One Hunkpapa eyewitness described the chaotic scene, "Sitting Bull was in front of [the] warriors . . . everyone [was] talking and could understand no one."[60] The spiritual leader escorted them to the western benchlands.

Reno stood in the middle of the line with his well-liked adjutant Benny Hodgson beside him. The jokester Hodgson was a West Pointer from Philadelphia who would miss the fireworks of the Centennial Celebration in his hometown. Instead, he was witness to the pop, pop, pop of the deadly fireworks in front of him, as bullets flew in his direction, he ordered the men to remain calm and fire low.[61]

Reno was notified of a horrifying development 10 minutes into the fight. The warriors, who could not reach their own ponies, were firing into the cavalry horses and holders from across the river.[62] The right end of the line, manned by Company G and dark-skinned Lieutenant Donald McIntosh, was a short distance from timber. Reno grabbed them and disappeared into the heavy brush to investigate.[63]

The skirmish line existed for approximately 15 minutes. Seeing their precarious situation on the exposed left, some of the allied scouts, as well as Bloody Knife, Varnum, Reynolds, Gerard, Herendeen, Dorman, Dr. DeWolf, and Dr. Porter quickly disappeared into the safety of the trees. Multitudes of Lakota poured out of the benchlands and attempted to flank the left end of the dangling soldier line. Ryan, along with his cool-headed commander French, held the ground as bullets zipped past their heads.

The two remaining companies found themselves on the firing line alone against warriors who fought more aggressively as the soldier numbers shrank. Reno's offensive thrust turned into a defensive fight for survival, within only 10 minutes. In the timber, Reno could see no support from Custer. He was heavily outnumbered and now feared being stranded without horses, along with a dwindling ammunition supply.[64]

Reno ordered the rest of the line to withdraw into the timber for protection.[65] Moylan's men marched into the timber, leaving French's company alone on the line trying to withdraw as well.[66] Private William Morris stated, "Some started for the timber. Captain French countermanded the order

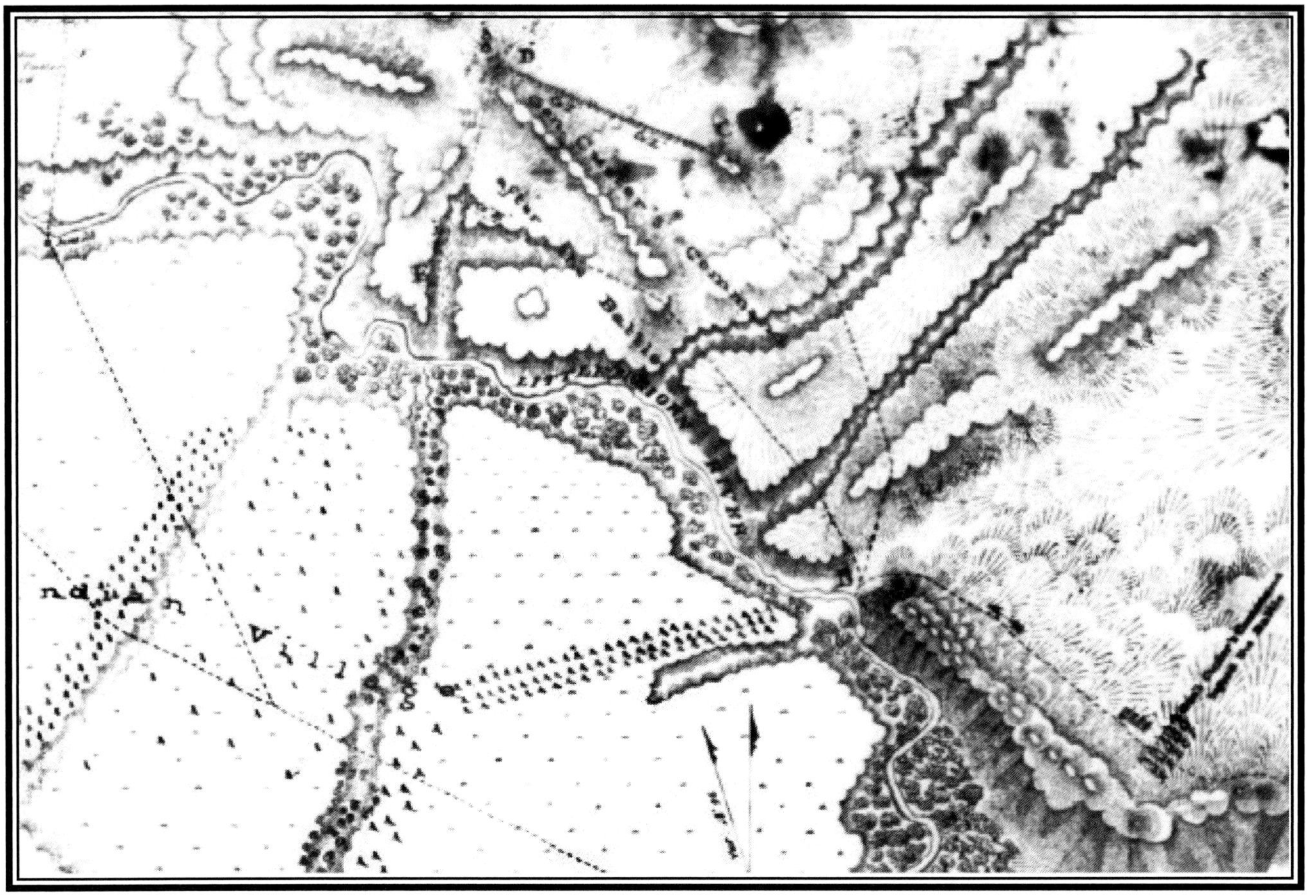

Fig. 64. Map: Edward Maguire map of the battlefield. Maguire's maps were heavily criticized for topographic errors by soldier eyewitnesses. He sketched Custer's five companies riding behind the edge of the bluffs before dividing into two separate divisions heading toward the battlefield [note dotted lines]. Note the large ravine which empties into the river, where Custer attempted to cross. Testimony by both warriors and soldiers indicated it was a difficult crossing due to boggy quicksand and may have been one of the determining factors for Custer's northward move.

Courtesy of National Archives

immediately: 'Steady, men! I will shoot the first man that turns his back to the enemy—fall back slowly. Keep up your fire!' I called to Sergeant [Miles] O'Hara, 'Come on, O'Hara! We are falling back to our horses.' He did not respond."[67]

O'Hara had left Fort Lincoln as a corporal and was promoted to sergeant only days before the battle. He was the solitary fatality on this line. Shot through the breast as the warriors pressed the line, he fell mortally wounded. As the rest of the men retreated, O'Hara screamed, "For God's sake don't leave me."[68]

At that moment, high above the valley fight, near Sharpshooter Ridge, a lone rider with a trumpet bouncing under his arm observed Reno's skirmish line withdrawing. He was traveling as fast as his exhausted and bloodied horse could carry him back toward Reno Creek. It was John Martin. He had been assigned to carry one of the most important messages in the history of Indian warfare, and it was stuck under his drooping, sweaty felt hat. He looked into the valley and was shocked by the images below him. The quiet scene of women and children leisurely going about their respective roles, only minutes

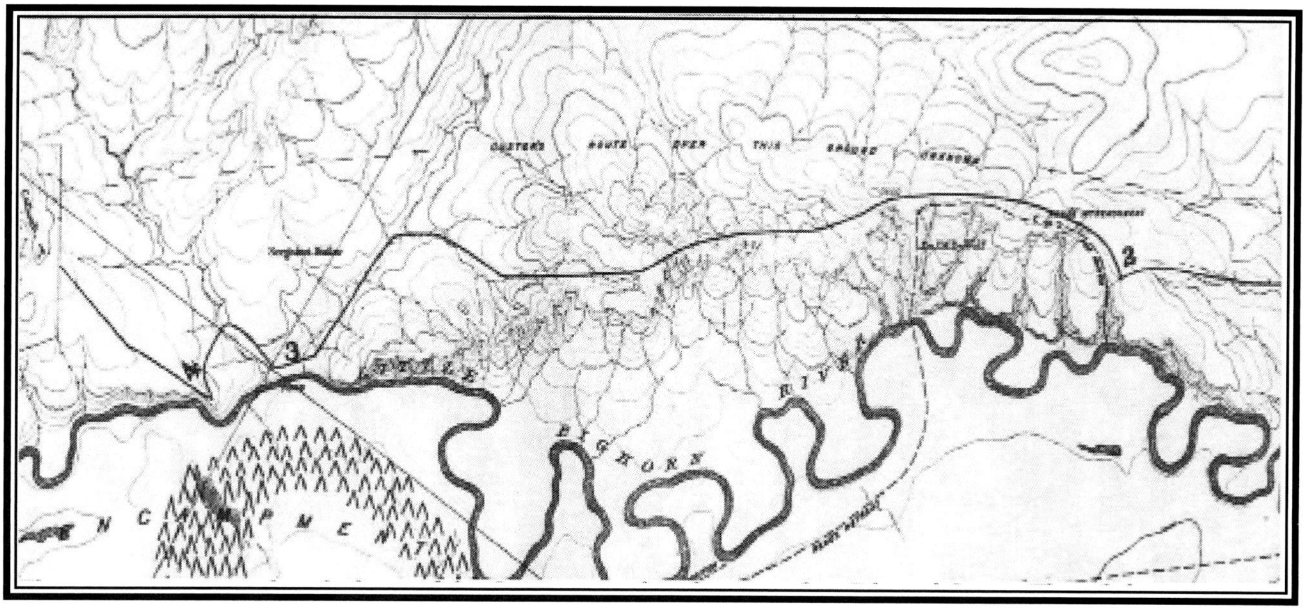

Fig. 65. Map: Edward Curtis battlefield map. This map was created with three of the Crow scouts by photographer, Edward Curtis. They did not remain as long as suggested by this map, but were eyewitnesses to the trail Custer took behind the bluffs in his advance toward the river. The trail goes over Weir Point [shown at the center of the map] and then down a ravine that led to Medicine Tail Coulee. The sharp left turn reinforces the testimony of John Martin as well as Gustave Korn and reinforces Maguire's map route of Custer's five companies.

Courtesy of Smithsonian Institution

earlier, had drastically changed into one of chaos and mayhem. Hundreds of Lakota and Cheyenne warriors were pressing Reno's faltering line.[69]

Only about 15 minutes had elapsed since he and Custer had left the bluffs. The five companies with Custer had galloped along the high ground toward the highest promontory on the battlefield, today named Weir Point.[70] The high ridge had a curved dip, that resembled a saddle, and the troops rode over this natural passage. It took only three minutes to gallop to Weir Point, where they had their first glimpse of the village at the south end of Sharpshooter Ridge. Martin evidently stayed in the 'saddle' of Weir Point as George, Tom, and Autie Reed rode to the summit just yards away.[71] Bouyer and the three remaining Crow scouts watched the valley charge of Reno from this area as well. Thompson, from atop his faltering gelding, watched as "the last of the companies had disappeared over the crest of the hill."[72] Later, Captain French made several references to the ridge, from which "None were ever seen alive again." While he does not state who saw the command along the bluffs, he wrote in a letter to the Cooke family, "The command was last seen crossing the saddle of the long ridge beyond which the terrible tragedy that robbed us of your dear one took place." He also noted that the gray horse troop was in the advance.[73]

Given the timing, both the Custers and the scouts would have seen the dismounting of Reno's troops and the villagers beginning to react to his presence. More importantly, it was their first view of the entire village, which had earlier been blocked by the hill on which they were now standing. After going over Weir Point, the troops could look to their left and see the whole village for the first time. Trumpeter Martin said, "About half way down to the Little Bighorn we came into full view of the village."[74]

The small group of F Company men, acting as a flanking unit, rode off toward the high ridges east of the troops, kicking up a huge cloud of dust.[75]

Tom and George Custer as well as Autie and Bouyer quickly joined the troops below. White Man Runs Him, Hairy Moccasin and Goes Ahead continued along the bluffs above the descending troops. The five companies reached a wide ravine that led to the river, today known as Medicine Tail Coulee.[76] Godfrey, who knew Custer well, believed, "He [Custer] must, then, have expected to find the squaws and children fleeing to the bluffs to the north, for in no other way do I account for his wide detour to the right."[77]

Medicine Tail Coulee, Montana Territory
Sunday, June 25, 1876, around noon

There was muffled silence, except for the clop of hoofs on the soft soil, hard-breathing horses, creaking leather, and tin cups banging against the saddles as the zealous soldiers dropped out of sight into a dry, descending ravine. Above them was a clear blue sky with few clouds. Custer was dressed in fringed buckskin pants that were tucked into his tall, knee-high boots with red tops. At age 36, a broad-brimmed hat covered his receding hairline, and he wore a blue-gray flannel shirt.[78] If only he had his band here to play *Garryowen* and inspire his men.

Keogh was anxious to lead his first attack against an Indian village as the senior ranking officer under Custer. Trumpet calls were sounded for movements, and officers shouted, "Column left!" as they turned sharply into Medicine Tail Coulee. Suddenly, they could see the cottonwood trees and brush along the river as well as the clustered tipis beyond.

Parties of panicked women and children, trying to escape Reno's attack from the south, crossed the river and ran up the ravine, until they spotted the approaching bluecoats. They quickly turned and fled back across the stream.[79] The wife of Spotted Horn Bull remembered, "From across the river I could hear the music of the bugle and could see the column of soldiers turn to the left, to march down the river to where the attack was to be made."[80]

Two small Cheyenne girls name Laura and Lenora Arapaho Chief saw the bluecoats coming toward the river. Laura asked her mother, "Who are those people?" She was told, "Those are the people who have come to kill us!" and young Laura was so frightened that she fainted.[81]

Custer realized that the lodges continued past the opening in the bluffs at the mouth of Medicine Tail Coulee; he still was not to the northern end of the camp. It truly was a big village. Custer had not seen the Reno fight, and there was no serious resistance to his own advance. Somewhere in the vicinity there were thousands of warriors who could descend on his approximately 151 men escorting the mules, and the 26,000 rounds of carbine ammunition at any time; that could be a truly disastrous situation.[82] At the Washita, Custer had feared an attack on his pack train and now the same applied. He had limited ammunition and needed to consolidate his scattered forces, so Custer finally decided on more urgent communication with Benteen and Mathey.

The command halted in Medicine Tail Coulee for a few minutes to allow the men to tighten their saddle girths, get drinks of water, and load their weapons. The long-whiskered Cooke turned in his saddle, and from his saddlebag, pulled out his stenographer-type notebook that was hinged at the top. He was in such a hurry that he wrote the note upside down on the page—it mattered little. In hurried, almost unreadable script, he jotted down a six-line order in pencil and hurriedly tore it from his notebook.

Benteen
Come on. Big
Village. Be quick.
Bring packs.
W.W. Cooke
& bring pacs [packs].[83]

Fig. 66. Adjutant W. W. Cooke was from an influential Canadian family and was considering a career change when killed at Little Bighorn. He was the writer of the last message and his long whiskers were too much of a temptation for a victorious warrior—one of them was scalped. This unpublished photograph was taken shortly after the Washita fight.

Courtesy of George Kush

Custer took the time to read it before he called for a messenger. An orderly stepped forward, but Custer said, "No, no, the other man," indicating Martin. The trumpeter recollected, "So I stepped out and was told to ride as hard as I could to Benteen and report to Custer again on Benteen's arrival."[84] He was to retrace their trail and to return quickly if he saw no danger.[85] Martin glanced back to see Custer and Bouyer in front of the command.[86]

Custer then rode to company commanders and ordered them to prepare for the attack. Each officer shouted, "Fours right, into line!" As the men lined up, Custer rode along the line, giving his final instructions to the company commanders. Several officers, including Captain Yates and Adjutant Cooke, had orderlies leading extra horses, and Custer realized that they would be unable to fight, as well as a hindrance, during the charge. Custer looked at Yates and said, "Better send that man back to the pack train with your horse. He can't accomplish anything in a charge, leading your horse."

Private Edwin Pickard, who was leading Captain Yates' extra horse, felt it was his bad luck since he hoped to collect some captured buffalo robes and beaded moccasins from the village. He encouraged an inexperienced trooper to take his place, but the soldier refused. Yates said in a kindly way, "Too bad Pickard, but I guess you will have to go back."[87] As Pickard rode back, some warriors in the area fired on him but missed.

Private Charles Banks, who was serving as an orderly for Cooke, was ordered to take the Canadian's extra horse back at the same time. Custer also wanted Banks to tell the cooks to get up quick and prepare a meal for the hungry men since they had not eaten lunch due to the hard riding. Banks survived by riding back up toward the reserve units. One of Custer's last thoughts before the battle was the comfort of his fellow soldiers.

It is clear from Custer's last comments that he believed this would be another rendition of the Washita fight. He would drive away or kill resisting warriors, capture nearby noncombatants, seize a portion of the village for destruction, and feed his men in the Indian camp. Banks and Pickard, along with Martin would survive immediate death due to their last-minute assignments.[88] Only one more soldier, Gustave Korn, riding next to Keogh, would depart after the short halt in Medicine Tail Coulee and survive to tell the tale.

John Martin's tired horse struggled up the steep incline following the trail he had just taken. He veered to a prominent knoll where he could see the entire area and heard gunfire. Martin noted:

> I looked behind and saw Custer's command over on the flat [an area next to the river] and Indians over in the village riding toward the river and waving buffalo hides. The battalion appeared at this time to be falling back from the river. This was the last I saw of the five companies alive and it was only a hasty glance.[89]

Martin also remembered, "Riding fast I soon reached the crest of the hill, and looking back, I could see that the Indians had already attacked, and our boys were acting very excitedly. I rested my horse on the brow of the hill a minute and sat watching the action in the distance."[90] Martin, in another account, explained what made him turn to look in Custer's direction. "I heard a volley and looked back and saw Custer retreating back from the river."[91]

Warriors, hidden in nearby ravines, aimed at the orderly and missed by inches. A bullet hit Martin's already struggling horse in the rear, right behind the saddle, spattering blood on the soldier's back.[92] As Martin rode over Weir Point, he met Boston and an unidentified Company C soldier. Boston was riding at a dead run, looking for his older brother. He shouted to Martin, "Where's the General?"

He also wanted to know if he had been attacked. Pointing back over the ridge, the Italian answered, "Right behind that next ridge you'll find him," and, "No," he had not been attacked. Martin, after hearing the firing behind him, warned Boston he had better look out, as there were Indians nearby. Boston said, "Well, I am going to join the command anyhow."[93]

Boston, who had waved at Benteen at the morass approximately 20 minutes before, cut through to his brothers, temporarily escaping the enveloping danger.[94] Such timing means that if Benteen had not watered his horses for such an extended time, he would have been involved in the beginning stages of the fight alongside Custer. His later suggestion, that Custer's five companies were dead by the time he received the important message, was unfounded and untrue.[95]

Neither Custer nor Reno knew what the other was doing. Almost simultaneously, Custer's command backed away from the Little Bighorn River, while Reno charged toward the same river, in a move that would be the bloodiest five-minute ride in 7th Cavalry history.

Chapter 12

Chases Along the Greasy Grass

As Custer and Reno both withdrew from different parts of the river, White Man Runs Him, Hairy Moccasin, and Goes Ahead sat on their horses looking down at the confusion among the tipis and at Custer's companies. Goes Ahead said, "[We] heard two volleys fired and saw the soldier horses standing back of the line in groups."[1] The Crow scouts sniped into the swirling mass of noncombatants and the warriors below, before retreating toward Reno Creek.[2]

Reno's attack preceded Custer's by only about 10 minutes, driving the noncombatants straight toward the five companies. According to Rain-in-the-Face, "When [the] fight commenced, Indian women and children left camp and crossed the river. They discovered Custer and recrossed and went into camp farther up the river."[3] A majority of the noncombatants took refuge in a ravine that emptied into the Little Bighorn River, aptly named Squaw Ravine or Chasing Creek. In the village the warning was sounded of the new, impending soldier threat. Excited warriors attempted to round up horses, while many on foot raced to the riverbanks to halt the bluecoats.[4] Lame White Man warned Cheyenne Roan Bear, Calf,

Fig. 67. Crow scouts White Man Runs Him, Hairy Moccasin, Curly, and Goes Ahead after the battle. The scouts gather around Custer's original burial spot marked here by a wooden cross. Curly supposedly left early before the battle and the others retreated shortly after. These four scouts were sought after for their testimony since they were among the last to see Custer alive. Circa 1913.
National Park Service, Little Bighorn Battlefield National Monument, LIBI_00019_00435

Bobtail Horse, and White Shield that there were too many bluecoats approaching, and any attempt to stop them would be in vain. They ignored him and rode to the riverbank anyway.

Warrior Shave Elk remembered, "All along the river there were many dismounted Indians and the banks were steep everywhere except at Medicine Tail Coulee. The soldiers came down the coulee toward the river and stopped just a little while, but not long, and the Indians crossed over and attacked them. There were a few soldiers ahead of the main body."[5] Bouyer, Custer, and Keogh led the column as it trotted over the broken ground toward the river.[6]

A young, hazel-eyed Prussian with a large moustache rode to the edge of the river. Gustave "Yankee" Korn was Keogh's orderly and near his commander as Company I approached the river. According to Korn, he was sent slightly ahead to check the river crossing. He said, "The entrance to the stream was not in good order, but the other shore was excellent." A closer inspection by other soldiers would prove his assessment incorrect. Korn stated:

> I was turning to go back to my comrades and report the condition of the stream, when immediately in my front the Indians opened fire from the tall grass where they had been concealed. I was not wounded but a rifle ball struck my horse's neck and rendered him unmanageable. I was carried away toward Maj. Reno's position.[7]

In a variation of the story, Korn stated:

> We shortly came in sight of the village. 'Charge!' was the order. My horse became frantic and plunged madly through the ranks straight for the Indian village. Gen. Custer called out to me as my horse galloped past him, not to go towards the village. This was the last time I ever saw Custer alive. I managed to rein my horse around and stopped him when about half way up the big bluff.[8]

Korn would be the last soldier to see Custer alive and survive the battle.

Eyewitness White Shield saw seven groups dismount near the river.[9] He was one of the dismounted Cheyenne firing across the river and saw Company E near the river as well. He remembered, "When the Gray Horse Company got pretty close to the river, they dismounted, and all soldiers back as far as I could see stopped and dismounted also."[10] "Only two companies charged on Bobtail Horse and [his] party," remembered the Cheyenne warrior.[11]

As Gustave Korn suggested the soldier side of the riverbank was cut up with thousands of hoof prints from horses and was called a 'watering place' because a large number of Indian ponies had occupied the area before the fight. Lame White Man and Horned Horse noted the difficult crossing at the mouth of Medicine Tail Coulee on their side of the river.[12] The bank on the village side of the river was 18 inches higher than the water and covered with a miry quicksand.[13] Dr. Charles Eastman, an instructor at the

Carlisle Indian School and collector of Indian oral histories, relayed stories told to him by participants in the fight. He recorded:

> Custer with his formidable column appeared on the slope of the ridge. They knew well he could not cross the river at that point. He must go down half a mile. The crossing therefore became at once of first importance . . . Custer appeared at the river bank. Having discovered that it was impossible to cross, he began to fire into the camp, while some of his men dismounted and were apparently examining the banks.[14]

Soldiers realized that the crossing was a problem as Cheyenne gunfire erupted from the brush. According to Cheyenne oral history, one soldier could not stop his horse and rode to his death in the village.[15] Maud Arapaho Chief told of another casualty:

> When the soldiers came across the river, it was right by the Cheyenne camp. Before they made a charge to go across, they were looking for a place to cross. Some got caught in a crossfire and one leader was shot . . . When that man got shot, two riders came on either side and one to the front and grabbed the reins and they kept the man on his horse, turned around, and went back across.[16]

Soldier William Philo Clark interrogated Oglala Horned Horse and reported:

> The Little Big Horn is a stream filled with dangerous quicksand, and cuts off the edges of the northern bluffs sharply near the point where Custer perished . . . He [Custer] made a dash to get across, but was met by such tremendous fire from the repeating rifles of the savages that the head of his command reeled back toward the bluffs after losing several men who tumbled into the water, which was there but 18 inches deep, and were swallowed up in the quicksand.[17]

Another warrior account noted, "An officer was killed where Custer made his first stand—nearest the river. The officer had a pair of field glasses and a compass—[wooden box]."[18]

Who were the early casualties at the river's edge? Some have suggested that George Custer was killed or seriously wounded at the river's edge. However, from the fact that the headquarters staff was found around his body and cartridges from his weapon were discovered under his body, this is highly unlikely. Even more convincing is the later battle movement toward the noncombatants that was a typical Custer tactic.[19]

While there can be no certainty, there is evidence that indicate some strong possibilities of who was killed or wounded near the river. Young Jack Sturgis, the son of the colonel of the 7th Cavalry, may have been killed at the river. He was attached to the gray horse company and was the acting engineer on the Reno scout. He most likely would have carried field glasses as well as a compass, like those recovered by the warriors there.[20]

Circumstantial evidence leads to Keogh being seriously wounded at the river's edge, and he was seen leading the troops by Korn. Keogh was itching to lead his first charge into an Indian village as senior captain. As the bullets poured into the advancing group, the dandy dressed in buckskin was likely shot through the knee. His shoe and bloody canvas legging were found near the riverbank. He was riding a government horse named Comanche that had a matching bullet hole in his flank where the officer's knee was situated. Keogh ended up on higher ground, probably in a hospital area.[21] Custer was holding on to his wounded as ordered by Terry.

Another soldier in the area that was probably killed in the initial charge was the buckskin-clad first lieutenant of Company I—James E. Porter. Captain Thomas French noted, "Lt. Porter was not found but his blood-stained blouse was discovered in the Indian camp alongside his horse, Fritz, and it is believed that he was killed there and his body thrown into the river."[22]

Sergeant James Bustard, under Keogh, may have been the soldier with stripes on his arm that rode across the river, although other accounts place him elsewhere. Daniel Knipe later stated, "[The] sergeant of Troop I was lying, he and his horse, across the ford in the Indian village or camp. He was not mutilated, [and] his horse was some 20 or 30 steps from him."[23] Another soldier recalled seeing a horse that Bustard had borrowed lying dead on the village side at the ford.[24]

Some of the casualties near the river were moved. Two Moons stated that, "Here [next to the river] some soldiers were killed and were afterward dragged into the village, dismembered and burned at [the] big dance that night."[25]

Custer had an immediate dilemma, and it was not the death of his lead soldiers that blunted his attack. He had found an impassable crossing, was taking warrior gunfire, and now had wounded men to care for. Every fourth soldier took the other three men's linked horses and led them behind the skirmish lines for protection. The tired horses bucked and reared due to the gunfire and war whoops from across the river. According to one warrior:

> Contrary Big Belly made a charge down in front of the Gray Horse Company and from where the Indians were, they saw that the horses of one other company began to get frightened and started to circle around the men who were holding them. When Contrary Big Belly got back, the companies began shooting fast.[26]

A Cheyenne named Soldier Wolf stated, "For some time [they] fought in the bottom, neither party giving back . . . and two soldiers were killed and left there, but soon the Indians overpowered the soldiers and they began to give way, retreating slowly, face to the front."[27] The statuesque Tall Bull relayed, "For quite a long time the troops stood their ground right there. They then began to back off, fighting all the time, for quite a distance, working up the hill."[28] He also recalled, "Some [soldiers were] mounted and some [were] on foot and not in very good order."[29] One of the three companies that rode bay horses pulled back from the river and were provided cover fire by the rest of the soldiers, as a warrior noted, "One company started to run when Custer was near the river and the rest [of the soldiers] fired on them and made them come back. This was the Bay Horse Company."[30]

The soldiers rode parallel to the river according to another warrior. Most of the warriors fighting near the river were on foot, and 'a galling fire' was coming from the other bank.[31] Angry warriors began to cross the river at the mouth of nearby Deep Coulee and flanked the soldiers. Survivor Korn witnessed this dangerous situation with his last glimpse, noting:

> When I looked around I could see that the troops had entered the ravine and the Indians were closing in upon them. One avenue of escape was still open, as at this time they were surrounded on three sides only. At this moment the Indians saw me and about 20 of them started after me and pursued me to Major Reno's camp [Reno Hill].[32]

Two Moons saw the soldiers in two wings fighting on foot with the horses between them. Custer was using the same exact formation that he had used on August 4, 1873, along the Yellowstone, to protect his horses.[33]

The five companies splintered into two battalions—one with two companies and the other with three. Since Keogh was incapacitated, Yates likely became the senior battalion commander and would have controlled three companies. Both battalions slowly began to withdraw toward the higher ground behind them. The fight near the river lasted about 20 minutes.[34] Bobtail Bull noted that two of the companies that fought near the river retreated and crossed the gulch [Deep Coulee] to the current battlefield.[35]

Archeology and Reno's later testimony of evidence in the area indicate that C and I companies likely angled across Deep Coulee.[36] Custer had been ordered to hold onto his wounded, and these incapacitated soldiers had to be placed in a safe area, free of Indian gunfire. It seems probable that the retreat of the two companies to the Calhoun Hill area provided such a destination. It has a protective area that could hide horses and wounded men.

Warrior Red Hawk arrived late and saw the other three companies, (likely E, F and L), some distance from the river riding to the high ridge overlooking the area.[37] (Today it is called Nye/Cartwright Ridge, after individuals who discovered soldier cartridges there.) The wary Cheyenne and arriving Lakota kept their distance, and many were on foot, unable to keep up with the mobile soldiers. Ten-year-old Hunkpapa youth, *Cankula*, later named Cannonball Woman, stated, "I looked up and saw soldiers on high ground over to the east. Saw soldiers going up a long hill in a direction away from the village, and there was no fighting until they got on top of it."[38]

Meanwhile, Reno's small force was fighting for their lives only a mile and half away. His retreating skirmish line reformed along a large crescent-shaped cut bank in the woods and one soldier stated, "When

Little Bighorn Battle Map 2

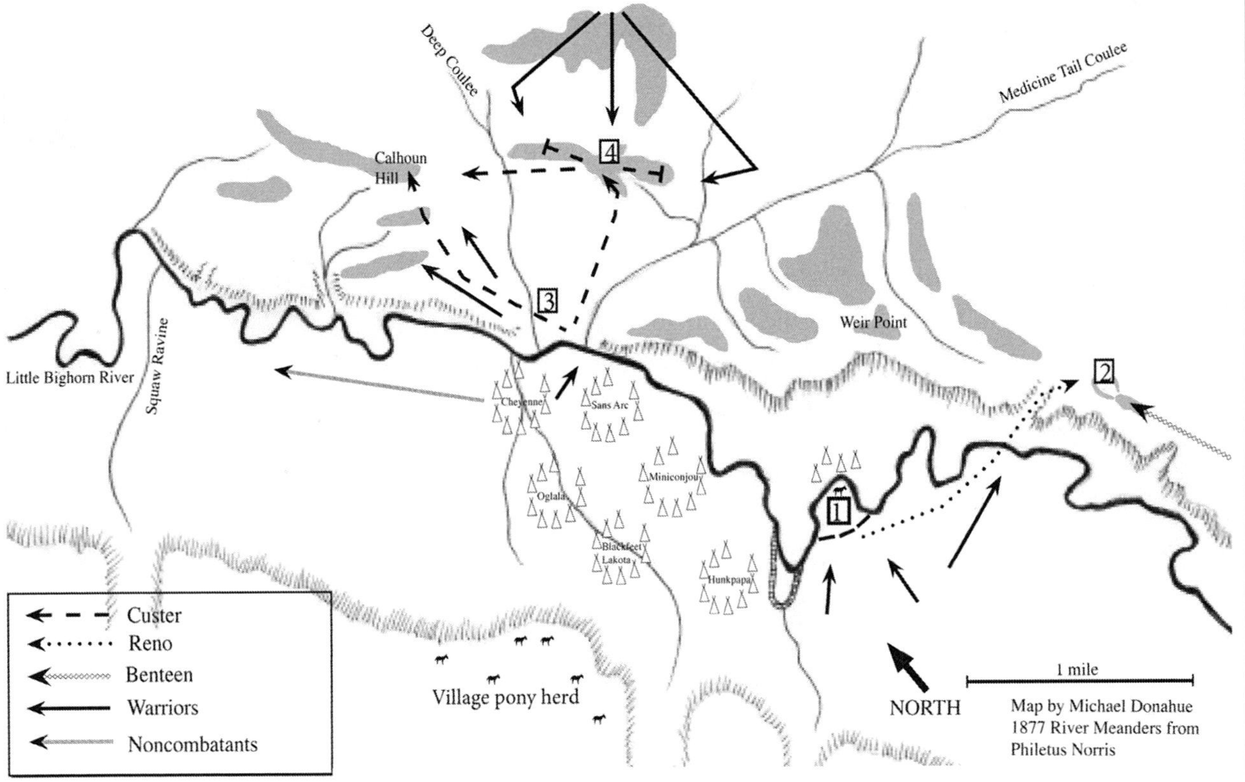

1 Reno's companies withdraw to the edge of the timber near their horses, and fire at surrounding warriors. After approximately 30 minutes of fighting in the valley, his panicked companies take heavy casualties as they gallop across the river to the bluffs.

2 Benteen, with companies D, H, and K, arrive ten minutes after Reno reaches the bluffs and reinforce the whipped soldiers.

3 Two companies withdraw to Calhoun Hill, followed by warriors. The other three companies, including E Company, withdraw to Nye Cartwright Ridge.

4 Wolf Tooth's party attacks the rear and front of the column. The companies deploy and skirmish for about 20 minutes before withdrawing to Calhoun Hill.

Fig. 68. Map: Little Bighorn Battle Map 2.

we reached the wood, we continued to fire for a short time."[39] Warriors tried to start fire to the grass to drive the soldiers out as they had done on August 4, 1873.

Wooden Leg was riding among the quickly growing multitude of warriors who were pressuring the bluecoats in the woods. The tall Cheyenne was a scary specimen. He had hastily painted a red and yellow circle across his face and tied his long black hair with a leather thong. Racing past the fleeing women and children, he noted, "I went on with a throng of Sioux until we got beyond and behind the white men. By this time, though, they had mounted their horses and were hiding themselves in the timber."[40]

In the woods, a frustrated Reno decided he did not have enough men to protect his horses and to cover the perimeter. The men were running out of ammunition and could see no promised support from Custer. A few horses had been killed, dismounting some of the soldiers in the timber. Reno later stated, "I knew I could not stay there unless I stayed there forever."[41]

Verbal orders were given to some of the company commanders to get to their horses, but many men did not hear the command, and in the confusion there were no trumpet calls to order men to mount. Those who were close enough to grab their horses wondered where they were going, and not even Reno could answer that question.[42] About 40 warriors, along with Crazy Horse, penetrated the brush as soon as the soldiers withdrew from the embankment.[43]

German immigrant George Lorentz fell off his horse after he was struck in the back of the neck by an Indian bullet that came out his mouth.[44] About the same time, another private was hit in the stomach. Private William Morris and another soldier were ordered by Captain French to dismount and help the man. The wounded soldier was propped against a tree and supposedly told Morris, "Go on, you cannot do me any good," to which he responded, "All right, if you say so, I will go."[45]

While mounted on his horse, Reno tried through sign language to ask Bloody Knife what the warriors might do next, and the fusillade instead hit the scout in the head, spraying blood and pink brain matter into Reno's face. The scout was later scalped as well as decapitated, and, according to one soldier, "Hacked to pieces, as you would cut up a hog."[46]

An unnamed soldier near Reno screamed, "Oh, my God! I have got it!" and fell from his saddle.[47] Herendeen recalled, "That volley and the man hollering seemed to startle everybody and they ran."[48] Reno shouted, "Dismount!" to help his men avoid becoming easy targets. Then realizing no other volley was eminent, he screamed, "Mount!"[49] Reno decided it was time to leave the timber, and while some heard orders to follow him, others did not. Perhaps suffering from battle shock, the blood-spattered major did not order his trumpeter to sound any calls to assemble the men.

Nearby, cool-headed Charley Reynolds asked, "What kind of damned move is this?"[50] Finding one's horseholder and horse in the brush became problematic as panic took hold of men scrambling for mounts. Morris recalled hearing someone shout, "Every man for himself!"[51] Wallace created more confusion by shouting for the men to grab any horse and get out of the timber.[52] Some men pulled their weapons and threatened those attempting to take another troopers' horse, while other men were left dismounted.[53] Varnum believed they were not in the timber more than 10 or 15 minutes, and the entire valley fight was a half an hour long. De Rudio thought the whole engagement lasted, "20 or 25 minutes, perhaps half an hour . . . I looked at my watch every 10 minutes."[54] Wooden Leg observed:

> Suddenly the hidden soldiers came tearing out on horseback, from the woods. I was around on that side where they came out. I whirled my horse and lashed it into a dash to escape them. All others of my companions did the same. But soon we discovered they

were not following us. They were running away from us . . . we whipped our ponies into swift pursuit.[55]

Hundreds of angry and excited warriors pursued the panicked soldiers as they raced to an unknown destination with Reno in the lead. He initially rode toward his first river crossing but soon realized he could not cut through the wall of warriors, veering straight toward the tall bluffs to his left.[56] The leading unit was fairly well organized, but the rear that included mostly Company G began to fragment and straggled into deadly chaos as warriors rode among them.

The troopers with their Colt revolvers had to use their six shots judiciously; otherwise, it was up to their brass spurs to save their lives. Ryan fired his revolver, hitting two warriors directly in their chests. He recalled, "As we cut through them, the fighting was hand-to-hand, and it was death to anyone who fell from his horse or was not able to keep up with the command."[57]

Charley Reynolds took a drink of whiskey to calm his nerves and then decided to make a break. Several warriors cut off Reynolds and shot his horse. The scout nicknamed, "Lucky Man," dropped his gun in the fall, and his leg was pinned under the horse, making him an easy victim. Other passing soldiers saw Reynolds recover his gun and fire several times before he was overrun. His prediction of death had become reality.[58]

Isaiah Dorman, a black man who had lived with the Hunkpapa, was now being chased by them. One of his friends, soldier Roman Rutten, trotted past Dorman who had his horse downed by an Indian bullet. Dorman was on one knee calmly firing his sporting rifle, and shouted out, "Goodbye Rutten."[59] According to a contradictory account by warrior Runs-the-Enemy, "We passed a black man in a soldier's uniform and we had him. He turned on his horse and shot an Indian right through the heart. Then the Indians fired at this man and riddled his horse with bullets. His horse fell over on his back and the black man could not get up."[60] The Hunkpapa saw him as a traitor and brutally tortured him on the spot.

Herendeen, peering from the nearby brush, saw Dorman being systematically shot in the lower legs with bullets only an inch or two apart. The women pounded him with stone clubs and ripped open his torso filling his coffee pot as well as his cup with his own blood. Other humiliations included stuffing his severed penis in his mouth, as well as hammering a twelve-inch long iron picket pin through his testicles. Another private noted that Isaiah's head had been cut off as well. Such was the example made of Isaiah Dorman, who had helped the soldiers and had used a dead warrior's body as fishing bait just a few days earlier.[61]

Irishman Thomas O'Neill mounted his horse in the timber only to have it shot from under him. He decided to try to follow the retreating command on foot and recalled that one large Indian rode right up behind him, but he quickly turned and shot the warrior off his horse. Changing his mind, O'Neill decided to return to the woods where he encountered scout Billy Jackson and interpreter Gerard, who still had their horses, along with a dismounted De Rudio. The small group stayed hidden all afternoon and attempted to follow the troops after dark.[62]

The withdrawal became a panicked rout, and numerous warriors used the analogy of a buffalo hunt in their descriptions. Cheyenne American Horse remembered, "The Indians rode right up to them [and] knocked some off their horses as they were running, and some fell off in the river. It was like chasing buffalo—a grand chase."[63] Wooden Leg taunted the fleeing bluecoats and lashed a soldier with his pony whip. The soldier turned and fired a bullet from his Colt revolver into the thigh of one Wooden Leg's friends who was riding nearby.

Wooden Leg angrily reacted by striking the soldier over the head with his pony quirt, stunning the man. He then grabbed the carbine that was strapped on the soldier and yanked him from the saddle. The soldier was trampled by the multitude of trailing warriors.[64]

Other soldiers attempted to return to Reno Creek and were chased and killed.[65] Some soldiers thrown off their horses approached the warriors with their hands up and were shot down.[66] Varnum remembered:

> The Indians surrounding on the flank with their Winchesters laying across their saddles and pumping them into us . . . I was soon at the head of the column and tried to check it saying we could not run away from the Indians. We must get down and fight. Maj. Reno was there, however, and informed me that he was in command. I subsided.[67]

Donald McIntosh could not hide his mixed Indian heritage as he led his scattered company from the timber. He had lost his horse, and a private offered his mount to the officer. As he got within 30 yards of the river, McIntosh was surrounded by a group of about 30 warriors. A fellow soldier stated, "[McIntosh] was calm as a deacon at service and he knew he had no chance."[68] The lieutenant fired his revolver at the warriors until he was killed. McIntosh was stripped of his clothing, with the exception of his neckband with an attached button. He and his borrowed horse lay near one another, both filled with arrows. He was scalped from his forehead to the back of his neck, and a nearby brush fire scorched his body.[69]

As the frightened soldiers reached the river, they had difficulty fording it since the banks were from four to five feet high on both sides. It quickly turned into an awful traffic jam of fleeing soldiers.[70] Cheyenne Brave Bear described it, "The worst of it was [that] the soldiers struck very high banks of the river and all went over. When they struck the water it sounded like [a] cannon going off. This was awful as the bank was awful high . . . I could see lots of blood in the water."[71] Black Elk recalled, "Soon the soldiers were all crowded into the river, and many Lakotas too; and I was in the water awhile. Men and horses were all mixed up and fighting in the water, and it was like hail falling in the river."[72]

The brave Thomas French attempted to cover the soldier retreat at the river and was able to place more scratches on the stock of his already notched rifle.[73] Reno's close friend and adjutant Benny Hodgson had his horse shot from under him near the river. Hodgson was shot in the back as he neared the river and grabbed a trooper's stirrup hoping to be pulled across the stream to safety. As soldiers raced past him, he screamed, "For God's sake, help me."[74] As he reached the other bank, he was shot a second time and killed.

Arikara scouts Bobtail Bull and Little Brave attempted to retreat along with the troopers but turned and crossed the river to Reno's left only to encounter warriors already occupying the east bank. Wooden Leg witnessed the death of fellow Cheyenne Whirlwind who charged and fired at Bobtail Bull.

He said, "The enemy Indian bravely charged also toward Whirlwind. The two men fired rifles at the same moment and both of them fell dead."[75] Little Brave was shot in the right shoulder, and blood soaked through his white shirt. Struggling at a slow trot across the river, he decided to make a stand behind a small knoll. He killed a Lakota named Elk Stands Above and continued to fight, even though he was outnumbered. Wooden Leg bashed Little Brave in the head with his rifle barrel. He remembered, "Others beat and stabbed him to death."[76]

The two Crow scouts who accompanied Reno into the valley wanted to stay in the valley and fight. White Swan was shot three times and lanced in the face, as well as being unhorsed near the Hunkpapa camp. His friend Half Yellow Face saved his life by dragging him into the bushes. White Swan finally retreated to the bluffs with the help of his friend, when the warriors vacated the area.[77]

In his first battle, Black Elk could not get close to the fleeing soldiers due to his small size. Near the riverbank, he was ordered by a warrior to scalp a wounded soldier so he fired his revolver and then took his first war trophy with his dull knife.[78]

Some tired soldiers attempted to ride their exhausted and wounded horses up the 100-foot bluffs; while others dismounted and pulled the reins of their mounts, urging them beyond all possible endurance. Dr. James DeWolf climbed the steep ravines along with several other soldiers; he perished with seven bullets in his body fired from warriors above them.[79] Dr. Henry Porter was the only surgeon who survived the carnage and reached the hilltop safely.[80] Climbing up the same ravine behind DeWolf was an Indian war veteran who participated in the fight along the Washita eight years earlier. Private Elihu Clear, who reputedly had killed a woman at the Washita, was shot to death as he climbed up a ravine.[81]

A hatless Reno, with his face covered with sweat, blood, and brain matter, peered over the edge to see his scattered troops climbing up the nearly perpendicular ravines. Riding on his Kentucky mare that had been shot through the jaw, Hare made the hilltop and shouted, "If we've got to die, let's die here like men. I'm a fighting son-of-a-bitch from Texas!" Hare then screamed at a panicked Moylan who was riding away, "Don't run off like a pack of whipped curs."[82] As the dust cleared, Reno had lost 32 men, seven wounded, and many others were missing in action.[83] Eighteen warriors and 10 non-combatants were killed in the valley.[84]

While Reno and Custer were fighting, Benteen and his battalion had been watering their horses at the cattail-filled morass for about 20 minutes. Some recalled hearing scattered gunfire from that position. Godfrey recalled:

> After waiting some time Capt. Weir came to me and expressed his impatience at our inaction saying we ought to hurry up, after waiting a few minutes he asked me to go with him to Benteen and urge him to go ahead. I told him I was only a lieutenant and would probably be told to mind my own business; with that he went back to his own troop. After waiting a while and the firing below becoming very heavy, Weir came to me in great agitation and said, 'I am going anyhow,' went back to his troop, mounted it and took up his march on the trail. Soon after, when Weir had marched some distance, Benteen gave the order to mount and took up the march.[85]

Fig. 69. Trumpeter John Martin (Giovani Martino) survived abandonment as a newborn baby and, later, death as Custer's messenger at the Little Bighorn. His eyewitness testimony is important in deciphering Custer's intentions and the early chronology of the battle.

Courtesy of John Doerner

As Weir took the advance, the men looked behind them to see 30 mules plunge into the morass in spite of their herders. The mule train had caught up with Benteen.[86] Messenger Knipe saw Benteen departing the watering hole and met him near the burning Lone Tipi. As he passed Benteen on his way to the mule train, "They want you up there as quick as you can get there—they have struck a big Indian camp."[87] Godfrey expounded, "As Knipe passed the column, he said to the men, 'We've got em, boys.' From this and other remarks we inferred that Custer had attacked and captured the village, [and] they are licking the stuffing out of them."[88]

Two miles farther, Martin rode up, pulled the sweaty message from under his hat and handed it to his captain. Benteen told the unaware but lucky Martin that his horse had been wounded. Benteen handed the paper to Weir who shared it with Edgerly.[89] The white-haired commander was unhappy with the messenger and his new orders. He had been promised the advance and now was stuck with an order to go back to the pack train. Benteen lamely justified his insubordination on Martin's comments that "the Indians were all 'skedaddling,' therefore there was less necessity for me going back for the packs."[90]

Edgerly overheard the excited Martin's conversation, not to Benteen but to his orderly, "He [Martin] was laughing and seemed very much elated, said it was the biggest village he ever saw, that they had found the Indians all asleep in their tepees, that Major Reno was charging it and killing everything, men, women, and children."[91]

Benteen came to the split in the trail and, rather than asking Martin which direction to go, he rode toward what he thought was a small party of warriors. The party of three Indians turned out to be three Crow scouts—White Man Runs Him, Hairy Moccasin, and Goes Ahead, who ended up joining him. As Benteen reached the lower bluffs above the river, he could see Reno's soldiers being killed by groups of warriors below him.[92] Soon Benteen could see other whipped soldiers racing up the bluffs as a group gathered on the high ground. Wounded and disheartened men lay prostrate near the top of the bluffs and had been on the hilltop for only five minutes when a nonchalant Benteen arrived on the hilltop wearing a straw hat.[93]

Reno pleaded, "For God's sake, Benteen, halt your command and help me. I've lost half my men and could do no better than I have done."[94] Reno had lost his composure and was angrily, but uselessly, firing his revolver at Indians in the valley floor a half-mile away. Benteen's first question to the major was, "Where is Custer?" Reno claimed not to have a clue as to the location of their commander.

Benteen pulled out Cooke's message and shared it. Benteen flashbacked to the Washita and queried Reno, "I wonder if this is to be another Maj. Elliott affair?"[95] Within minutes, Benteen had convinced himself that Custer had abandoned them.

Wooden Leg rode up the bluffs chasing the soldiers, until he reached Sharpshooter Ridge that overlooked the disorganized soldiers and he fired a couple of shots from a great distance. Many other warriors, including White Bull, were also gathered on the side of the hill watching the bluecoats. Wooden Leg testified that someone said, "Look! Yonder are other soldiers!"[96] They could see soldiers riding across Nye/Cartwright Ridge behind them. There was a new threat from the north about to capture their women and children. The confident and victorious Lakota and Cheyenne warriors who had protected their families from Reno's attack had pushed them straight toward other enemies. Their collective demeanor changed, according to Runs the Enemy—"I looked into their eyes and they [Sioux] looked different—they were filled with fear."[97] Little did they know that 'Creeping Panther Custer' was chasing the women and children and watching them through the lens of past experience as if he was looking under a magnifying glass.

Chapter 13

Death on a Pale Horse

Wooden Leg and other adrenalin-filled warriors fired their weapons from behind some rocks on Sharpshooter Ridge into the blue-coated mass on Reno Hill. Weir took the initiative and ordered a skirmish line thrown out to drive away the warriors.[1] Private Jacob Hetler, from Weir's company, charged up the hill on foot toward the rocks with the rest of Company D. It was an easy task, since the warriors themselves were disappearing to the north toward the new threat. From atop Sharpshooter Ridge, Hetler could see what they had witnessed. The former carpenter from Ohio recorded something unidentified by other soldiers. Hetler reported:

> I saw six *vedettes* [signal men] stationed on a bluff, and next I saw Custer and a group of men going out of sight over the hills. This was the last time I ever saw my leader alive. For an hour and a half after that, Captain Thomas B. Weir of my company asked Major Reno for permission to go after Custer, and this was granted. We heard firing and knew there was trouble.[2]

Hetler likely saw the movements of the Company F detail [*vedettes*] and one battalion of soldiers riding over Nye/Cartwright Ridge—it was the same sight that Wolf Tooth's party of Cheyenne were simultaneously witnessing from a different vantage point. The fight along the ridge top had not yet begun. The vedettes would have witnessed the troops driving warriors from the area near Reno Hill and likely communicated the news to Custer that Benteen had finally arrived in the area. Custer's troops moved northward with the intelligence that Benteen's troops were on Sharpshooter Ridge and only minutes away. It would become a fatal move to the north, because Benteen would stop to render aid to Reno.

Dr. Henry Porter tended to wounded soldiers on Reno Hill, and about 15 minutes later, Luther Hare was sent by Reno to direct the pack train to the bluffs. It would be another half hour before the first mules of the straggling train arrived.[3]

One sergeant began to hear gunfire to the north and stated, "We could hear the firing. At first it was a couple of volleys, very heavy, afterwards it was lighter and appeared more distant. Varnum made the remark that Custer was hotly engaged or was 'giving it to the Indians hot,' or words to that effect, and a few minutes after, Major Reno came up."[4]

According to one private, the firing lasted an hour and a half.[5] Men hiding in the valley timber heard the volleys as well.[6] Back on the hilltop, Sergeant Culbertson was ordered by the brain-speckled Reno to accompany him to find adjutant Hodgson's body, and they disappeared into a narrow ravine toward the river.[7] Reno, who was suffering from battle shock and may have been somewhat inebriated, had only one thing on his mind—to locate Benny. According to the major, "I suppose I was gone for half an hour, Captain Benteen was the senior officer in command when I went down, and he was a man in whom I had the greatest confidence."[8] Finding Hodgson dead by the edge of the river, Reno removed the ring from his hand and a bunch of keys from his pocket.[9] While Reno was away, soldiers on top of the hill discovered Custer's trail, which led behind the hilltop they occupied; they pointed it out to Reno upon his return.[10]

Neither Reno nor Benteen seemed concerned about the well-being of the five companies or in following the newly-discovered trail. Both officers knew Custer's men had only 124 rounds of ammunition each, and Reno's men had expended over half of their ammunition in 20 minutes of fighting. They immediately, instead, saw themselves as the victims of abandonment by their leader. Reno later recorded, "It was supposed he could take care of himself as well as we could."[11] Benteen, likewise, thought, "I supposed General Custer was able to take care of himself."[12] George Wallace recalled, when they got on top of the hill, "I heard a great deal of swearing about General Custer running off and leaving us."[13]

Korn, who had escaped from Custer's command only minutes earlier, remembered, "I immediately reported the terrible situation of the [Custer] troops to Captain Vere [Weir] and he to Major Reno, who could render no assistance as many of his troops were wounded and unfit for service."[14] Only Weir seemed worried about Custer and his whereabouts after receiving the orders to, "Be quick." However, even Weir would wait for at least 35 minutes to move.

When Thomas McDougall finally arrived with the slow-moving pack mules to the bluff top, he could hear faint volley firing.[15] McDougall told Godfrey, "I think we ought to be down there with him [Custer]," and went to find Reno. McDougall found Reno and Benteen talking, and he expressed his desire to ride to Custer, but according to McDougall, Reno did not appear to regard the seriousness of the situation, and the captain [Benteen] just walked away.[16] The dazed and traumatized Reno had returned from getting his adjutant's belongings and was still obsessing over Hodgson. All he could mumble was, "Captain, I lost your lieutenant, he is lying down there."[17]

In Reno's report, immediately after the battle, he stated, "We had heard firing in that direction and knew it could only be Custer."[18] Three years later, he altered his previous statement, "If I had heard the firing, as they represent the firing, volley firing, I should have known he [Custer] was engaged while I was on the hill but I heard no such firing."[19] Benteen claimed he heard no volleys but only 15 or 20 shots, which he believed came from near the river crossing. Three years after the battle, Reno and Benteen's 'memory' of gunfire was negligible, and it is clear that their comments were a result of a self-protective cover-up.[20] Winfield Edgerly stated that almost immediately after arriving on the hill he heard firing:

> I remarked [about] it, heavy firing by volleys down by the creek. Captain Weir came to me and said General Custer was engaged and we ought to go down. I said I thought so, too. He went away, walking up and down rather anxiously. I heard the fire plainly. The first

> sergeant came up then and I saw a large cloud of dust and thought it must be a charge and said, 'There must be General Custer. I guess he is getting away with them.' He said, 'Yes, sir, and I think we ought to go there.' I did not answer him.[21]

Weir walked over to his towering lieutenant who ran the everyday activities of Company D and asked, "Edgerly will you go down with the troop if I will?" Edgerly agreed, and Weir said, "Well, I will go and get permission."[22]

Private John Fox was in a position to overhear the animated conversation between Reno and Weir. Weir remarked that "Custer must be around here somewhere and we ought to go to him." Reno replied, "We are surrounded by Indians and we ought to remain here." Weir demanded, "Well if no one else goes to Custer, I will go." Reno replied, "No you can not go. For if you try to do it you will get killed and your Company with you." Fox thought Reno was "intoxicated or partially so."[23]

More importantly, the private stated that both Benteen and Moylan stood by and overheard the conversation but did not seem to approve of Weir going either "and talked as though to discourage him. Finally, Weir said that he was going anyhow, and Reno did not object."[24]

Martin witnessed the exchange between Weir and Reno and "Could tell by the way he [Weir] was acting that he was excited and angry. He waved his arms and gestured and pointed down river. Then we heard some volleys, and Captain Weir jumped on his horse and started down the river all alone."[25]

Weir rode past Edgerly who assumed he had received permission. Edgerly mounted the men of Company D and followed.[26] A miffed Benteen later testified, "Captain Weir sallied out in a fit of bravado, I think, without orders." Benteen said he did not see him leave, but guessed it was about 30 minutes after they arrived on the hill.[27] Reno noted that Weir left, "On his own hook."[28]

Company D rode toward the highest promontory in the area that today bears his name, Weir Point, where they could see miles to the north. The five companies under Custer had been fighting for their lives for over an hour and were less than four miles away from Reno Hill—a trot of only 12 minutes.

Meanwhile, the five companies with Custer had withdrawn from the river after about 20 minutes of fighting. Two companies, likely C and I, trotted northward along the edge of the bluffs to keep warriors from circling them. Eventually, the soldiers sheared away from the bluff's edge and rode toward a ridge that predominated the area. It would later be called Finckle/Finley Ridge. Then they advanced to an even higher elevation called Calhoun Hill. It seems from warrior testimony that most of the pressure was on these two companies early in the fight.[29]

The other three companies nonchalantly withdrew to the high ridge between Medicine Tail and Deep Coulees, a little over a half-mile from the river.[30] Most of the warriors that had resisted at the river were dismounted and were quickly outdistanced by the troopers. The 40 to 50 Cheyenne warriors under Wolf Tooth and Big Foot, who had snuck out of the village early that morning in search of soldiers, were riding along a small creek two miles north of the same ridge. They saw the line of soldiers riding unmolested along the top of Nye/Cartwright Ridge.[31] Wolf Tooth and Big Foot raced toward the bluecoats and split into two groups—one attacked the leading soldiers and the others harassed the rear of the troops. Wolf Tooth told Stands in Timber, "Custer['s soldiers] split up and started shooting on one side and then the other." The fight lasted about half an hour.[32] Not having the long-distance firepower of the soldiers, the harassing warriors kept their distance. Soldiers came off the higher ridge, dismounted, and fired at the Cheyenne party that had attacked the rear as well as the advance unit of the soldiers.[33]

An interviewer of Oglala warrior Red Hawk recorded:

> Red Hawk saw the fight as well, stating, 'All this time the soldiers fought bravely.' He adds that 'he never saw soldiers fight like they did.' He says, 'several Indians were killed now, but no soldiers were killed. The officers tried their utmost to keep the soldiers together at this point, but their horses were unmanageable; they would rear up and fall backward with their riders; some would get away. The Indians forced the troopers back to where the first stand was made on Calhoun Hill and the ridge [Finckle/Finley] running therefrom towards the river.'[34]

The wary warriors finally pulled back to the northern hills, out of range of the Springfield carbines.[35] Wolf Tooth and Two Moons saw the soldiers form two wings or lines, with the horses protected between them, and it was the same tactic used by Custer along the Yellowstone River in 1873, when attacked on open ground. The battalion had some real issues. There was no place to protect the horses on the open terrain. More devastating was that some men had lost their mounts and would have to ride double with a fellow soldier.

One of the three companies in the battalion on Nye/Cartwright Ridge was Company E. "The gray horse soldiers acted as rear guards for about half an hour to hold off Indians attacking from the south; then they followed the other soldiers along the Nye-Cartwright Ridge and on to Custer Battlefield," noted historian Stands in Timber.[36] The three companies, that probably included E, F and L, had to find a safer position for the protection of their horses.[37] John Two Moons saw the soldiers leaving Nye/Cartwright Ridge and stated, "They were on a lope and Indians were behind them, but they [soldiers] paid no attention to them."[38]

The five companies reunited on Calhoun Hill where they stopped for a little while.[39] There was a trumpet call, and some of the companies dismounted.[40] Calhoun Hill was the southeastern end of a narrow, half-mile ridge shaped like a fishhook. Known today as Horse Holder's Ravine, the bend of the hook formed a ravine that was perfectly situated for hiding the horses. It was the appropriate spot to make a defensive stand, drop off dismounted soldiers, or create a hospital for any wounded.

Custer had a real dilemma: it was not the quickly multiplying numbers of warriors that bothered him. Disciplined firepower and offensive thrusts could fend off huge numbers of warriors, as was learned from his past experience. Currently, the real problem was his limited ammunition supply, soldiers without horses, and perhaps several wounded men. He had to "hold on to his wounded" as had been ordered by Terry. It must have puzzled Custer that after the fight on Nye/Cartwright Ridge, Benteen had not appeared even though he was in the vicinity. The initial plan to capture the women and children, as he had suggested earlier, was still an option. He could see them escaping to the north as he sat watching, and he

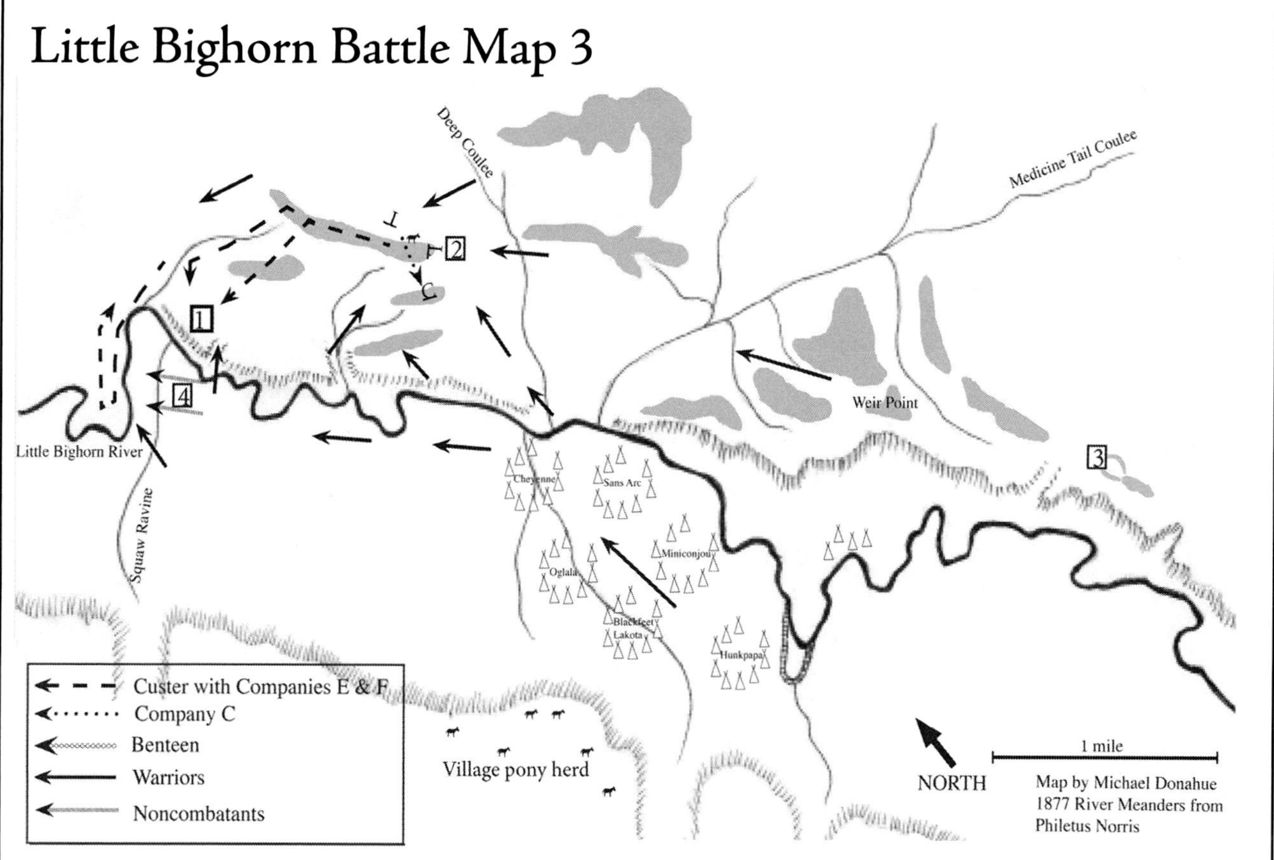

1. Custer's headquarters, along with Companies E & F, ride behind battleridge before separating near Custer Hill. E & F approach the river near the noncombatants. Custer's headquarters reunites after riding farther north. One contingency rides into the valley before they halt for about 20 minutes. Some warriors attempt to halt the capture of the noncombatants.

2. Meanwhile, Companies I, C, & L hold Calhoun Hill as warriors leaving Reno converge on the position. Company C rides to a lower ridge to drive warriors back and are quickly overwhelmed. The warriors' charge continues wiping out most of the men from L an I as well.

3. Reno and Benteen wait for the pack mules to arrive, and men are listening to the gunfire to the north. Weir is impatient and encourages movement in the direction of northern gunfire.

4. Several thousand noncombatants converge along Chasing Creek (also known as Squaw Creek) where their progress is stopped.

Fig. 70. Map: Little Bighorn Battle Map 3.

simply could not allow it. Their capture could neutralize the aggressiveness of the warriors. Victory was across the river, but it was quickly escaping from his grasp.

Three companies [C, I and L] under Harrington and Calhoun, along with any wounded or dismounted soldiers, were left defending Calhoun Hill. It would stretch Custer's companies thin, but as a gambler, he was willing to risk it. Custer's headquarters staff, along with Companies E and F, rode along the northern backside slope of a long ridge, today known as 'Battle Ridge.' Flying Hawk saw the division of troops on Calhoun Hill, stating, "[Custer] went up onto Calhoun Hill, leaving a detachment there, and he went right over to Custer Hill."[41]

As the soldiers looked in front of them, it looked like a green, velveteen ocean of hills and gullies capable of hiding hundreds of warriors or soldiers. As the two companies and staff approached the other end of the ridge, they rode past a small hogback knoll, later called Custer Hill. There were several ravines and an Indian trail that led to the river toward the noncombatants.[42]

It seems likely that Custer and his staff rode another 100 yards along the ridge, where they could see down the valley for miles with the hope that some sign of the Terry/Gibbon command would materialize. They observed two things: Terry was not in the vicinity and there were no more camp circles spread down the valley. It was important and strategy-changing intelligence. Sergeant Major William Sharrow, who had been a messenger earlier, was almost certainly called into action again. He rode to the north in the supposed direction of Terry's column. Flying Hawk recalled that a bluecoat rode north before the soldiers were surrounded, only to be killed some distance from the battlefield.[43] The two companies under Smith and Yates followed the Indian trail toward the river, while the headquarters group dropped into another nearby ravine to meet up with them.[44]

Some warriors quickly surmised what the troops were doing, and the numerous accounts are compiled here for the first time. It seems clear that Custer's objective was the noncombatants. Lakota Charles Corn shared, "While they were fighting, the women folks move[d] on the other side of the Little Big Horn and the soldiers were on the other side."[45] One warrior said, "The report came to us that another party [Custer] was coming to attack us. We could not see them from where we were. The report was that they were coming to head off the women and children from the way they were going."[46] Another stated, "The women ran back and someone rode to

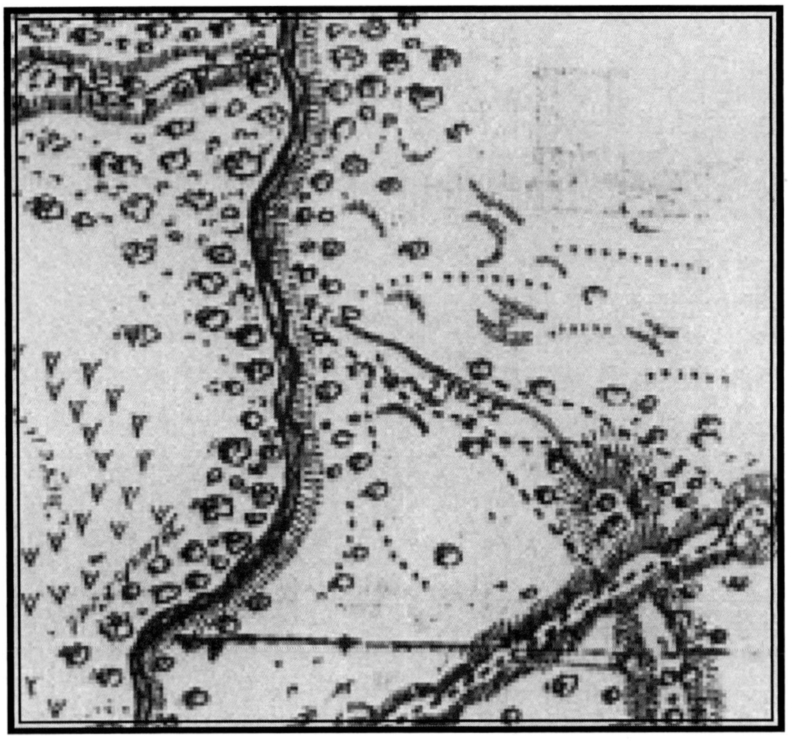

Fig. 71. This map was created by William Clark based on an unnamed warrior's map. The drawing depicts the same area as the Bell map (adjacent page) and also notes a troop division. The ravine at upper left where noncombatants hid and the lodgepole trail mentioned by Benteen at the Reno Court of Inquiry are included. The dashed lines adjacent to the 'trail' are Custer's movements toward the river. The dotted lines indicate warriors.

Courtesy of the National Archives and Records Administration.

where the men were fighting Reno and told them that more soldiers [Custer] were coming below. Then all the men rushed down the creek again to where the women were."[47] Runs the Enemy stated, "They yelled to us that the genuine stuff was coming, and they were going to get our women and children."[48] White Bull said, "The women and children were about a half a mile [away], and Custer stopped about straight across from the camp. . . . The Indians see [saw] the women folks, and they stay[ed] between the troops and the women."[49] Red Horse recollected:

> The women and children ran down the Little Bighorn River a short distance, into a ravine . . . a different party of soldiers [Custer] had all the women and children prisoners. Like a whirlwind the word went around, and the Sioux all heard it and left the soldiers on the hill and went to save the women and children.[50]

Lakota Flying Hawk remembered, "When Custer reached the Custer Hill, he was right above the women who had collected down by the river."[51] Two Moons stated, "Messengers came saying more soldiers were going to kill the women."[52]

Crazy Horse, who was unfamiliar with Custer's history of hostage taking, mistakenly believed that his actions were accidental, and he said:

> Women, papooses, children, and in fact all that were not fighters, made a stampede in a northerly direction. Custer, seeing so numerous a bodies, mistook them for the main body of Indians retreating and abandoning their village, and immediately gave pursuit. The warriors in the village, seeing this, divided their forces into two parts, one intercepting Custer between their noncombatants and him.[53]

The soldiers were directly across the river from the ravine filled with thousands of noncombatants.[54] Recent protests to change the name of the creek due to the insensitive nomenclature—Squaw Creek—may be politically incorrect,

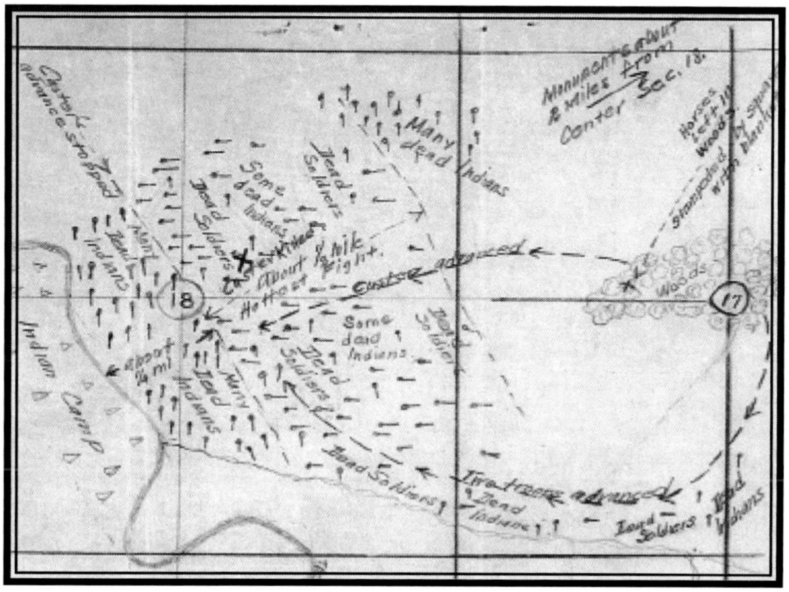

Fig. 72. This map was supposedly made by a trooper named James Bell based on Indian testimony. The area depicted is near the area near the noncombatants in Squaw Ravine/ Chasing Creek. While exaggerated in casualties, it notes a division of soldiers advancing toward the river and an area where horses were left in woods and stampeded by 'squaws' with blankets.

Courtesy of Yale Collection of Western Americana, Beinecke Rare Book and Manuscript Library, Yale University but it disavows the origin and name of this important place. Lakota Respects Nothing had another, even more descriptive name for the creek. He recalled, "The women fled down the river, toward the mouth of *Chasing Creek*."[author's emphasis][55]

Hanging Wolf and other Cheyenne were hiding in the river brush and fired on the soldiers in the valley, turning them back.[56] Whether this was a small detachment, the two companies, or the headquarters staff cannot be determined from the warrior accounts. The soldiers remained near the river where they waited a long time—20 minutes or more.[57] Someone wearing buckskin may have been wounded or killed in the area, as noted by some warrior sketches.[58]

Two Moons and Gall noted that the soldiers: "Followed the deep ravine thence to the river, supposing they could make a crossing there. When he [Custer] reached the outlet of the ravine he found it was several feet to the water and his horses could not get down."[59] According to Gall, his warriors met the soldiers there and drove them back with some soldiers being killed. More intriguing is an account by Red Hawk, who returned to the battlefield a month later and noted, "At the mouth of a ravine they found eight dead soldiers lying with their uniforms on and their guns and ammunition and everything by them. Red Hawk said he could not understand how these soldiers were killed down there."[60] Moving Robe Woman was likely in the area and revealed:

> Custer's men got to the river above the beaver dam where the water was very deep. [The] soldiers dismounted and fired. [The] soldiers' horses got loose and ran to the river. Their packs got loose and floated down the river. [The] horses wanted water. [The] packs were fished out by the squaws. [The] mules' packs got loose.[61] [She also witnessed a tactical move that was all Custer and stated:] With my father and other youthful warriors I rode in that direction. We crossed the Greasy Grass below a beaver dam (the water was not so deep there) and came upon many horses. One soldier was holding the reins of eight or 10 horses. An Indian waved his blanket and scared all the horses. They got away from the men [troopers]. On a ridge just north of us, I saw blueclad men running up a ravine, firing as they ran.[62]

Custer had ordered fewer men to hold more horses during the skirmishes along the Yellowstone, and in his Civil War battle at Trevilian Station, when he needed more firepower and was outnumbered. Now he repeated the same tactic.[63]

Correspondent Mark Kellogg lost his mule as well as his notebook and was killed in the area. The *Bismarck Tribune* would later mention him, not as the story-telling correspondent of the battle, but instead as a casualty.[64] The warrior tactic that the soldiers feared most became reality. Some stampeded horses galloped off with any remaining ammunition in the saddlebags, leaving soldiers stranded. Suddenly, the two companies were on the defensive, yet it is clear from Indian participants that the soldiers maintained their discipline and did not panic. The soldiers withdrew to the flat, open ground above them. As the soldiers worked their way to the high ground, they could hear intense gunfire from Calhoun Hill. Things had turned desperate for the outnumbered soldiers left behind.

A little more than a half a mile away, Calhoun's skirmish line of 37 men knelt in a thin blue line at right angles and fired at the oncoming warriors. It is likely that Company I was firing from atop the battle ridge as well.[65] The inexperienced Harrington with Company C was nearby.

Many determined warriors lined up behind a long ridge next to the river; the ridge was called 'Greasy Grass Ridge,' and it was a perfect staging area for the ever-increasing numbers of Lakota and Cheyenne.[66] Wooden Leg observed that the warriors approached the battlefield, each choosing his own personal course.[67] The tall Cheyenne found himself watching the battle from Greasy Grass Ridge with other Cheyenne and Oglala warriors as well as Kate Bighead. The 29-year-old Cheyenne woman wanted to sing strongheart songs for her young nephew Noisy Walking fighting in the battle.[68] The fighting was slow and took a long time, according to Wooden Leg.

Two Moons testified, "The Sioux rode up the ridge on all sides, riding very fast. The Cheyennes went up the left way [Greasy Grass Ridge]. We circled all around them—swirling like water round a stone."[69] Gall, Two Moons, Rain-in-the-Face, Crow King and other warriors arrived from the Reno fight and ascended up the large ravine known as Deep Coulee.[70] Crazy Horse likely crossed the river at Deep Ravine and followed the gully toward the soldiers.[71]

The warriors crawled to another ridge closer to the bluecoats in order to fire on Calhoun Hill. It was decided to send Company C, under Harrington, to drive the warriors from the deadly position.[72] Calhoun had befriended the young officer and had an affinity for the lad: they were proudly photographed together. At this point, Harrington was ordered off the hill to offer protection for Calhoun and his men.

The non-commissioned officers Bobo, Finckle, and Finley got their men on their sorrel horses, and the men in white hats rode toward the lower ridge driving the warriors before them. Just like the Washita and Yellowstone fights, the warriors, when confronted by a soldier charge, ran away. Kate Bighead stated:

> The soldiers on the ridge mounted their horses and came riding in a gallop down the broad coulee toward the river . . . The Indians hidden there got back quickly into the deepest parts of the gulch . . . The soldiers who had come galloping stopped and got off their horses along another ridge, a low one just north of the deep gulch.[73]

Finckle, Finley, Bobo, and Harrington screamed at the men to hold their line. Frightened and excited horses bucked behind the firing line, with the helpless horse holders doing everything within their power to restrain them. Nearby, Crazy Horse shot the mounted horseholders as fast as he could load his gun. Flying Horse stated, "They fell off their horses as fast as he could shoot. When they found they were being killed so fast, the ones that were left broke and ran as fast as their horses could go to some other soldiers that were further along the ridge toward Custer."[74] Corporal John Foley was in the unenviable position of trying to hold the company's colors from his wooden staff, and the banner was like a beacon for gunfire. They were on deadly open ground. The soldiers fired several volleys, forcing the warriors to take cover.[75]

While other warriors scampered back over the sagebrush and tall grass of Greasy Grass Ridge, the fearless Cheyenne, Lame White Man, stayed hidden in the area below where the small band of soldiers dismounted. Interrupted from a sweat, his hair was loose, and he was adorned in only a blanket. He called to the young men, and they began creeping and dodging back to him. The Oglala Lakota chiefs also called to their young men, and these also returned to the fight.[76] Yellow Nose, Contrary Big Belly, and others joined in the counterattack.

Lame White Man screamed, "Young men, come with me and show yourselves to be brave!"[77] Then Lame White Man called out, "Come. We can kill all of them."[78] Kate Bighead remembered, "On all sides of this band of soldiers the Indians jumped up . . . I think there were about 20 Indians to every soldier there."[79] Wooden Leg related," All the white men went crazy . . . the Indians took the guns of these soldiers and used them for shooting the soldiers on the high ridge."[80] Iron Hawk remembered, "Custer's men in the beginning shot straight, but later they shot like drunken men, firing into the ground, into the air, wildly in every way."[81]

Fig. 73. This unpublished image of August Finckle reinforces that he was one of the tallest men in the 7th Cavalry. Fellow soldiers chided him about his strong German accent. He died with Company C during Lame White Man's attack. His body was found near his dead horse and was recognized by Daniel Knipe, Edward Godfrey, and Samuel Alcott.

Courtesy of George Kush

The chaos lasted only minutes, and the panicked men of shattered Company C were found all over the battlefield. Today, the ridge where most of the company disintegrated is named after the two sergeants found there: Finckle/Finley Ridge. Young Harrington's body was never recognized. Finckle and Finley were killed next to their horses. The bald-headed Finley was later found with 20 of his cartridges next to him, indicating not only that he resisted the onslaught, but also that he fought there for some time.[82] From the saddlebags of Finley's horse, the warriors reclaimed the dried, red-headed scalp, taken from Sitting Bull's Sundance site, as they now collected fresh scalps. Some soldier survivors raced on foot toward Calhoun Hill while others fled on horseback. Evidence seems clear that Bobo and Foley were able to flee the chaos, arriving on Calhoun Hill.[83]

Lame White Man was also killed during the charge. He was shot through the breast with the bullet coming out his back, and in the bloodlust, was scalped by another warrior who thought he was an enemy scout. White Bull was moved by the apparent recklessness of the inspirational elder, stating that he had never seen a man "throw his life away."[84]

Crazy Horse was also involved in the melee with the men on sorrel horses. With a lightning bolt painted across his face and unbraided hair whipping behind his head, he rode ahead of the small group of warriors, which included his friend Red Hawk. A Company C soldier attacked Crazy Horse and fired his revolver at him but missed. Crazy Horse shot his horse, and as the soldier fled on foot, Red Hawk attempted to stick him in the back with his lance. The bluecoat quickly turned and fired two more shots from his Colt revolver at Red Hawk. The shots missed, but the range was so close that the powder burned the young warrior's face. Red Hawk ended the episode by striking the soldier on the side of the head with his Spencer carbine.[85]

On the hilltop, Calhoun and Crittenden each stood behind platoons of dying but determined men. Crittenden, already handicapped with his glass eye, could see little beyond the smoke and dust. Soldiers began to crumple in death along the angled skirmish line. All at once, charging warriors were on top of them before the remainder could react. Flying Hawk remembered, "On Calhoun Hill, a part of the soldiers stood and gave battle . . . finally, the line on Calhoun Hill was broken and the soldiers fell back toward Custer Hill, fighting as they went."[86] Calhoun and Crittenden died near one another and Custer's brother-in-law kept his promise, "If the time comes you will not find me wanting."[87]

Gall witnessed the stampede of the horses and death of the horseholders in the ravine behind Calhoun Hill. He noted, "The warriors directed a special fire against the troopers who held the horses, while others fought. As soon as the holder was killed, by moving blankets and great shouting, the horses were stampeded, which made it impossible for the soldiers to escape."[88]

John Stands in Timber said that Yellow Nose seized the striped guidon from Company L and noted, "Just over on the other side [of the hill] many soldiers were still fighting . . . he . . . used that flag stick and hit the soldier and counted coup."[89]

Red Hawk and his cousin decided to count coup on the soldiers as they fled the hilltop. Red Hawk saw a soldier with a big hat pointing his gun at them, and he dropped to the ground, screaming a warning to his cousin that was too late. His dead relative fell right into his lap. As the soldier reloaded, the angry Red Hawk shot the bluecoat in the chest as another warrior counted coup on the wounded enemy. The stubborn soldier attempted to get up and Red Hawk rushed in and killed him with the soldier's revolver.[90]

Miniconjou warrior Turtle Rib saw the confusion near Calhoun Hill and remembered, "I saw a running fight with some of the soldiers on foot. Those who kept their horses seem stampeded. Some were going toward the monument and some were trying to ride back the way they came."[91] Today, we know the name of one of the men who attempted to ride back. New Yorker James Butler from Company L was easily recognized by his huge, bushy sideburns that covered his jawline, and somehow in the dust and confusion, Butler was able to get into the lower recesses of Deep Coulee on top of a small knoll. According to warrior accounts, the man with three yellow stripes on his sleeve fought desperately over a half an hour, some distance from the ongoing battle, before he was killed.[92]

Keogh's Company I had chosen a deadly spot of low ground to defend. There may have been a feeble attempt to bring along the wounded including Keogh. Most of them died around Keogh's body in a disorganized heap near Horse Holder's Ravine.[93] White Bull pulled a soldier off his horse and counted several coups.[94] The combat became hand-to-hand with soldiers swinging their carbines like clubs and warriors striking soldiers with stone war clubs and tomahawks. Soldier survivors began to run toward the other end of the ridge—Custer Hill.[95]

Wooden Leg was positioned north of the Keogh fight and stopped shooting since he was afraid he might hit one of his own warriors in the close fighting. One bullet had hit a young, unnamed Lakota boy. Wooden Leg said:

> I saw one Sioux walking slowly toward the gulch, going away from where were the soldiers. He wobbled dizzily as he moved along. He fell down, got up, fell down again, and got up again. As he passed near to where I was I saw that his whole lower jaw was shot away. The sight of him made me sick. I had to vomit.[96]

Stands in Timber recollected, "He was shot through both sides of the jaw and it was hanging down on his breast, the teeth were showing like a necklace. He came down to [the] washout and they met him and tore up a cloth and tied his jaw back on top of his head. He took care of himself from there."[97]

Crazy Horse, who, like Custer, had an uncanny ability of knowing when to successfully charge the enemy, attacked on horseback and cut the soldier line in two. An Arapaho warrior noted, "Crazy Horse, the Sioux Chief, was the bravest man I ever saw. He rode closest to the soldiers, yelling to his warriors. All the soldiers were shooting at him, but he was never hit."[98] Another warrior stated that it looked like a stampede of buffalo.[99]

Atop the high knoll named Custer Hill, warrior snipers shot at the soldiers below them from behind the grass and sagebrush. Custer's headquarters staff, along with Companies E and F, fired from skirmish lines as they took turns providing cover fire for each other and taking the hill from the warriors. Two Moons remembered, "I could not break the line at the bunch of the gray horses and I wheeled and went to the left down the valley with the line of soldiers facing me as I went, firing at me, and my men were firing at the soldiers. . . At last we saw that Custer and his men were grouped on the side of the hill, and we commenced to circle round and round."[100] Some soldiers left their gray horses with their holders in a natural depression below Custer Hill, while some of them made it to the hilltop.

Foley and his guidon had been luckier than most, at least temporarily. Mounted on his fast sorrel, he was able get to the other end of the ridge and Custer Hill. About 19 other soldiers, including at least six men from Company C, escaped the bloodshed at the southeastern end of the ridge and made it to their comrades a half-mile away.[101] White Bull reinforced that many of the horses on Custer Hill were the sorrels, bays, and whites from the other end of the ridge.[102] Red Hawk, Crazy Horse, and others raced toward the last soldiers gathering on a distant hill. A volley of gunfire from Custer Hill repulsed the warriors that were chasing the soldiers running from the other end of the ridge. Red Hawk dropped to the ground, and bullets hit so close to him that they kicked up dust in his mouth. Crazy Horse asked him if he was hurt and Red Hawk replied "no," but that his cousin had been killed.[103]

When Custer reached the hilltop, the commander of the 7th did not like what he saw. There was tremendous residue from gun smoke and dust kicked up by fleeing soldiers being chased by hundreds of warriors toward his position. Their panicked comrades were riding or running on foot through a bloody gauntlet of death in their direction. It was time to make a stand.

Chapter 14

Last Stands

The exhausted soldiers on Custer Hill fired an intense volley at the warriors chasing their comrades along the ridge, and the Lakota and Cheyenne pulled back. As the troopers looked down at their homemade leather belt loops, the copper cartridges were few in number. The isolated horseholders grappling with the gray horses below Custer Hill were in a terrible situation on the low ground. It was almost impossible to hold the rearing horses and fire a weapon. Their comrades had taken the high hill above them, but they were threatened by an ever-increasing number of warriors from every direction. Two Moons mentioned, "The Sioux and Cheyenne came up the valley swarming like ants toward the bunch of gray horses where Long Hair stood."[1] Gall remembered, "They were fighting good. The men were loading and firing, but they could not hit the warriors in the gully and the ravine. The dust and smoke was [were] black as evening."[2] Warriors also began to fire from a nearby ridge that overlooked the horseholders and Custer Hill. Something had to be done to drive them back.[3]

A trumpet call ordered the men who owned the gray horses off Custer Hill. According to Two Moons:

> At last about 100 men and five horsemen stood on the hill, all bunched together. All along the bugler kept blowing his commands. He was brave, too . . . then the five horsemen and the bunch of men, may be so 40, [sic] started toward the river.[4]

It is clear from body locations found after the battle that these were men from Company E. Other warriors noted, "While Custer was making his last stand on the hill, about 40 of his men, letting their horses loose, made a break from the command and endeavored to reach the timber on the Little Big Horn, but none succeeded [and] all were killed."[5] While they saw it as an attempt to escape, it was a half-hearted offensive movement meant to dislodge warriors. It would eventually become an escape attempt, but not for some time.[6] Warrior Little Hawk noted that on Custer Hill, "The soldiers, having tied their horses in fours, let them go and they scattered, most of them running toward the Little Horn. One company of soldiers went down toward the Little Horn, and all but one man dismounted."[7] A man on a sorrel horse exited the hill alone in the direction they had come from. It was flag carrier Corporal John Foley.

Company E seized the ridge above the horseholders and formed a skirmish line there.[8] Runs the Enemy was more successful in getting through that line and was more definitive in location, noting, "From the point that juts out just below where the monument stands [Custer Hill], about 30 of us got through the line, firing as we went, and captured a lot of Custer's horses and drove them to the river."[9] Stands in Timber likely refers to the same episode and stated that young men called suicide warriors "also rode to this area where the museum now is. Some turned and stampeded the gray horses of the soldiers. By then they were mostly loose, the ones that had not been shot. The rest charged right at the place where the soldiers were making their stand, and the others followed them as soon as they got the horses away."[10]

It was no use—the men holding the gray horses could die holding them or flee. Some mounted and tried to get to Custer Hill, while others fled toward the river.[11] Dead gray horses lay in and all around the Deep Ravine that emptied into the river; one belonged to Lieutenant Algernon Smith. The incapacitated officer who needed help to mount and dismount his horse had somehow managed to get to the top of the hill. Unable to raise his almost useless left arm or load a weapon, he had stayed on the hilltop when his company departed. This meant that Jack Sturgis or German First Sergeant Frederick Hohmeyer and his noncommissioned officers were running the company.[12]

Another pale horse rider was straight-laced George C. Brown. Born in Baltimore, Maryland, the 25-year-old wanted to leave his mundane life as a candymaker for the adventurous life of a soldier. He was soon disillusioned and noted, "Our life is not very romantic." His honesty and moral character led to the promotion to corporal only three months before the campaign. Brown needed to be able to keep time, so he ordered an expensive pocket watch right before the campaign and had it on his body.[13] Irishman Corporal Thomas Hagen was also from the gray horse company. He wrote a prophetic letter before he departed:

> *Dear Sister*
> *. . . the old chief Sitting Bull says that he will not make peace with the whites as long as he has a man to fight . . . as soon as I got [sic] back of the campaign I will rite [sic] you. That is if I do not get my hair lifted by some Indian.*[14]

One warrior remembered that the soldiers fought most stubbornly at Custer Hill.[15] George and Tom were near one another and appeared similar in appearance. One difference was a colorful tattoo on Tom's forearm with the initials 'TWC.' Shouting orders nearby was Tom's bachelor counterpart W. W. Cooke. The Canadian had serious thoughts about his lack of promotion and had considered a transfer, and as he stood facing death on the barren hillside, probably wondered why he had not done so.[16] He had exhibited an unusual coolness toward Custer on the campaign over some personal matters, but that mattered little as they all fought for their lives.[17]

Another Canadian stood in the proximity of Cooke—Sergeant John Vickory. He had the prestigious job of carrying the blue regimental standard, but since it had been left behind, he was probably fighting.[18] Another flag bearer was a tall Irishman, Sergeant Robert Hughes, who toted Custer's personal flag with its red and blue background and white crossed sabers. He had the difficult task of chasing Custer around the field, and now he found himself on the highest ground that had a panoramic view of the surrounding country.[19]

Yates, along with his Company F, the Bandbox Company, was nearby. He had expressed in 1873, "It is only a question of time until Custer will get us into a hole from which we will not escape."[20] He had predicted his own demise. It had been Yates who killed the Cheyenne ponies along the banks of the Washita and now the screaming cycle had come full circle. His men were ordered to shoot cavalry horses to serve as breastworks, in an attempt to survive this hot summer afternoon.[21] Custer knew that warriors did not like to attack a fixed position that could be defended, and he wrote about the Beecher Island fight during which soldiers survived by hiding behind their dead animals against overwhelming odds. It was a last resort until help arrived.[22] Five or six sorrel horses that escaped the other end of the ridge were shot in a crude circle. Custer and other men gathered and lay low behind the dead mounts on the top of the hill.[23] Bullets plowed into the horseflesh and iron-tipped arrows fell like rain inside the small circle.

Fig. 74. Canadian John Vickory was responsible for remaining near Custer as part of the headquarters staff. The regimental flag that he normally carried was left behind with the pack train and survived. George Custer's body was buried next to Vickory on the side of Custer Hill. (Previously unpublished)

Courtesy of George Kush

Nearby was one of the most dashingly handsome bachelors in the 7th Cavalry, William Van Wyck Reily. His normally slicked-back hair was ruffled and he had attempted to be on a ship rather than a horse. At age 16, he had been a poor student in math and science and had secured special influence from his stepfather to gain entry into the Naval Academy. Resigning, due to his scholastic deficiencies, he found himself far from any ocean.[24]

Dr. George Lord had to forget his own illness as he yanked bandages from his leather pocket surgical kit. He tended to the wounded and dying, but it was a hopeless cause as he looked through his sweaty, wire-rimmed spectacles. Men fell around him faster than he could react. One officer who knew him commented on his beautiful, soft hands. Now they were being used to help those hands that were covered with hard calluses—the dying enlisted men.[25]

Boston and Autie Reed were nearby. It seems highly likely that their caretakers, George and Tom, had them hiding behind dead horses and lying low. The older brothers did not want to have to explain their deaths to Libbie, Maggie, or other relatives.[26]

Trumpeter Voss had bullied Martin on the journey and forced him to serve as Custer's orderly. At this point, the Italian was four miles away listening to their gunfire. Voss may have been blowing the calls on Custer Hill, or perhaps it was one of the trumpeters of Company F. *Officers' Call* and *Rally* pierced the noise of gunfire for a few brief moments. The first was a command for any officers nearby to gather, and the second for the men to continue to regroup. These two trumpet calls indicated clearly that things were falling apart.[27]

Warriors who had defeated the soldiers at the other end of the ridge slowly surrounded the last bluecoats. Wooden Leg climbed to the top of a hillside northeast of Custer Hill. From his vantage point 50 yards away, he could see soldiers hiding behind dead horses. He fired his newly-captured carbine at the desperate bluecoats.

Several of the men who were on top of the hill were excellent shots and Civil War veterans. It seemed to be somewhat of a stalemate since the soldiers could not be seen behind their horses. It is impossible to know how much ammunition was left, but the dwindling supply must have preoccupied their thoughts and actions. Kate Bighead watched the fight on the hill from a distance and remembered:

> At that time there must have been hundreds of warriors for every soldier left alive . . . Many hundreds of Indian heads were being popped up for a quick look and then jerked down again for a movement forward a little farther. The soldiers must have seen many of these heads, but they must have been puzzled as to which ones to shoot at. The remaining soldiers were keeping themselves behind their dead horses. The Indians could get only glimpses of the white men, but it was easy enough to see where they were.[28]

The Cheyenne woman also saw a soldier who had been knocked unconscious along the battle ridge. She recalled:

> He was just sitting there and rubbing his head as if he did not know where he was or what was going on in the world. While I was watching him, three Sioux men ran to him and seized him. They stretched him out on his back. They went at this slowly, and I wondered what they were going to do. Pretty soon I found out. Two of them held his arms while the third man cut off his head with a sheath-knife.[29]

Benteen had penned accusatory words about Custer's abandonment of Elliott after the Washita. He had written:

> Who can describe the feeling of that brave band, as with anxious beating hearts, they strained their yearning eyes in the direction whence help should come? What must have been the despair that, when all hopes of succor died out, nerved their stout arms to do and die? Round and round rush the red fiends, smaller and smaller the circle, but the aim of that devoted, gallant knot of heroes is steadier than ever, and the death howl of the murderous redskin is more frequent.[30]

The condemnation and responsibility for death had come full circle as Benteen stood on a hillside—unmolested—four miles away and listened to the dwindling gunfire.

The small knot of soldiers behind their dead horses on Custer Hill looked longingly to the southeast for guidons and galloping bluecoats to come over the hilltop and rescue them. Surely 'Custer's luck' would kick in. Suddenly, as had happened at the Washita and the Yellowstone, soldiers on their horses appeared next to a single guidon on a hilltop overlooking the scene. They were three miles away on the highest ground—Weir Point. Captain Weir and his company were attempting to make sense of the chaos from their vantage point.

Before the horses were killed, and the circle of warriors had surrounded Custer Hill, one man had been able to ride away. All hope had rested on him—it was Corporal John Foley. He had escaped the collapse of his own company and had made it to the chaotic knoll. He was probably ordered to ride back through the gauntlet of warriors to bring help to the desperately shrinking number of soldiers.

Foley had removed his colors from the wooden staff and stuffed his company guidon into his dirty blue blouse. His strong and fast sorrel was still unharmed and had saved his life earlier. The bald-headed soldier with blue eyes and a strong Massachusetts accent took off as fast as he could ride through the dust and smoke along the battle ridge.[31]

Stunned warriors watched Foley gallop past them. Many warrior stories mentioned the man on the sorrel horse—he and his ride became something of a legend. Little Hawk remembered:

> One company of soldiers, [Company E], went down toward the Little Horn, and all but one man dismounted. The man who did not dismount rode away. He was riding a sorrel horse and Indians began to shoot at him, but could not hit him or overtake him.[32]

White Shield remembered that when the Calhoun survivors reached Custer Hill, the man on a sorrel horse took off. Warriors Old Bear and Kills in the Night chased the man on the sorrel horse and another Lakota fired a shot at this soldier but missed.[33] Foley miraculously got to Calhoun Hill and disappeared into Deep Coulee. The party of warriors chasing the soldier grew—Cheyenne Low Dog, Little Son, Old Bear, two unnamed Santee boys, and even the famous Crazy Horse. This would account for his absence from warrior accounts in the fight on the deadly knoll where Custer died. The great warrior given credit for over a century of stopping Custer dead in his tracks on Custer Hill—was not even there.[34]

Like so much of the testimony of the battle, the death of Foley is steeped in mystery and contradiction. Several warrior accounts revealed that he was outpacing his pursuers when, suddenly, he pulled his revolver and committed suicide. Another possibility that may have been misinterpreted was that Foley was whipping his sorrel with his revolver when it accidently discharged, killing the corporal.[35] A more likely scenario is that an Indian bullet killed him, which was noted by at least four warriors. Crazy Horse added another detail that others neglected to mention: when Foley fell from his horse, his foot was caught up in the stirrup and the horse dragged him some distance.[36]

The pursuing warriors were not the only ones to witness the frantic chase. Foley had spotted Weir and his men on the highest hill overlooking the battlefield. The corporal was riding hell-bent-for-leather toward the small group of soldiers with their guidon fluttering on top of Weir Point, only one mile away.

Sergeant James Flanagan peered through his field glasses and saw something the others did not. A single soldier was riding as fast as he could toward them, going up and down over the hilly terrain. As the soldier came on top of a rise a mile away, warriors cut him off and killed him.[37] The balding Foley with small tufts of black hair was left alone. He was not scalped or stripped of his clothing. His sorrel ran off, leaving the soldier lying on a grassy rise overlooking Medicine Tail Coulee. As life faded from his body, his blood seeped into the red stripes of the company colors still hidden under his dirty blouse.[38]

The anxious Weir and Edgerly had advanced without permission to a high peak overlooking the Custer fight. Reno now had second thoughts and decided to send Hare toward Weir with orders to open up communications with Custer.[39] From Weir Point, the soldiers looked three miles away toward the dust, and Edgerly saw, "Indians shooting as though at objects on the ground."[40] The soldiers stood closely beside one another attempting to make sense of the distant chaos. Flannigan had field glasses and could see more clearly than the others. Another soldier could see many horsemen on the distant ridge with guidons flying. Weir said, "That is Custer over there." The men mounted up, then Flanagan stated, "Here, Captain, you had better take a look through the glasses. I think those are Indians." Weir looked and changed his mind. The men dismounted and horses were led behind Weir Point.[41]

Weir stayed on the high ground as Edgerly took the company down in the vicinity of Medicine Tail Coulee. Weir, from his higher vantage point, eventually saw warriors coming toward them, so he waved at Edgerly to return.[42] The lieutenant, who went farther than any other soldier looking toward Custer, revealed in a later report, "I saw no Indians on the bluff during our advance until near the lowest point, where we could see the bodies of Custer's men and horses with swarms of Indians."[43] He also stated, "One part of [the] hill on Custer battlefield was black with Indians and squaws standing there."[44]

Thirty minutes later, three other companies joined Weir. French with Company M, Godfrey with Company K, and finally, Benteen with Company H arrived at the highest peak on the battlefield.[45] Hare was there long enough to surmise, "The Indians were thick over on Custer Ridge and were firing and at that moment . . . thought Custer was fighting them."[46] One private shared, "At first when [he] looked over toward Custer Ridge, the Indians were firing from a big circle, but it gradually closed until they seemed to converge into a large black mass on the side hill toward the river and all along the ridge. [He] thinks what they saw was the last stages of the fight."[47] Godfrey wrote:

> The air was full of dust. We could see stationary groups of horsemen, and individual horsemen moving about. From the grouping and the manner in which they sat their horses we knew they were Indians . . . Looking toward Custer's field, on a hill two miles away we saw a large assemblage. At first our command did not appear to attract their attention, although there was some commotion observable among those near to our

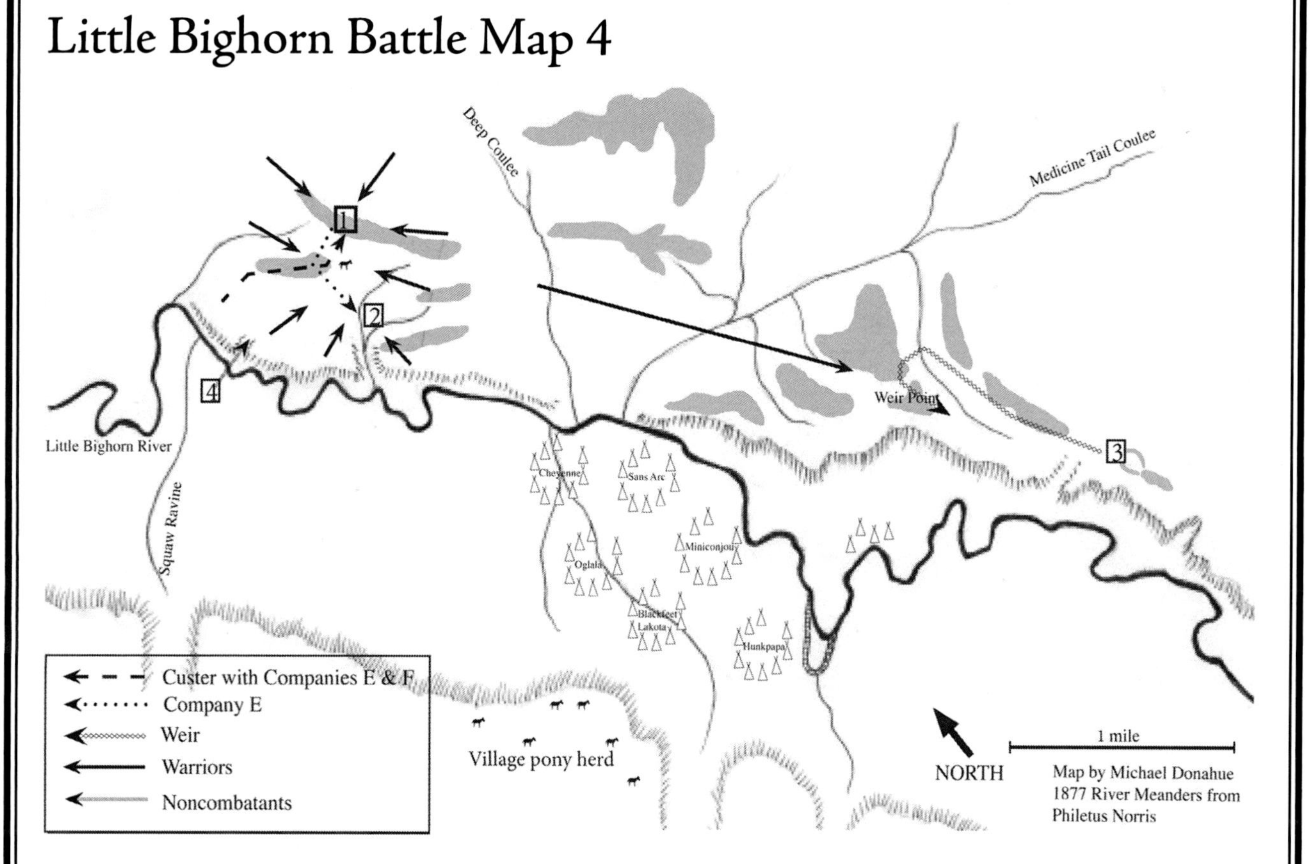

1 Custer's headquarters along with Companies E & F withdraw to Custer Hill leaving the gray horses below them. Warriors begin to surround the entire hill where the soldiers are gathered.

2 Company E is ordered off Custer Hill as warriors pressure the gray horses from below. Warriors stampede the last horses and kill the men on Custer Hill and the surviving members of Company E flee into the Deep Ravine where they are killed, ending the battle for the five companies.

3 Weir, with Company D, attempts to reach Custer and is witness to the end of the Custer fight, but is unaware that the entire command has perished. They are joined by Companies M,K, and H, who quickly retreat to Reno Hill with Weir's company behind them.

4 The noncombatants cross the river and help strip and mutilate the dead soldiers. The warriors spot the other soldiers and race off toward Weir Point driving them back to Reno Hill where they are surrounded. A 24-hour siege begins.

Fig. 75. Map: Little Bighorn Battle Map 4.

position. We heard occasional shots, most of which seemed to be a great distance off, beyond the large groups on the hill. While watching this group, the conclusion was arrived that Custer had been repulsed, and the firing we heard was the parting shots of the rear guard.[48]

Fig. 76. Oglala warrior Red Hawk accompanied Crazy Horse during the Custer fight. Recalling that he bravely fought soldiers riding sorrel horses, it seems clear that he was involved in the destruction of Company C that rode that color of horse.

Courtesy of the author

John Ryan's comments are more enlightening and point to a truth that some did not wish to consider. "We saw at a distance from a mile and a half to two miles parties, whom we supposed were Indians, riding back and forth and firing scattering shots. We thought they were disposing of Custer's wounded men, which afterwards proved to be true."[49] They stayed there long enough to see the huge group of Indians head back toward their camp and others ride in their direction.

There was no place to hide. Hundreds of lead bullets from repeating weapons and captured soldier carbines thudded into the ground and dead horse flesh on Custer Hill. Men lying behind the crude circle could not avoid the iron-tipped arrows that fell like hail from the blue sky. There was no way to avoid being hit by the heavy barrage. A few lucky men lay motionless and unharmed. A short distance away on another hill, the remaining men who had lost their gray horses watched the fight rage above them.

Stands in Timber noted, "At the end it was quite a mess. They could not tell which was this man or that man, they were so mixed up. Horses were running over the soldiers, and over each other. The fighting was really close, and they were shooting almost any way without taking aim."[50]

The question for many must have been—do I fire my last few cartridges or attempt to save them? For some it would be a choice of when to fire the last round into your head to avoid capture and torture. Wooden Leg remembered that eventually on Custer Hill:

> The shots quit coming from the soldiers. Warriors began to call out that all the white men were dead. All the Indians jumped up and rushed forward. All the boys and old men on their horses came tearing into the crowd. The air was full of dust and smoke. Everybody was greatly excited. It looked like thousands of dogs might look if all of them were mixed together in a fight.[51]

Standing Bear was there on the hill near the end. Warriors were yelling:

> *Hoka Hey!* Hurry! Hurry! Then we all went up, and it got dark with dust and smoke. I could see warriors flying all about me like shadows, and the noise of all those hoofs and

guns and cries was so loud it seemed quiet in there and the voices seemed to be on top of the cloud. It was like a bad dream.[52]

Red Hawk remembered, "an Indian yell and bone whistles, and later, a stampede."[53] He shot a soldier aiming his weapon and as the Oglala got closer, he noticed another bluecoat with ornaments on his breast, likely an officer. The soldier fired a bullet that whizzed past Red Hawk who immediately dropped the enemy with a shot from his carbine. The soldier attempted to get to his knees, but Red Hawk pushed him to the ground with his weapon. The soldier then tackled him around the legs until the warrior buried his knife into the bluecoat's chest.[54]

As warriors tumbled over the northeast side of the ridge, surviving soldiers began to run to escape death back along the battle ridge or toward the river. The breakout party likely included Bouyer, Boston, and Autie Reed. None of them made it far from the hill of death. White Bull saw 10 men come running off the hill toward him, led by two soldiers out in front. He killed one soldier and then was hit by a spent bullet in the ankle. He crawled into a ravine until the soldiers were all dead.[55]

Fig. 77. Mitch Bouyer. The Lakota and French mixed-blood Mitch Bouyer was one of the most knowledgeable scouts of the region and Custer borrowed him from John Gibbon's command. While apprehensive of the large numbers indicated by Indian trails, he followed Custer into the fight anyway. A skull found below Custer Hill by a visitor after the 1983 fire indicated where his body was buried.

Courtesy of the author.

Mitch Bouyer, wearing a calfskin jacket, ended up on a small knoll 200 yards below Custer Hill. The warriors had placed a bounty on his head and he knew what lay in store for him if caught. Oglala Eagle Bear recalled seeing a wounded soldier in or near a ravine. The man could speak Lakota and pleaded for his life saying, "Do not kill me. I have a wife and child at home waiting for me." Eagle Bear said, "If that is true, then why are you here trying to harm our wives and children?"[56] The warrior ended the debate with a bullet into Bouyer's head.[57]

Soldiers lay wounded and dead all over the broken ground with a concentration of men and horses on the western slope of Custer Hill. Soon the moaning of wounded men was silenced with the sound of stone mallets crushing skulls and the slashing of foreheads by tomahawks. Wooden Leg saw that a plump-cheeked soldier had been knocked unconscious, then woke up, propping himself on his left elbow as he looked over his left shoulder. "His expression was wild, as if his mind was all tangled up and he was wondering what was going on."[58] Warriors jumped back, thinking he had come back from the dead, but a Lakota quickly wrenched the six-gun from his hand and shot the soldier in the head. Other warriors rushed up, counted coup, and stabbed him.

Young Cheyenne Big Beaver captured a new soldier revolver and saw the warriors rushing the hill so he joined them. He saw no soldiers standing up, but some were still firing or on the ground sitting up. The warriors quickly killed them.[59]

Exactly when Custer and his fellow officers went down will never be known for sure. Before the dust and gun smoke had enveloped the small band, the men had a spectacular view of the snow-topped Bighorn Mountains and the beautiful valley below them. It was a beautiful place to die. It looked like the Albert Bierstadt paintings that encouraged the pursuit of 'manifest destiny'—the landscapes that Custer had recently admired in the artist's studio.

Even Sitting Bull, who watched the battle from the hills on the other side of the valley, supposedly heard of his adversary's death. The great spiritual leader did not know who had attacked his camp, but learned who their short-haired leader was after the battle. Whether any of the Lakota knew who Custer was is questionable, but Sitting Bull was poetically told by his warriors:

> Where the last stand was made, the Long Hair [Custer] stood like a sheaf of corn with all the ears fallen around him. . . . He killed a man when he fell. He laughed . . . he had fired his last shot . . . he rose up on his hands and tried another shot, but his pistol would not go off."[60]

From the top of Custer Hill, Big Beaver remembered:

> There were still some men fighting of the White Horse Troop. E Troop down near where the Nat.[ional] Cemetery is now. About 15 or 20 men got up and started to run towards the river down the ravine. He says these men were afraid as they did not fire back and that some mounted Indians pursued them and killed them all. The last men to be killed are those next to the river down this draw. The reason they did not fire back when they got up to run is evident as they had run out of ammunition.[61]

The last men from Company E watched as hundreds of warriors swamped their commander and his small band of bluecoats on top on the high ground, 1,000 feet away. The terrified soldiers began to run toward Deep Ravine that led to the river timber.[62] They were the last soldiers alive on the field, and running out of ammunition for their carbines, tossed them to the ground. Two Moons later complimented the troop. "The White Horse Troops fought with signal [single] desperation. If the others had not given up, but had fought with equal stubbornness of the White Horse Troop, Custer would have driven the Indians from the field."[63]

Thirty-year-old Private John Darris was a Civil War veteran and had been a fireman before his enlistment. Somehow he had been able to hold on to his gray horse and took off in the direction the soldiers had come from. He raced his frightened horse up and down the ravines and was able to reach Finckle/Finley Ridge. Making it about 100 yards past the dead body of Finley, one warrior recalled,

"One soldier on a white horse almost got away—but we kept shooting at him, and before long we knocked him off his horse."[64] Warrior Flying By mentioned, "The last man killed was one who rode a horse and who fell over near Sgt. Finley."[65]

First Sergeant Hohmeyer and former candymaker George Brown, who joined the army for a life of adventure, made it into Deep Ravine, where there was more excitement than he ever could have imagined. Brown's expensive silver pocket watch bounced in his pocket as he ran for his life. Deep Ravine became a place of last resistance and survival. If the men could get to the river and the timber, there was a chance to hide in the thick brush. Iron Hawk was near the gully and he recalled:

> Nearly all of them were afoot, and I think they were so scared that they did not know what they were doing. They were making their arms go as though they were running very fast, but they were only walking. Some of them shot their guns in the air. We all yelled, *Hoka Hey*! and charged toward them, riding all around them in the twilight that had fallen on us.[66]

The Indians followed and killed them with war clubs of stone and wood, some of the latter having sharp blades embedded in them.[67]

The tallest man in the company was William Rees and the 6'1" private, a former laborer, slid down the steep walls of the ravine.[68] Once in the curving ravine, there was no way to get out except to run toward the river that was 400 yards away. The ravine curved toward the south and it was impossible to see around the bend. There were hundreds of warriors chasing them. Above the scared soldiers were two warriors with Winchester or Henry rifles that began to pour lead into the soldiers rounding the bend. It was like shooting fish in a barrel.[69]

As dead and wounded men began to pile up at the bend, others halted and decided to try to climb out of the ravine. Most of the soldiers were shot in the side attempting to climb out of the death trap.[70] Some soldiers ran down one side of the ravine and unsuccessfully tried to scramble up the other.[71] One Lakota warrior followed the soldiers in the ravine too closely, and the soldiers killed him. The overexcited warriors, when entering the ravine, mutilated his body mistaking him for an enemy scout.[72] One German private, William Brown from Company F, escaped the carnage of Custer Hill with his horse and rode into Deep Ravine. Through stealth, timing, or speed, he bypassed the killing in the ravine and was able to get across the river into the brush where he was finally caught and killed.[73]

Deep Ravine was the end of the five companies. Death had taken men on both sides. Custer with 209 of his men perished in less than two hours. There is no consensus on the number of warrior dead and many probably died later from wounds. The Indian memorial that exists today on the battlefield lists approximately 60 warriors killed in the battle, and many others were mortally wounded.[74]

While Custer's units during the Civil War had some of the highest casualty rates in the Union army, his Indian war losses had always been negligible. He had 21 killed and 18 wounded in his first three Indian fights, with 20 of those killed at the Washita.[75] Even when outnumbered, in Custer's mind, the possibility of large numbers being killed under his command did not exist. There was nothing in his past experiences to lead him to that conclusion.

He had underestimated one major factor that had not played out in his other Indian fights. It was not that the soldiers had single shot weapons—they had those in 1873. It was not that the Lakota and Cheyenne had repeating weapons and plenty of them—they had those as well along the Yellowstone. It was not because soldiers were tired—they were exhausted and freezing before the Washita.

The difference was an intangible—motivated anger combined with fear. The warriors' families were right across the river and the men would fight with a ferocity Custer had witnessed only once in the early moments of the Washita fight, where his overwhelming numbers worked in his favor. The warriors' families were in imminent danger, and there was no time for a delaying action. It became obvious to the warriors who the cavalry targets were and they tactically whipped the soldiers. Guerilla-style fighting had defeated disciplined, controlled lines of fire. The warriors' brilliant use of terrain, the strategy of stampeding horses, and the capture of soldier weapons and ammunition meant defeat for the bluecoats. The strategy of stretching the troops thin and away from additional support had almost worked along the Rosebud, it was perfected at Little Bighorn.

Those very women and children who Custer had hoped to capture wreaked vengeance on the bluecoats. Standing Bear witnessed:

> The women [were] coming over in a swarm and they were all making the *tremolo* [a victory sound rather than a song].[76] To make sure soldiers were not playing possum, they slashed bodies with knives. Kate Bighead saw several soldiers who were not dead and "the Indians cut off arms or legs or feet of these, the same as was done for those entirely dead.[77]

A chief named Wopide lost two brothers in the fight and his sisters came up to the hill with axes and knocked the brains out of some of the wounded soldiers.[78] Their enemies would return in the next life in this condition. Big Head found her nephew Noisy Walking below the battle ridge shot through the body and stabbed several times. Lame White Man, who suffered similar wounds, was also found nearby.[79]

Women brought travois pulled by ponies to take the dead and wounded warriors across the river. The camp circles moved downstream in the verdant valley below the battle ridge. Many tipis were not set up in the new camp as the inhabitants erected small dome shelters from hides or slept under the open stars instead. Lakota dead were placed in trees near Deep Ravine or in burial tipis. The Cheyenne dead were buried in hillside caves and rocks away from the camp that afternoon.[80]

Black Elk saw wounded bluecoats wiggling on the ground. He rode around with his young friends shooting arrows into the bodies.[81] He found a pocket watch and wore it as necklace, not understanding its use. Eagle Bear recalled that the boys were much more cruel to the dead and dying bluecoats than the warriors.[82] Some soldier corpses were propped up and arrows shot into their backs and buttocks.[83] Black Elk recollected, "I could smell nothing but blood, and I got sick of it . . . I was not sorry at all. I was a happy boy. Those *Washicus* had come to kill our mothers and fathers and us, and it was our country."[84]

Oglala Julia Face stated, "The dead soldiers were quite plain as the Indians would strip them and their skins would shine in the sunlight. They would rob the soldiers . . . and keep the silver and gold [coins], but would throw the paper money away."[85] A seven-year-old Cheyenne noted that as he stripped the bodies he found green picture paper in the soldier trouser waistbands and thought it was pretty. Later, when playing along the banks of the Greasy Grass River, they modeled riders and ponies from mud and used the dollar bills as saddle blankets.[86] Several warriors found tobacco on the soldiers and whiskey in their canteens. Two Cheyenne participants remembered, "There was lots of fussing and quarreling among [the] Sioux after the fight over horses and guns that were captured. [The] Sioux very often quarrel over the plunder."[87]

The warriors and their women ignored Sitting Bull's vision warning them to leave the soldier belongings alone. The white man's goods were too much of a temptation to his people, and the predicted bad omen would materialize into reality: they would never win another victory over the white man.

One Cheyenne warrior slowly picked up four carbines and as many revolvers from the dead bluecoats. He had been wounded twice in the Rosebud fight with Crook's bluecoats, a week earlier. He was unable to fight in the battle but was proud of his fellow Cheyenne. Some of these same soldiers had wounded him in 1868 along the Washita River and had killed many of his friends there. Magpie witnessed the death of soldiers he had fought before and he had survived once again.[88]

"I did not like to hear the weeping of the women. My heart that had been made glad because of the victory was made sad by thoughts of our own dead and dying men and their mourning relatives left behind," remembered Wooden Leg.[89]

Warriors who had been intensely involved in the fight began to realize that other bluecoats were watching them from a high hill three miles away. It appeared that the soldiers they had chased to the hills earlier were now coming their direction. Pumped with the confidence of victory, they began to gallop toward the bluecoats.

Weir stood on the high hill that today bears his name with only 55 soldiers. One of them was a redheaded man with a name that sounded strangely like two first names—Vincent Charley. The 27-year-old former farmer-turned-farrier took good care of the company horses. The hills he observed were miniscule and pitiful in comparison to the beautiful Alps he was familiar with in his home country of Switzerland. He had been more than lucky not to be a part of Custer's five companies or Reno's three companies decimated in the valley. He had been on a wild goose chase with Benteen for most of the day. That would soon change.[90]

The soldiers watched as hundreds, if not thousands, of warriors rushed toward them. One soldier said the numbers of rushing warriors, "Looked as thick as grasshoppers in a harvest field."[91] Godfrey said, "The firing ceased, the groups dispersed, clouds of dust arose from all parts of the field, and the horsemen converged toward our position."[92] The soldiers on the hilltop dismounted for a fight.

Chapter 15

The Rescue of the 7th Cavalry

Thirty minutes after Weir's departure, a hesitant Benteen joined him. Benteen recorded, "There was no sign of any troops or any fighting going on, nothing of the kind could be seen. We had not been more than two or three minutes at that high point before the gorge was filled with Indians rushing toward us, then we fell back to where we were corralled."[1] His tardiness had not allowed him to see what the soldiers had witnessed earlier. He jammed his company guidon between some rocks on the high ground to let Custer know where he was. If he was out there, it was his job to come to Benteen. He exclaimed, "This is a hell of a place to fight Indians. I am going to see Reno and propose that we go back to where we lay before starting out here."[2]

The white-haired captain, in a self-protective mode, testified later, "Some of the officers say that the battlefield was in sight, but I know positively that it was not, having gone over it two or three times since."[3] Such a statement is blatantly false, as anyone who climbs Weir Point today can easily see the area. Benteen was the last to arrive and first to leave after only minutes.[4]

Reno had moved only a few hundred yards with the wounded carried on blankets. Benteen convinced him to take refuge at the crater-shaped formation where they had met. Francis Gibson recalled, "Reno did not know which end he was standing on, and Benteen just took management of affairs in his own hands, and it was very fortunate that he did."[5] Weir left his company with Edgerly and rode back to see what was going on. Soon, an order to retreat was given to the men on Weir Point. According to Gibson, the huge numbers of warriors coming toward Weir had tempered his initiative. He recounted, "[Weir] returned saying he could find no signs of Custer's command and that there were enough Indians over there to eat up his company a hundred times over."[6]

On top of Weir Point, French suddenly hollered to withdraw. Edgerly, who was along the ridge top overlooking the river, could see a few warriors gathering on a lower ridge nearby. When ordered to retreat, the young lieutenant did so reluctantly. As soon as the soldiers vacated the high point, the Lakota and Cheyenne seized the ground and delivered a withering fire.[7] Farrier Vincent Charley was shot through the hips and fell off his horse. With blood also streaming from his head, he desperately struggled to stand but could not. Edgerly told the farrier to hide in a nearby ravine and assured him that he would get some men to rescue him.[8]

The warriors raced toward the stragglers and advanced within 15 feet of Edgerly. Cutting their way through the warriors, they escaped under a hail of gunfire.[9] One soldier looked back and saw a

Fig. 78. Dr. Henry Porter was the only surviving surgeon of the three that accompanied the 7th Calvary to Little Bighorn. He established a field hospital among the horses and mules atop Reno Hill.
The Denver Public Library, Western History Collection [B606]

desperate Charley crawling on his hands and knees. According to Edgerly, Weir refused to allow more men to risk their lives to rescue the farrier. The warriors not only killed the Swiss native, but also rammed a stick down his throat. The military mantra of not leaving a fellow soldier behind was not in the vocabulary of the 1876 soldier on this hot day.[10]

French, who was supposed to cover the retreat, took off, and Edgerly was right behind him. A perceptive Godfrey realized that the panicked retreat "would throw the whole command into confusion, and perhaps, prove disastrous."[11] Godfrey ordered his men to dismount, and with the help of Hare, delivered several volleys at the warriors chasing the troops. Godfrey threatened to shoot any man who did not keep his designated interval, and his bold stance drove the warriors back. Just as he had done at the Washita, his coolness under fire helped save his unit as well as the other troops that had raced past him.[12]

The soldiers formed a line around the rim of what could be described as a skillet with an edge missing. Companies lined the edge of the depression, and Benteen commanded the handle: a high ridge-like spur that was exposed from three directions. The horses and mules were herded into the lower part of the depression, and the wounded from the valley fight were placed among them for protection from the gunfire. Dr. Henry Porter, the only medical staff that survived, busily attended to the numerous wounded.[13] The warriors quickly surrounded the position, firing from distant hills and nearby knolls. Edgerly remembered, "From that time, about 5:00 p.m. until nine or 10 o'clock, the Indians kept up desultory fighting . . . most of the firing coming from bluffs a thousand or more yards away."[14]

The men under Weir and Moylan defended an area that had no natural protection, so boxes of hardtack, packsaddles, and dead horses were piled up along the northeastern side of the depression to stop incoming bullets. Other soldiers quickly threw up mounds of dirt, some using tin cups, forks, knives, and spoons.[15]

Mathey was relieved that the mules were within the defensive area and relaxed his responsibility of watching them. Several of the animals smelled the water below and took off toward the river. Benteen rounded up several of them personally; however, one ammunition-laden mule reached the Indian position below. How many soldiers would killed as a direct result of this careless act will never be known.[16]

Benteen took this opportunity to further chastise Mathey, who had reported the captain's poor packing along the Rosebud. Benteen stated:

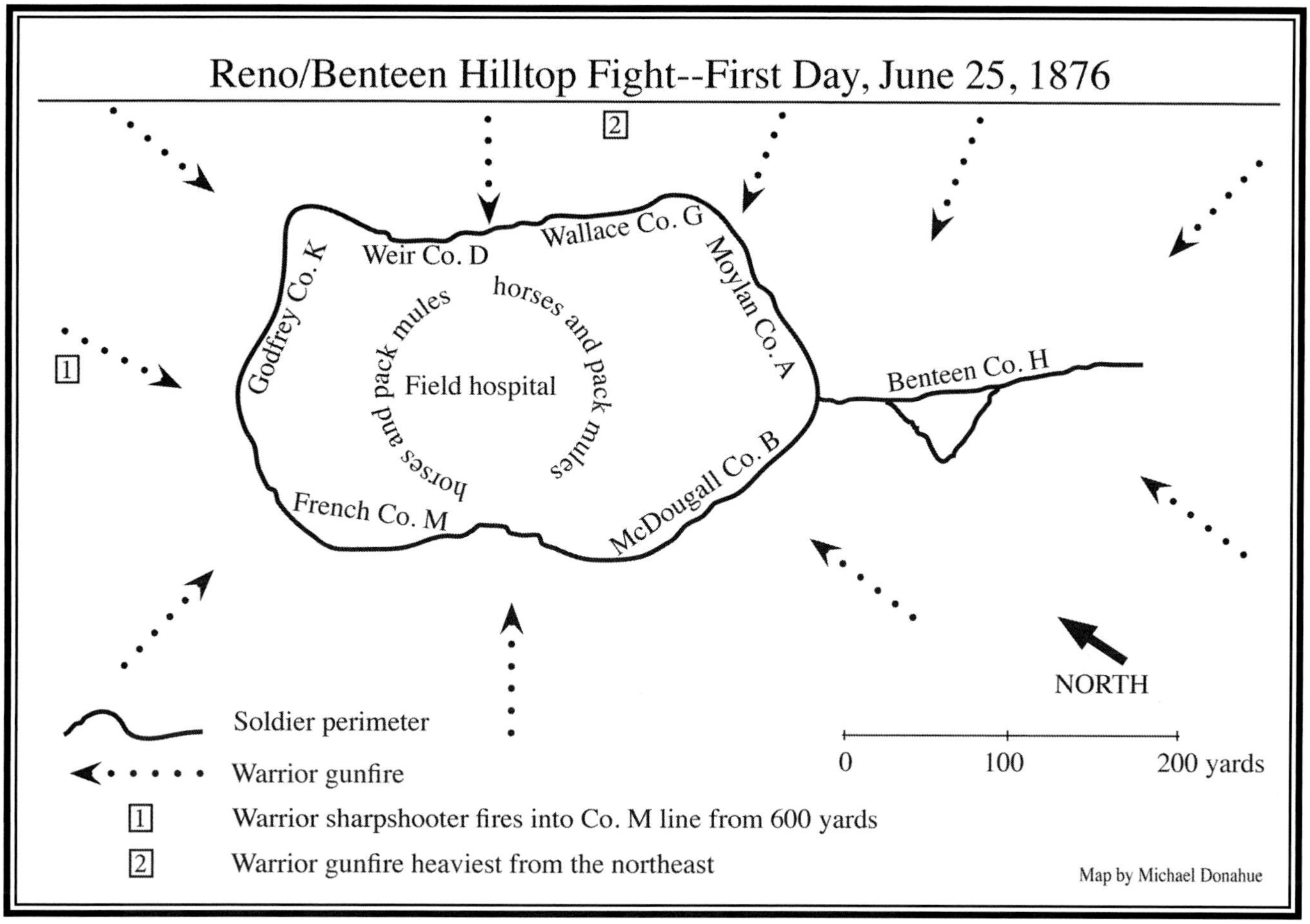

Fig. 79. Map: Reno/Benteen Hilltop fight. First Day, June 25, 1876.

> I don't suppose there was ever an officer of the army that got such a 'cussing out' as I gave Mathey on the eve of June 25th, and before crowds of enlisted men, officers and 'packs.' Personally, I brought three of the mules of the train back—the mules being loaded with ammunition, and had gotten quite a long way down toward the water, for which they were heading.[17]

Soon after the mule incident, Thompson and Watson, whose horses had given out, came wandering up from the ravine where Dr. DeWolf had been killed.[18] They had spent over an hour and a half hidden in the nearby ravines.

The soldier horses neighed and mules brayed as bullets from Sharpshooter Ridge pierced them. A soldier came over to find a horseholder, suffering from battle shock, holding the reins of five horses—three of them were dead.[19] Animals were not the only targets that warriors were hitting. John Ryan and Thomas French lay in a line in the tall grass looking over the ravine they had just climbed. A dead-eyed warrior on Sharpshooter Ridge killed or wounded four men lying near them. The soldiers pivoted and fired a volley that ended the sniping from the warrior behind the rocks 600 yards away.[20]

As Benteen walked up and down the line, an Indian bullet tore off the heel of his boot. He turned around, looked down, and taunted the warriors, "Pretty close call—try again."[21] The gunfire was close on both sides. One warrior recalled, "The soldiers were all sharpshooters, and the moment we put our heads up they fired at us and nearly hit us."[22]

Corporal Dan Nealon of Company H became known as "Cracker Box Dan" because he would not move from behind his box of hardtack. Moylan, who had performed bravely in August of 1873 against some of the same warriors, lost his nerve in the deadly race from the valley. Not only had he not wanted to go toward Custer, but he also found an Aparejo packsaddle and hunkered down, weeping. He later received the name "Aparejo Mickie."[23] Benteen, who saw Moylan upon his arrival on the hilltop, wrote that the gallantly-mustached captain was "blubbering like a whipped urchin, tears coursing down his cheeks."[24] Francis Gibson ordered one of his privates to hand over his carbine. The same soldiers later discovered Gibson lying flat on the ground frozen in fear and having tossed the weapon aside. The soldier remembered that Gibson acted cowardly all through the battle. When charging the warriors, Benteen ordered his men to step over his frightened lieutenant.[25]

White Swan and Half Yellow Face had left the valley after Reno, but they made it to the hilltop before it was surrounded. During one of the charges by the Lakota on the hilltop, unable to crawl due his wounded legs, White Swan grabbed the grass to pull himself forward to fire at the enemy. The soldiers would drag him backward; only to later, find he had crawled back to the firing line. It was a sight to inspire the more timid bluecoats.[26] Hairy Moccasin, Goes Ahead, and White Man Runs Him escaped down Reno Creek before the warriors completely surrounded the site.[27]

A 15-year-old Lakota boy named Breech Cloth circled close, and as he neared the flat ground below Moylan and Benteen's troops, his horse tripped, flipping end over end and throwing the youth. Breech Cloth was shot dead and could not be removed from the field until darkness settled.[28]

Crazy Horse believed the soldiers would attempt to retreat down Reno Creek. He and a group of warriors went to the bluffs above the burned Lone Tipi to set up an ambush. As they waited, Crazy Horse etched a drawing into the sandstone rock. The image was a horse with a lightning bolt on its flank and a large rattlesnake that arced over the horse. Tiring of waiting for the uncooperative soldiers to escape, they eventually returned.[29] As darkness covered their movements, groups of warriors raced forward, unleashing scores of arrows into the soldier circle.[30] The strategy to starve out the soldiers rather than risk warriors recklessly was initiated.[31]

The bluecoats prayed that darkness would soon envelop them and hide them as targets. It had been a deadly evening—eight soldiers were killed and 21 wounded.[32] The sun had created a symbolic red dawn that morning, and it dipped behind the hills to the west through the dust and smoke; Wallace saw it as a 'big red ball,' and Windolph, 'a ball of fire.'[33] It seemed a fitting end to a red Sunday. Windolph wrote:

> Welcome as the darkness was, it brought a penetrating feeling of fear and uncertainty of what tomorrow might bring. We felt terribly alone on that dangerous hilltop. We were a million miles from nowhere. And death was all around us.[34]

Soldiers had quickly emptied their canteens and were desperate for water. Some men, in vain, tried to place hardtack, grass, or pebbles in their mouths to create saliva. One soldier went insane with thirst and had to be tied up.[35] Nature, in a cruel tease, dripped rain all around the battlefield, but very little hit the valley or the hilltop. Some of the exhausted men could no longer keep their eyes open and, no matter how hard the officers kicked them, kept snoring. Charles Windolph reminisced:

> Across the river and down in the valley of the Little Horn, where so many of our men lay dead, we could see great fires and hear the steady rhythm of Indian tomtoms beating their wild victory dances . . . now and again you'd hear what sounded like the high-pitched wailing of the squaws, crying to the Great Spirits for their departed dead. All through the short, black night the orgy went on.[36]

Godfrey saw fires all up and down the valley floor, recording:

> The long shadows of the hills and the refracted light gave a supernatural aspect to the surrounding country, which may account for the illusions of those who imagined they could see columns of troops . . . All night long they continued their frantic revels: beating tomtoms, dancing, whooping, yelling with demoniacal screams, and discharging firearms. We knew they were having a scalp dance.[37]

Black Elk was in the camp and remembered that everyone was up all night. The warriors had decided to watch the soldiers in shifts. Those watching up on the hill would be relieved when the sun came up.[38] At sundown, Wooden Leg went to visit his mortally wounded friend, Noisy Walking. The young man was lying under a shelter dome on buffalo hides. Wooden Leg asked him how he was doing and he said, "Good, only I want some water."[39] His father had refused his wish suggesting it would kill him. Wooden Leg felt helpless and, realizing his friend would die, told him how bravely he had fought. Noisy Walking died that night.

Hunkpapa Little Knife admitted that a soldier with stripes on his sleeves was killed that night during a dance. Rain-in-the-Face captured the bluecoat who was bound hand and foot, then placed in a lodge. Little Knife stated, "A wild dance followed in the night after the battle, and a few men, drunk with the excitement of the results of the conflict, sought out the lodge in which the captive was held and killed him with a knife."[40]

Benteen came off his ridge, and Reno suggested they destroy all the surplus equipment, abandon the position and leave the wounded men who could not ride. A clear-headed Benteen quickly nixed the proposal, but the rumor somehow seemed to have eventually reached the wounded in the hospital, which caused great consternation.[41] John Burkman was ordered to perform orderly duty for Reno. He remembered, "I had to guard Reno's tent. He had a keg in thar [there] and he was drinkin' considerable. I couldn't see in but I could hear . . . I heard him say to another officer that was with him, "Wall' [Well], he says,

'I wonder whar [where] the Murat of the American army is by this time! And then I heerd [heard] them both laugh.'"[42] The statement that went over Burkman's head was a reference to Marshal Joachim Murat who was one of Napoleon's cavalry officers. Known for military success at a young age, Murat was daring, flamboyant and a French version of Custer.

As it got colder during the night, civilian mule packers John Frett and Benjamin F. Churchill went hunting for blankets and something to eat. Sometime shortly after 9:00 p.m. Frett passed Reno in the darkness and said, "Good evening," after saluting him.

Reno asked Frett what he wanted and was told he was looking for something to eat. Reno, with a slurred tongue, inquired, "Are the mules tight?"

Not understanding the question, Frett asked, "Tight? What do you mean by tight?"

Reno, with whiskey bottle in hand, shouted, "Tight, God Damn you!"

Not only was the picket line taunt, but Reno was tight as well. The frustrated major slapped the packer in the face, sloshing alcohol over him. Propping himself against a mule, an angry Reno leveled his carbine at Frett with the threat, "I will shoot you."[43]

Churchill pulled his friend away and they disappeared into the darkness. Both men thought that Reno was under the influence, and they did not see him for the remainder of the night. The threats within the soldier perimeter would soon become secondary to those coming from the warriors.

At around 4:00 a.m. a pinkish-yellow light began to illuminate the eastern horizon. Two shots from the Indian lines pierced the silence—it was June 26. According to George Wallace, the day was cloudy—a respite from the day before.[44] At 5:00 a.m. John Martin sounded *Reveille* and, as the last note echoed across the hills into the valley, there was an indecipherable response from the warriors blowing captured trumpets in the dawn.[45] During the night, a large number of warriors had used the cover of darkness to move closer to Benteen's dangling line. A deep ravine led to the exposed end of the ridge, and the warriors were close enough to pelt the bluecoats with dirt clods.

Benteen had not ordered his men to create breastworks since he believed the warriors would leave. His men began to take casualties and around 7:00 a.m., he decided to charge the Indians threatening his line. The offensive move had a limited effect, and soon, the warriors were pushing back. As soon as the soldiers withdrew, the warriors retook their former positions.[46] Benteen decided he needed reinforcements and left the hill to get help. The warriors pressed the timid Gibson, who was left in charge, and fear swept the ridge with some men running from the scene. Someone shouted to get the 'old man' back up there.

Benteen requested part of French's company to join him. His men, who had labored at their entrenchments, were not happy to be leaving their breastworks to the exposed position on the top of the ridge.[47] The white-haired captain seemed to dare the Indians to hit him as he walked about with his shirttail hanging out of his pants drawing the Indian gunfire. Such a bold face inspired his men, and he told them, "When the bullet is cast to kill me, it'll kill me, that's all."[48] One bullet nicked Benteen's thumb and came painfully close to doing so. Windolph lay at his cool captain's feet and was ordered to stand up

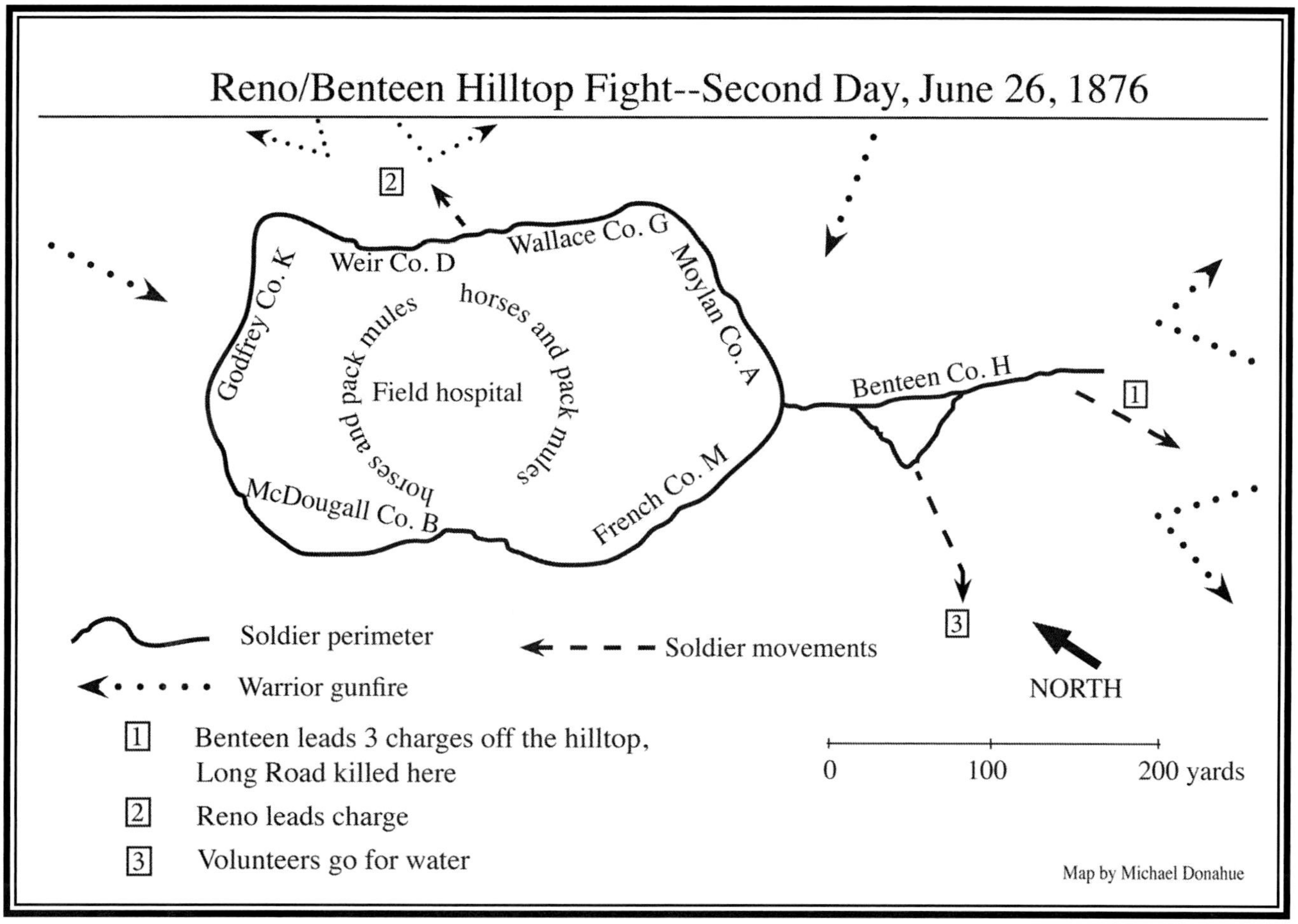

Fig. 80. Map: Reno/Benteen Hilltop Fight. Second Day, June 26, 1876.

as bullets whizzed past them. The young German private reluctantly asked him if it was necessary "to get up in the face of such a murderous fire." Benteen replied, "If you do, and ever get out of here alive, you will be able to write and tell the Old Folks how many Indians we had to fight today, as from here you can get a splendid view of their numbers and position."[49]

One of the nearby warriors was the Sans Arc warrior Long Road who had lost an older brother in the Rosebud fight. Private Edward Pigford was one of the unhappy transplanted men who had joined Benteen's company. He saw the warrior boldly creep forward popping up, shooting and getting down again. He was only 75 feet from the soldier line. Pigford sighted his carbine at the upper part of Long Road's chest, he pulled the trigger, and the warrior tumbled into the dust. Fellow warriors could not retrieve his body, and Gerard later scalped the youth.[50]

Benteen, reinforced with French's men, decided on another charge. Windolph elicited, "Yelling and firing, we went at the 'double quick' and the Indians broke and ran. When we had cleaned them out for a hundred yards ahead of us, we hustled back to our holes."[51] In the charge, Private Tanner of French's company was wounded and left exposed down on the hillside. John Ryan, with three men, ran down and wrapped him in a blanket. Bullets ripped through the clothing of Ryan and the other men but miraculously none were hit; however, Tanner died a short time later.[52]

Windolph noted a third and last charge by Benteen around 11:00 a.m. Benteen's account of one of his charges is humorously descriptive. Leaving the timid Gibson behind, Benteen stated,

> I walked along the front of my troop and told them I was getting mad, and I wanted them to charge down the ravines with me when I gave the yell: then, each to yell as if provided with a thousand throats...we dashed into the unsuspecting savages...and to say that 'twas a surprise to them, is a mild form, for they somersaulted and vaulted as so many trained acrobats, having no order getting down those ravines, but quickly getting; de'il [devil] take the hindmost![53]

The warriors also pressed the lower position defended by Reno. Civilian packer Churchill, who had witnessed the clash between his friend and Reno hours earlier, saw the major lying with his head on his carbine behind some packsaddles and hardtack boxes.[54] Benteen walked over to his commander and encouraged him to drive the warriors back. A timid Reno jumped up and led the rush as the men ran 15 or 20 yards. A critical Varnum recalled that it was a rush and not 'a charge.' The result was successful as the men saw the "whole outfit [of warriors] skipping out to the hills beyond."[55] The only casualty was a crying soldier who refused to leave his hole, and when the soldiers came back, they found him shot dead.

As more wounded men were brought to the field hospital, Dr. Porter warned the commanding officers that water was needed or he could not save some of them. Desperate men cried from their wounds and were shielded from the sun by a large piece of officer's canvas propped up by sticks. Weir's trumpeter had been shot through the bowels. Screaming in tremendous pain, he begged his fellow soldiers to put him out of his misery.[56] One man who could not be saved was Corporal George Lell who was shot while on Benteen's line. He was dragged to the hospital and, realizing he was dying, asked his fellow soldiers, "Lift me up boys, I want to see the boys before I go." Windolph recollected, "A smile came over his face as he saw the beautiful fight the Seventh was making. Then they laid him down and he died soon after."[57]

As the gunfire slackened around noon, a call for volunteers to obtain water went out. Benteen, who had driven the warriors from the ravines leading from the river, decided to take advantage of his success. Nineteen men offered to take the deadly risk. Thompson, who was wounded in the hand, made more than one trip for water. Fifteen soldiers, along with four sharpshooters, would eventually earn the Medal of Honor for their efforts.[58]

Saddler Michael Madden filled two kettles and, as he was returning, bullets punctured both kettles and his lower right leg. He had to be carried back up the hill by fellow soldiers, and Dr. Porter tended to his wound. While the soldier was given a field promotion from private to sergeant, losing his leg was not enough to earn him the prestigious Medal of Honor.[59]

Somewhere northeast of Reno's line, Dog's Back Bone shouted to fellow warriors, "Be careful . . . it's a long way from here but their bullets are coming fierce."[60] The Miniconjou warrior caught a bullet above his eyebrow and through his forehead, killing him instantly. The gunfire from the warriors slackened around noon; however, it was only a ploy to trick the soldiers into thinking the fight was over. At around 2 p.m., the warriors fired a massive volley once again.[61] The warriors then disappeared into the valley.

Down in the makeshift village, an exhausted White Bull slept until noon, when his father saw an unusual gathering of warriors and sent him to investigate. A scout had come to warn the village that more bluecoats were coming toward their village from the north. Wooden Leg noted that a council of chiefs was held, and the young men wanted to fight the new soldiers. It was finally decided that they would follow their original course of moving away from the soldiers. White Bull and a few other warriors decided to ride in the direction of the walking enemy.[62]

At the same moment, and only miles away, a group of over 400 exhausted soldiers under Terry and Gibbon tromped down the valley floor. The lack of communication from Custer puzzled Terry as he pressed along the Bighorn River to the mouth of the Little Bighorn River. Early on June 25th, Terry had sent Bradley and his Crow scouts to search Tullock Creek, but they discovered nothing.[63] Grant Marsh had been ordered to navigate the *Far West* with its supplies up the Bighorn to the mouth of the Little Bighorn.[64] Terry's column had gotten lost and struggled over terrible terrain with his infantry who were footsore and had run out of water for a short time.[65] The non-swiveling Gatling guns were a constant problem, and soon Terry realized Custer's wisdom in leaving them behind. The frustrated general had been unable to cover as much territory as he had hoped when the officers mapped their routes to the Little Bighorn Valley.[66]

On June 26th, Bradley's scouts stumbled upon White Man Runs Him, Hairy Moccasin, and Goes Ahead who delivered the shocking news of a 7th Cavalry defeat. The three scouts spoke of disaster, even though they had not personally witnessed it. Bradley trotted back to report the news to Terry. Beside him was an ill Gibbon, who had partially recuperated on the *Far West* and had decided to join the command that morning.[67]

A reserved Terry tried to comprehend the staggering news. Other officers scoffed at the claim, and one infantry officer bragged, "I don't believe he [Custer] thought that enough Indians could be brought together to whip his regiment—that is with himself in command. We soldiers thought that Custer could not be beaten."[68] Bradley remembered, "Terry took no part in these criticisms, but sat on his horse silent and thoughtful, biting his lower lip and looking at me as though he by no means shared the wholesale skepticism of the flippant members of his staff."[69]

By mid-afternoon, the soldiers crossed the Little Bighorn River and rode down into the valley.[70] The Crow scouts abandoned the command and rode back to their agency.[71] Smoke rose in dark columns into the valley sky to their front. Some believed Custer was burning the village, when in reality, they were seeing smoke rising from the burning grass lighted by departing warriors who were following their own scorched earth policy. Soon, a scout galloped back, having been fired on by warriors. He said excitedly, "You have been looking for Indians all summer; you'll find all you want there," pointing to the benchlands on their right.[72]

Terry ordered Lieutenant Charles Roe and his cavalry to the benchlands to investigate. A distant party in blue uniforms on the higher terrain maneuvered like a cavalry unit 'breaking into column,' and

Fig. 81. John Ryan was a Civil War veteran who also fought in the Washita fight where he killed and scalped a warrior. The sergeant vividly described the fight on Reno Hill as well as the fact that he was ordered to reinforce Benteen's faltering defense on the hilltop.
Courtesy of Sandy Barnard

Roe thought they were Custer's men. One first sergeant placed a handkerchief on his carbine barrel and raced forward waving it at the unidentified group only to receive a volley of gunfire. Soon, the advance party heard gunfire and war whoops as they raced back to Roe's troop. Terry's column tightened their formation and prepared for an attack. From the higher ground, Roe could see a huge procession going in the other direction toward the Bighorn Mountains.[73] A decision to halt the command was made, and it was an uneasy night as men slept with their weapons beside them. They were camped only one mile from Custer Hill.[74]

Sitting Bull and the other elders felt it was time to leave the area rather than fight again. He had resisted the urge to fight the soldiers at the Rosebud and the Greasy Grass, but he had been given little choice along the latter. He did not want to endanger his women and children again, and the council decided to move southward toward the White Mountains [Bighorn Mountains]. Flying Hawk recounted, "We packed up and took care of the wounded the best we could . . . we could have killed the men that got into the holes on the hill, but they were glad to let us alone, and so we let them alone, too."[75]

From the hilltop, Reno's soldiers began to notice a commotion in the valley, and then the grass below them was set afire. A huge wall of smoke rose to conceal the movement of the tribes as well as to keep the soldier horses from grazing on the grass near the river. John Ryan watched the "largest body [of Indians] that I ever saw" move below him and took his rifle firing a spiteful shot into the moving crowd. Ryan claimed to have fired the first and last shot of the Battle of Little Bighorn.[76] Charles Windolph recorded:

> The heavy smoke seemed to lift for a few moments, and there in the valley below we caught glimpses of thousands of Indians on foot and horseback, with their pony herds and travois, dogs and pack animals, and all the trappings of a great camp, slowly moving southward. It was like some Biblical exodus, the Israelites moving into Egypt; a mighty tribe on the march.[77]

Reno's relieved men cheered the departing warrior procession as the major ordered his men to move away from the stench of the dead soldiers and livestock. Windolph stated, "Gently, we buried our dead in the shallow trenches we had dug for the living."[78] Men leisurely walked the horses and mules down the steep ravines to be watered, and the soldiers filled their canteens as well as kettles during the night. Many of the men who had not slept in over two days quickly passed out in the hilltop darkness, while an unlucky few were ordered to sentry duty.[79] After dark, scout Billy Jackson, Gerard, De Rudio and O'Neill, who had been hiding in the timber for over a day, rejoined their comrades on the hilltop.[80]

The Indian procession rode past a buffalo skull with an empty hardtack box on top of it in the darkness. Only two months before, a Crow scout, along with 97 soldiers, had left the warning for the Lakota. Using charcoal, the Crow drew pictographs on the box to communicate how powerful the whites were. It was a warning and a message to any Lakota who might stumble upon it. They also stuffed a burned stick and grass into the cracks in the box to indicate that they were at war; the grass symbolized 'that they would be cleaned out this summer.' The scene left behind them on a hillside several miles away spoke louder than any wooden crate. They had left their own warning in the pale, grossly mutilated bodies of over 200 bluecoats.[81]

As the sun rose on the morning of June 27, Terry and Gibbon's soldiers ate a hurried breakfast of hardtack, bacon, and coffee.[82] By 7:00 a.m. the exhausted column was on the move. Bradley was in the advance with mounted infantry, in lieu of his Crow scouts who had abandoned him. An eerie silence prevailed as the men neared a narrow passageway in the valley. Bradley and his contingency of mounted infantry soldiers decided to explore the east side of the river where he had spotted strange, dark forms on the hillside the evening before.

As Terry and Gibbon crossed the river, a wide vista of the valley floor opened before them and two light colored lodges could be seen to their left. They headed toward the tipis and a single horse near the Deep Ravine. A soldier in the valley said, "About this time Lieut. Bradley . . . sent word to Gen. Terry that he had found the bodies of 25 cavalrymen in a ravine."[83] Bradley had stumbled upon the bloated bodies in the Deep Ravine before exploring the rest of the field around him. Bradley pulled out a piece of paper and tallied the number of bodies scattered for over a half mile—soon he was atop Custer Hill.

In the flat valley floor, Captain Walter Clifford and his infantrymen walked up to the two lonely tipis. A circle of dead horses had been laid around the lodges to help the warrior in the next life. An officer with a knife sliced the lodges open to view five Lakota warriors bundled in one and three others bundled together in the second.[84] In the rush to depart, other deceased warriors were hidden in ravines and the river brush.[85] Before the day had ended, the soldiers had discovered about 25 dead warriors.[86] Second cavalryman Charlie Schofield remembered:

> The bodies of 20 odd Indians were found and the ground was strewn with Indian camp equipment and dead horses, showing that the Indians had left in a hurry . . . the bodies of all the soldiers & officers were stripped and horribly mutilated. I tell you it was a sickening sight & I never want to see the like again.[87]

For the next three miles, the valley was littered with debris. Scattered around the flattened grass were piles of lodge poles, buffalo robes, saddles, cooking utensils, china dishes, coffeemills, new spades, axes, horn spoons and wooden bowls. A stack of new milk pans, unused tin cups on lines of string, tin plates in original packages, along with cook stoves were discovered. Photographs and love letters were strewn on the ground.[88] Double-barreled shotguns, rifles, knives, and pistols of unique patterns were left behind in what looked like a disorganized stampede.[89] Infantryman Charles Booth wrote:

> The Sioux abandoned their camp very hastily, as they left large quantities of meat and flour, blankets and camp utensils, besides a number of their dead chiefs . . . Our men found in the camp new camp kettles and coffee pots fresh from the Agencies that had never been used; sacks of flour with the Ft. Peck brand; bales of blankets marked 'Standing Rock Agency,' plainly showing that these were Agency Indians, and that they had recently come from their Agencies bringing their supplies of all kinds with them: many of them armed with the new pattern of Winchester Rifle.[90]

Soon, something even more ominous was discovered—cavalry saddles, an officer's bloody pants and a bloody coat with a bullet hole through the right breast marked 'Lieut. Porter, 7th Cavalry.' Assistant Surgeon Holmes Paulding found the buckskin shirt stripped of its hide. He also found a pair of gauntlets marked 'Yates, 7th Cavalry,' as well as the underclothing and spurs of Jack Sturgis.[91]

Sergeant Homer Coon of the 7th Infantry saw "a dead cavalry horse which had been cut open, and a dead, naked soldier was forced head foremost into the horse's belly."[92] White men's bodies were found in the camp, in addition to severed heads that had been drug around the camp by thongs. Charles Roe recalled, "We found a white man's head in the village which had been dragged around with a lariat tied to it, until the head was pulled from the body." He also found a human heart with a lariat around it.[93]

Paulding thought, "the bodies may have been of wounded or captured men tortured."[94] Three heads were found near a campfire, and one had the hair completely burned off. Souvenir hunting soldiers stepped out of ranks and pulled arrows from the naked bodies of soldiers.[95] A second messenger from Bradley arrived with an upgraded figure of 50 dead. Almost before the man could ride off, a third messenger arrived with news that almost 200 had been discovered.[96] According to Gibbon, Bradley rode up to Terry and Gibbon and with a trembling voice said, "I have a very sad report to make. I have counted 100 and 97 bodies lying in the hills!" Someone asked, "White men?" and Bradley answered, "Yes, white men."[97]

The number of bodies eventually swelled to 204.[98] That the 7th had suffered a tremendous defeat at the hands of Sitting Bull's warriors was now obvious, but where were the rest of the regiment?

At 9:00 a.m., on Reno Hill, the shattered remnants of the 7th watched as a dust cloud rose from the valley floor and moved toward them.[99] The two parties soon spotted one another, and a reconnaissance of Arikara scouts from Reno Hill identified Terry's troops. A relieved Reno then ordered Wallace and Hare to descend the heights and guide the column to the hilltop. Terry questioned the soldiers about the battle details, and Wallace related the story of the attack by Reno's troops and the survival of most of seven companies that dug in on the bluffs. Quickly, Hare and Wallace inquired about Custer's whereabouts. Their shock and disbelief were obvious when Terry stated that he believed that Custer, along with his five companies, had all been killed.[100]

Gibbon remained in the valley and proceeded to select a new campsite, while Terry ascended the bluffs.[101] The relief party was shocked to the see the 7th Cavalry soldiers with their torn and dirty clothing. Some wore gunnysacks on their feet since their boots had worn out.[102] The relieved soldiers gave Terry several hearty cheers.[103] Burkman observed, "We waved flags and hats and blankets at 'em. We run [ran] back and forth like we was crazy. Some of the men broke down and cried, jist [just] 'cause they was so tickled. We never stopped to think it was funny we didn't git [get] no answerin' cheer from those advancing."[104] Terry had tears running down his cheeks.[105]

Benteen was the first man to confront Terry with the question of where Custer had gone. While he had inspired bravery on the hilltop and had held the command together, the ghosts of the Washita raised their ugly heads again.

Terry soberly responded, "To the best of my knowledge and belief, he lies on the ridge about four miles below here with all his command killed."

The disbelieving Benteen quipped, "I can hardly believe it. I think he is somewhere down the Little Big Horn grazing his horses. At the Battle of the Washita, he went off and left part of the command, and I think he would do it again."

Terry, who did not want to argue the point with Benteen, sternly ordered him to mount his horse. "I think you are mistaken, and you will take your company and go down where the dead are lying and investigate for yourself."[106]

Gibbon noted, upon his arrival on top of the hill, that Custer's trail had skirted the hill they were on and passed to the right of the sharp peak [Weir Point].[107] With Bradley leading, Benteen took about 15 of his men on a two-hour exploration of the battlefield. Weir, Varnum, Edgerly and De Rudio were allowed to join the party, as well as Custer's last two surviving messengers Knipe and Martin.[108] Martin showed Benteen where he had turned around with the message about 600 yards from the river.[109] Benteen looked for evidence of fighting near the river, but noted only a wounded gray horse and a pair of enlisted

Fig. 82. Edwin Bobo marker. Researcher Walter Camp took this photograph near the marker of Edwin Bobo in the early 1900s. From left to right is Peter Thompson, Mrs. Daniel Knipe, Daniel Knipe and Curly. Knipe married Bobo's widow and brought her to the battlefield to see where her husband's body was discovered.

Courtesy of the author

man's trousers there, along with ground that had been cut up by horse hooves.[110] About 150 yards east of the mouth of Medicine Tail Coulee they found a clothed body that was not scalped.[111]

Following across numerous ravines, they came upon the high ground where Knipe recognized the corpses of his fellow sergeants, Finckle and Finley.[112] Soon, the bodies of Calhoun and Crittenden were found near one another on the higher hill nearby. Keogh's stripped but unmutilated body was discovered among a pile of soldiers.[113] The men of Company H began to scatter out in a desperate search for Custer.

Private Jacob Adams recollected, "I saw a little knoll covered with dead white horses. I rode forward to it and there discovered the mortal remains of the gallant Custer. I said, 'Captain, here's General Custer.' 'That surely is Custer,' " responded Benteen."[114] Adams later wrote that Benteen dismounted and walked over and said, "By God, that is him," then he mounted and rode away and never said another word.[115]

Perhaps Adams was being protective of his captain, as others disputed the recollection. Years later, General Nelson Miles quoted an anonymous officer who had been there to overhear Benteen's comments. The statement has a ring of truth, knowing his hatred of Custer. Benteen exclaimed, "There he is, God damn him; he will never fight any more."[116] Windolph, who was also nearby, contradicted Adams as well:

> I stood six feet away holding Captain Benteen's horse while he identified the General . . . he looked almost as if he had been peacefully sleeping. . . . Captain Benteen found a bit of wood, hollowed out a hole, found an empty shell, wrote Custer's name on a bit of paper and placed it in the shell, and shoved it deep in the hole in the piece of wood. Then he pushed this into the ground at Custer's head. It would make sure that the burial party would identify Custer's body.[117]

Benteen had offered his commander a small weapon to take his own life if things turned desperate when entering a Cheyenne village in 1869. Custer had respectfully declined. Benteen had suicide on his mind again. He, Weir, Varnum, and Edgerly stood looking down at Custer's head with dried blood from the temple wound. Benteen looked at Weir and asked him if he believed Custer had shot himself. All Weir could do was grunt.[118] Private Theodore Goldin and others carefully looked at the bullet holes in Custer's body and remembered, "[They] were made by a gun of smaller caliber than the .45, and we calculated it may have been from a Winchester or Henry. There were no powder marks about the wounds at all."[119]

Benteen identified and marked other officers' bodies in the same manner.[120]

He, years later, would write with little remorse, "That had Gen. Custer been alone killed, that in my opinion the U.S. Government would have been a big winner by the operation."[121] A similar statement was evidently made on top of the hill when Benteen returned and told engineer Edward Maguire, "By the Lord Harry, old man, 'twas a ghastly sight; but what a winner the U.S. Govt. would have been if only Custer and his gang could have been taken!"[122] Benteen later proclaimed to another soldier:

> The Lord, in His own good time had at last rounded the scoundrels up, taking, however, many good and innocent men with them. You know enough of me to know I'd have gone

> through to him [Custer] had it been possible to do so. At the same time, I'm only too proud to say that I despised him as a murderer, thief, and a liar.[123]

Knipe recognized Boston, Cooke, and Smith, but was puzzled that he did not see Tom. He had walked right past the horribly mutilated body of his company commander. Knipe separated from the other troops and rode toward the Deep Ravine where he counted 28 bodies. He thought that he recognized the body of Mitch Bouyer in the bottom, but was not certain.[124] The men were shaken by what they had seen and rode back toward Reno Hill. When a pale and somber Benteen rode up, he said, "We found them, but I did not expect that we would."[125]

Engineer Edward Maguire had something else on his mind—he had to map the battlefield. Lieutenant Edward McClernand, who had been mapping most of Gibbon's journey, knew there would be some recognition in creating the first map of the battlefield and started the task on June 27. However, Maguire pulled rank on McClernand. The angry subordinate turned over his effort to Maguire who spent the evening and most of the next day collecting measurements, drawing the topography, and marking death sites, including those of the officers. Maguire's efforts would eventually gain him notoriety for creating the first map of the field but also recognition for its poor depiction of the terrain.[126]

The rest of the day was spent making stretchers for Reno's wounded and preparing to move them across the river into the Terry/Gibbon camp. Dr. Porter, who was equipped with morphine, amputated the wounded right leg of Madden below the knee.[127] Those seriously injured were carried by men down the steep ravine and through the waist-deep water of the river. Dead soldiers and horses killed near the river were hauled away.[128] As darkness descended over the men in the valley, there was still fear that the warriors might return. Shapes in the darkness looked like warriors, and imaginations produced fears in the night. Captain Walter Clifford, one of the infantry officers, wrote:

> A great round moon, bright as burnished silver, rolls slowly over our sorrowing heads. By its uncertain light a motionless black object can be seen at no great distance, which upon a nearer approach, proves to be a dead cavalry horse, and beside it the body of the rider, naked. Both are swollen almost to bursting. The legs of the horse are sticking straight out front of the body, while the skin of the sleeping rider gleams in the moonlight like polished white marble . . . But the polluted air is here also, and one is forced to lie with face close to the water to be rid of the deadly poison that is permeating the clothing and filling the lungs with every respiration. A little delay on this death-stricken ground and we will remain forever. Let us hide our slain comrades from sight and resume, with quickened footsteps, our pursuit of their butchers.[129]

The next day, they would indeed get a chance to 'hide their slain comrades,' but it would be a difficult task and one never forgotten. The gruesomeness of their camp would be no match for what lay on the ridges only four miles downstream.

Chapter 16

Reno's Skirmish Line of Horror

As the sun rose on June 28th, Captain Edward Ball and one company of the 2nd Cavalry followed the Indian trail down the valley of the Little Bighorn. As they crossed ground torched to discourage pursuit, they found bloody rags and dead warriors in the ravines near a deeply rutted travois trail. Ball traversed the trail for almost 12 miles and located an Indian night campsite near present day Lodge Grass, Montana, that had been vacated only a day earlier.[1]

This ground had been the scene of a victory dance on the night of June 26th, where the warriors decided to parade their spoils and talk about their brave deeds. There, the Cheyenne painted their faces black as a symbol of victory and danced around a pole decorated with the scalp and hand of one of the Arikara scouts.[2] Stands in Timber recorded, "They were just having fun, showing off that they had done a great thing."[3] The Lakota were riding 10 of the 15 captured gray horses and displayed a soldier flag on a wooden pole. The villagers were hungry and hunters brought meat into the camp that quickly disappeared. They could not feed all the people, and it was safer to split up into smaller groups. Ammunition for another fight was scarce. From the camp, the huge trail divided. The Lakota went south to the Bighorn Mountains, but a few went with the Cheyennes up to the Wolf Mountains.[4]

Ball and his small contingent had no real desire to pursue the massive village, especially with what they had witnessed the day before. Intelligence demanded that an effort be made to determine where the tribes had gone. He quickly and nervously headed back with his information.[5]

Meanwhile, Reno's exhausted men ate a quick breakfast in the camp; they had relocated above the river and away from the stench of their dead. He surmised, "I thought that was a proper duty for the 7th, that we should take care of the wounded and go and bury our comrades because we would be best able to recognize them."[6] A little after 5:00 a.m., the troops numbered off in groups of four. Just as in battle, the number four men would hold horses, while the other three men buried their friends. The companies formed in rows of two soldiers abreast and moved slowly toward the scene of carnage, four miles away.[7]

Other soldiers were left behind to destroy anything in the camp and prepare to move the wounded. Private John Burkman desperately wanted to go to see Custer. A vengeful Burkman daydreamed of Phil Sheridan being buried on the plains rather than his hero. An officer told Burkman that he had to stay on Reno Hill to work, and he recalled, "I wanted to go right over to the battlefield . . . seemed like if I could see him layin' thar [there], not smilin' anymore, not speakin' to me, then I could believe it was so."[8] Burkman and others piled damaged guns and cut up saddle and horse gear, then set them on fire. Nothing useful would be allowed to fall into the hands of the Indians.

Charles Varnum stated, "We started on the 28th to go down and bury the dead, and in going down I was on a trail which I supposed was Gen. Custer's, and when we got to a high hill that had a pile of stones and Indian medicine bags and other things on it, I went to see what they were, and rode off the trail."[9] Varnum was then detailed to sentry duty to watch for Indians. The companies rode to the edge of the river and looked at the watering place for evidence of fighting.[10] A trumpet call sounded *attention* and the men received their instructions from Reno. The men were to move in parallel columns, "and each company, as it moved along, would bury the dead it found."[11] One hundred-yard intervals were created between the companies.[12] After they had completed the duty, the number that each company commander had buried was to be reported to George Wallace who was to record their tallies.[13] The fatigued columns veered from Custer's trail near the river and cut straight across to the battlefield.[14]

The few spades available allowed for about one per company, while others used broken canteen halves, tin cups, butcher knives and mess kits as burial tools. On a small rise near the river lay a body that was neither scalped nor stripped of clothing. The body had several arrows in it.[15] A corporal recognized him and exclaimed, "There lies Foley of C!" According to Sergeant Samuel Alcott, speculation among the troops was that Foley had been a late casualty of the battle and believed it had "no doubt saved him from the tomahawk and scalping knife."[16] Sergeant Ferdinand Culbertson dismounted to examine the corpse. He discovered the bloodstained guidon underneath Foley's blouse and kept it as a souvenir.[17]

Almost simultaneously, Butler and his dead horse were discovered 200 yards away.[18] A shallow hole was dug next to the body. Godfrey remembered:

> The first dead body we came upon was within my sector, [it] was the body of Sergeant Butler. He was not entirely stripped; he had several wounds, and he was scalped and otherwise mutilated. In fact, I thought he was nearly in the same position as when he died. . . . When we lifted his body to place it in the shallow grave, we found underneath it a large number of empty cartridge shells. This fact and his several wounds gave evidence that he had sold his life dearly.[19]

As they climbed the rising terrain, the burial skirmish line dismounted. The morning sun rose over their right shoulders and angled toward the ground in front of them on what looked like white stones lying in the knee-high grass. Long shadows were cast by the hills into the ravines. Godfrey, who had not been to the battlefield the day before, asked, "What are they? Are those rocks?" Most of the others thought the same thing until an officer raised his field glasses and gasped, "My God! They are the bodies of the dead!" Weir was only several feet away, and even though he also had seen them the day before, they looked different in the morning sun's raking light. "Oh, how white they look! How white they look!"[20]

These soldiers had tanned arms, faces, and necks, but the remainders of their bodies were rarely exposed to the sun. The bodies looked like marble from a distance, but closer inspection revealed the darkened dried-on blood. Exposure to the near 100-degree temperatures, and three days of decomposition had taken its toll on the bodies. Where blood had pooled in the bottom parts of the corpses, some appeared black. Horses and bodies looked like overfilled balloons, with the legs and limbs protruding straight out in unnatural positions, due to swelling.

Nineteen empty soldier cartridges were found next to one bloated corpse discovered on the slope of Finkle/Finley Ridge. It is with a great deal of certainty that it was Private John Darris who had attempted to escape the carnage on his gray horse.[21] The body of Corporal John Briody was found in the area. He had the fastest horse in the company, leading one soldier to believe that the soldier had tried to escape or had been sent as a messenger. His right leg was cut off and laid under his head like a pillow.[22] Another soldier remembered the first body in his sector:

> All of the bodies we heaped mounds of earth over were mutilated more or less, mostly more. Will give a description of the first we buried: first he was scalped; the skull was bare to the ears, the crown of the head was chopped out, his cap put into the cavity. The body was nude between the waist and throat. There were 12 or 15 places where no doubt a spear or knife had been thrust to the hollow. Blunt arrows were driven in and left in the wounds.[23]

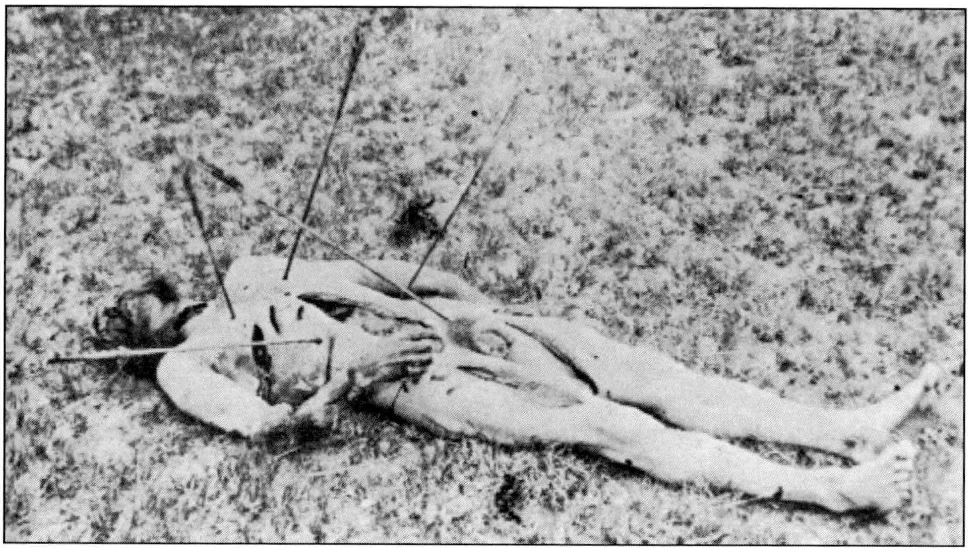

Fig. 83. Seventh Calvary Sergeant Frederick Wyllyams, was killed near Fort Wallace in June 1867. His mutilated body was typical of those discovered on the Little Bighorn Battlefield. White Bull, Sitting Bull's nephew, said that many times this was done out of anger, while others noted it had religious implications in the next life.
Courtesy of Kansas State Historical Society

Jeremiah Finley looked like a human pincushion with 12 arrows protruding from his corpse. He lay next to the head of his horse Carlo.[24] Dead horses lay thick on Finckle/Finley Ridge.[25] Only about 30 yards from Finley lay the other sergeant of Company C—6-foot tall August Finckle. A photograph bears witness to his stature, his clear eyes and dark drooping moustache. Finckle's horse, that had almost saved his life by collapsing before the battle, lay next to his lanky corpse.[26]

The odor was nauseating and men covered their mouths with bandanas. It was impossible to avoid the gases that were being expelled from the ever-swelling corpses. Men gagged and choked and vomited.

The physical discomfort of losing your breath to the stench of death was compounded by the psychological realization of how close each had come to being in the same state. "The smell was so bad we had to be relieved every five minutes," wrote one soldier.[27] The honorable idea of digging a hole, then carefully lifting the body into a shallow grave was quickly abandoned. The skin would slip and joints were dislocated by the act of lifting. Dirt was tossed on top of the corpses. Other hardened men, during a lunch break, sat near mangled bodies munching their hardtack and bacon. Godfrey noted:

> All the bodies, except a few, were stripped of their clothing; according to my recollection, nearly all were scalped or mutilated . . . many faces had a pained, almost terrified expression . . . other bodies were mutilated in a disgusting manner.[28]

As the burial parties advanced, some found few bodies in their sectors, while others were overloaded with more than they could bury.[29] On Calhoun Hill, an organized skirmish line of dead soldiers still lay in position. Moylan found 28 cartridges around one soldier and around another, 40.[30] Officers recognized John Crittenden and James Calhoun behind the line. Crittenden's body was filled with arrows bristling in the morning sunlight.[31] Luther Hare wrote, "I buried the men near Calhoun. Jim Calhoun and John Crittenden lay close together. An arrow had been shot in Crittenden's [glass] eye."[32] Myles Moylan requested notification of Calhoun's discovery and stated:

> I at once rode to the spot; when I arrived, the body was partially covered with earth—from the neck to the waist was covered. I examined the exposed parts of the body carefully. It was all that remained of your husband. The face was not disfigured in any way, nor were his limbs mutilated, but like all the bodies that I saw, except Gen. Custer, he was scalped. I could not tell where he was shot on account of the body being covered up. Mr. Crittenden and Jim [Calhoun] were not buried in the same grave.[33]

The tame picture painted by Moylan of his brother-in-law to his wife Maggie was tempered, as Varnum relayed the once handsome officer was identified by a unique filling in his mouth.[34]

As the men moved beyond Calhoun Hill, they descended into 'Horse Holder Ravine,' where bodies were strung out in the grass and sagebrush. Sergeant Bobo's horse had been shot along the ridge top and had slid into the ravine. At the opening of the ravine, a large cluster of bodies surrounded the remains of Keogh.

Evidence indicates that Keogh was likely shot while on the back of his horse Comanche. The bullet wound in Comanche's flank matched where Keogh's knee would have been when he was mounted.[35] Strangely, Keogh was lying almost untouched, stripped of his clothing except for his socks. The warriors had taken his pocket watch from his body, along with a photograph of McDougall's sister spattered with the officer's blood. A Catholic medal was still suspended from his neck as if it were some potent medicine respected by the warriors.[36] Corporal John Patton, Keogh's trumpeter, was found lying across his body.[37] The corpse of First Sergeant Frank Varden completed the cluster of dead men.[38] According to one witness, there were approximately 38 dead in that group. It was clear that a surge of warriors had killed Keogh's staff in a consolidated effort. Edgerly, as he went through the Keogh sector, saw evidences of a fighting line, but it had disintegrated into a pile of dead bodies.[39] Nearby, bald-headed Edwin Bobo was discovered. He had not died with his company and seems to have fled on foot once his horse was killed.[40] Reno was unnerved by the scene and remembered:

> The scene at Custer's battlefield was beyond description, and it filled us with horror and anguish. The dead had been mutilated in the most savage way, and they lay as they had fallen, scattered in the wildest confusion over the ground, in groups of two or three, or piled up in an indiscriminate mass of men and horses. . . . Many of their skulls had been crushed in, eyes had been torn from their sockets, and hands, feet, arms, legs and noses had been wrenched off. Many had their flesh cut in strips the entire length of their bodies, and there were others whose limbs were closely perforated with bullets, showing that the torture had been inflicted while the wretched victims were still alive.[41]

A critical Benteen stated, "I arrived at the conclusion then, as I have now, that it was a rout, a panic, till the last man was killed . . . there was no line on the battlefield. You can take a handful of corn and scatter it over the floor and make just such lines, there were none."[42] Men tried to identify blood-encrusted heads that had no body nearby. Feet and hands had been severed and slung around the field. Private parts were cut off and placed in a dead soldier's mouth. Men had been gutted like antelope and buffalo. The Lakota often slashed the thigh muscles of their enemies as a way of identifying their kill, and so limbs could not be used in the next life. Other corpses showed evidence of warriors' lances being plunged into the bodies.[43]

One scout said, "The bodies [were] swollen and blistered in the sun, and each hideous gaping wound filled with a host of buzzing, quarrelling blue flies."[44] Godfrey described the battlefield as a "scene of sickening, ghastly horror."[45] One soldier reminisced, "We piled brush over the slain soldiers, weighting it down with dirt."[46]

A corporal was assigned to kill all the wounded horses and single gunshots intermittently shattered the silence.[47] As Thomas McDougall's company crossed the river at the mouth of Deep Ravine to look for burial implements, they stumbled upon a seriously wounded horse in the bushes. It was the big buckskin ridden into battle by Keogh—Comanche. The gelding stood watching the troops with large black eyes, unable to communicate the horror he had witnessed. One of the leather cheek straps holding up his bridle had broken, and the Shoemaker bit hung below his mouth. Keogh's saddle hung upside down under Comanche's belly, while his pad and blanket were gone. The big bloody horse had two severe wounds along with four minor injuries. McDougall detailed one of his men to watch the animal until he reported back to Reno. Rather than put the horse down, it was decided to take the gelding back to Fort Lincoln, initiating the myth of the horse as the 'only survivor of the battle.'[48]

McDougall had commanded Company E for five years before the battle, and Reno thought it best that he attempt to identify as many of his former company as possible near Deep Ravine. He found about half of the company in the ravine and the other half outside.[49] The lack of a moving breeze in the depth of the ravine multiplied the stench of death. The burial party began to gag and vomit as soldiers counted at least 28 bodies. The men raced down the ravine to get water from the river. A new strategy was devised as men got out their knives and took shovels to break off edges of the ravine from above, which toppled on top of the bodies. Other men kicked off the crumbling soft dirt with their boot heels.[50]

Sergeant Hohmeyer was identified by his name found on his left sock.[51] The tall body of William Rees was in the ravine as well.[52] Another soldier recognized Corporal Albert H. Meyer, a German and former bartender, and nearby, he saw another German, Private William Huber, who had been a gunsmith before joining the 7th.[53] Private John Kimm found his good friend George C. Brown with a gunshot through the heart. He removed an arrow from his head to send back to Brown's brother in Ohio and noticed that someone had cut off his little finger to remove a ring.[54] Kimm also found another Company E man, John Ogden, near the river and supposedly scratched his name in a tree near his burial site.[55] Not all the men in the ravine were from the gray horse company. The body of Private Francis Hughes of Company L ended up in the bottom of Deep Ravine.[56] McDougall also recognized Custer's personal flag bearer, Sergeant Robert Hughes. Somehow he had managed to flee the hill.[57] A number of dead gray horses were found near Deep Ravine, including that of Algernon Smith.[58]

Officers could not locate Sturgis, Porter, or Henry Harrington. Sturgis' bloody underwear, and Porter's coat, with two bullet holes, that were found in the village, testified to their fates. The only circumstantial evidence of Henry Harrington's death was his gold pocket watch, later purchased from the Indians in Canada.[59] Officers finally resorted to feeling the palms of men, in an effort to find the soft, uncalloused hands of officers who were not used to doing manual labor.[60]

A small broken line of men extended up the sloping terrain toward Custer Hill. One officer remembered seeing only nine or ten men.[61] One of these men was Private Weston Harrington who had been a farmer or laborer before enlisting. According to a fellow soldier, he had been friendly to the Lakota before he enlisted and they must have recognized him. He was not mutilated and, strangely, a blanket had been carefully laid over the body.[62]

The man in the 'calf-skin vest', Mitch Bouyer, was found on top of a little knoll. He was wearing cavalry boots and trousers as well as a civilian shirt with mother of pearl buttons. He likely suffered a blunt force trauma to the head although other reports said he had been shot.[63]

About 150 yards below Custer Hill lay the body of Autie Reed. He would not get a chance to show off the souvenirs that he had pilfered from an Indian grave; in another ironic twist, warriors did not take his clothing in death. His youthful body was strangely still fully clothed, and it was hard to determine the manner of his wounds or death. Nearby, was Boston Custer. He was mutilated and shot in the body several times. According to Moylan, "Bos Custer was about 75 yards from his brother [George] and Autie Reed about the same distance from him."[64] Boston's luck had run out. He had been encouraged to stay on the *Far West*, but could not resist tagging along with his older brothers.

On the southwestern side of Custer Hill, there were approximately 30-35 dead men scattered among dead horses. It seemed that their comrades had put up a chaotic and desperate fight. About 14 bodies from the bandbox company were near their commander, George Yates.[65] One of the burial party stated, "The bodies were mutilated in every conceivable way. One body had its leg cut off with a sharp knife at the hip joint. It was done so carefully that the bowels had not come out. Some bodies were set up on their knees and elbows and their hind parts had been shot full of arrows."[66]

Two Company F soldiers lay near one another: Gustav Klein and William Teeman. Both soldiers had checkered pasts. Klein was a hell-raiser from Germany who had been busted from corporal to private. Both had dealings with Rain-in-the-Face. Klein was the corporal of the guardhouse at Fort Lincoln when Rain escaped.[67] The warrior's cellmate was the Corporal Teeman from Denmark. They had suffered quite different fates. Klein had his skull crushed in by a large flat rock.[68] Teeman lay nearby, and someone, perhaps Rain, had created a strange memorial to the Dane. His blouse had been removed and placed over his face. The sleeve of the jacket had been removed, and the double chevron insignia was carefully resting on top of the wool jacket 'in a peculiar manner.'[69]

Tom "Boss" Tweed, a printer from Ohio, was found among the men on Custer Hill. For some reason, the warriors had been brutal to the bluecoat. A bloody axe lay near his body, and his crotch had been split with an axe. One of his legs was thrown over his shoulder. He was shot with arrows in both eyes and a third arrow was sticking from his face.[70] One officer, angry about the axes left around the field, stated, "The squaws finished the wounded by knocking them in the head with axes furnished by the United States government."[71]

George Yates was found on Custer Hill, and the Indians had cut off one of his fingers to get his ring off.[72] His handsome First Lieutenant W. V. Reily was near him, full of arrows, but his body was not disfigured. Reily's arms were crossed over his face and his gold ring with a griffin carved in jasper stone was missing.[73] Both officers' bodies were found a little lower than the hilltop. Near the top of the hill was the lifeless body of W. W. Cooke lying between two horses. He was missing one of his huge side-whiskers and had gashes in his thighs in several places.[74]

Nearby, Cooke's close friend, Tom Custer, was unrecognizable. He was found on the highest point on the knoll. The warriors or noncombatants had severely abused the corpse. The body lay on its face and a stone hammer had pulverized his head to the thickness of man's hand. An arrow was shot into the head but could not be removed as the metal point was bent like a fishhook. The muscles of the arms and legs had been slashed. Red dirt had been thrown into the wounds.[75] The body 'bristled with arrows,' and his throat was cut. His eyes and tongue had been cut out and his genitals removed. His body was sliced open with a cut up the torso and one across the abdomen—disemboweled, with his intestines spilling onto the ground. It was only when the burial party rolled the body over that they recognized a dark tattoo on the bloody skin of one arm. It was the Goddess of Liberty with an American flag next to the initials T. W. C. 'Thomas Ward Custer.' The two young bachelors, Cooke and Tom Custer, who were good friends and had wooed young girls back at Fort Lincoln, had died atop the knoll together.[76]

Also, on the slope near Cooke's body, were the remains of Dr. George Lord. Although many men claimed they did not recognize the body of the sick doctor, an officer spotted him only 20 feet from George Custer.[77] Private Charles McCarthy of Philadelphia had been detailed to do work for the headquarters group all the way from Fort Lincoln. He had previously gone AWOL and resisted arrest, leading to a 25-pound ball being attached to his ankle and being confined to hard labor for six months as punishment. His body lay with headquarters staff—it was his final punishment.[78] Another of the Custer clan lay on the hill—Algernon "Fresh" Smith. He was one of the only Company E men found there and situated far from his company in the Deep Ravine below. The warriors had left his trousers on his body and Moylan recovered a gold watch fob that had been overlooked and left in his back pocket.[79]

Twenty yards from the top of the hill were the nude bodies of two noncommissioned officers whose duties required that they remain near Custer. One was Chief Trumpeter Henry Voss. His trumpet was gone, and his corpse lay across the head of another soldier—Sergeant John Vickory.[80] The regimental color-bearer Vickory did not carry the regimental flag on this day. He stared face toward heaven with Voss on top of him. Vickory's right arm was severed at the shoulder, and he would be unable to hold a standard in the next life. Someone had taken the huge amount of cash he had sown in his boot.[81]

On the highest ground in the area, there was a ring of dead sorrel horses about 30 feet in diameter.[82] Near a white horse, inside the ring, lay the body of George Custer. The hilltop offered a panoramic view of the entire region. He was only 20 feet away from his brother Tom. Some, romantically, said that his body looked like he was asleep, and his face showed no signs of the horrific grimaces on the faces of surrounding soldiers.[83] Custer was just as bloated and grotesque as the men under him. When found, he lay on top of two other soldiers with only the small of his back touching the ground. His head rested on his right arm as if he were napping. The warriors, unable to get hold of his short hair, had not scalped him. His body was nude with the exception of two woolen socks and the instep of one of his tall boots. The bullet that had killed him entered his left temple and exited the other side of his face. Blood had oozed down his face into his mouth along into his moustache. Godfrey checked for the black powder residue as a sign of suicide and found none.

Another bullet wound was almost impossible to see. It had entered his left breast near the heart and had not bled at all, probably making it a post-mortem wound. The bullet had exited on the right side of the body near the lower ribs.[84] Jacob Adams projected, "I think the wound in the temple killed him as it had bled and run down his face, and the other wound had not bled any."[85] A slash wound four to six inches long had been inflicted upon his thigh. It had been inflicted to keep him from mounting a horse in the next life. There was also an arrow shoved into his penis.[86]

Two soldiers did a thorough job of covering up the body of Vickory and had created a good size depression in the ground next to the body. George Wallace said, "That would be a good place to bury the General."[87] He wrote Custer's name on a piece of paper and placed it inside an empty cartridge case. The grave was one of the deepest on the field—18 inches.[88] Custer's body was carefully picked up and placed on a blanket and some tent canvas. A soldier, taking needle and thread, sewed the body into this make-shift wrapping. As his body was picked up, the soldiers found some of Custer's empty Remington shell casings. Such evidence seems to indicate that Custer was alive and fighting at the end of the battle.[89]

Dr. Henry Porter collected locks of hair from officers to return to family members and other officers did the same.[90] John Ryan ordered a soldier to take a lock of hair from his commander to give to Mrs. Custer. Ironically, the souvenir hunters were slowly removing more of Custer's hair to give to his wife, who had urged him to get it cut so he would not lose it. While the Indians had not taken his hair, his fellow soldiers were inclined to do so.

Custer's wrapped body was taken 20 yards down the slope and placed in the hole next to Vickory. Tom's mangled body was placed next to his older brother's left side. According to Moylan, "All the bodies of officers were buried where they were found except those on the hill with Genl. Custer. They were buried in a row [—] Genl. Custer on the right."[91] On top of the grave, a basket-like affair from an Indian travois was pinned to the ground with stones and wooden pins to keep the wolves from digging up the body. Such care and attention was like erecting a sign saying—'Dig here'.[92]

Reno's skirmish line of horror--June 28, 1876 & Indian casualty sites

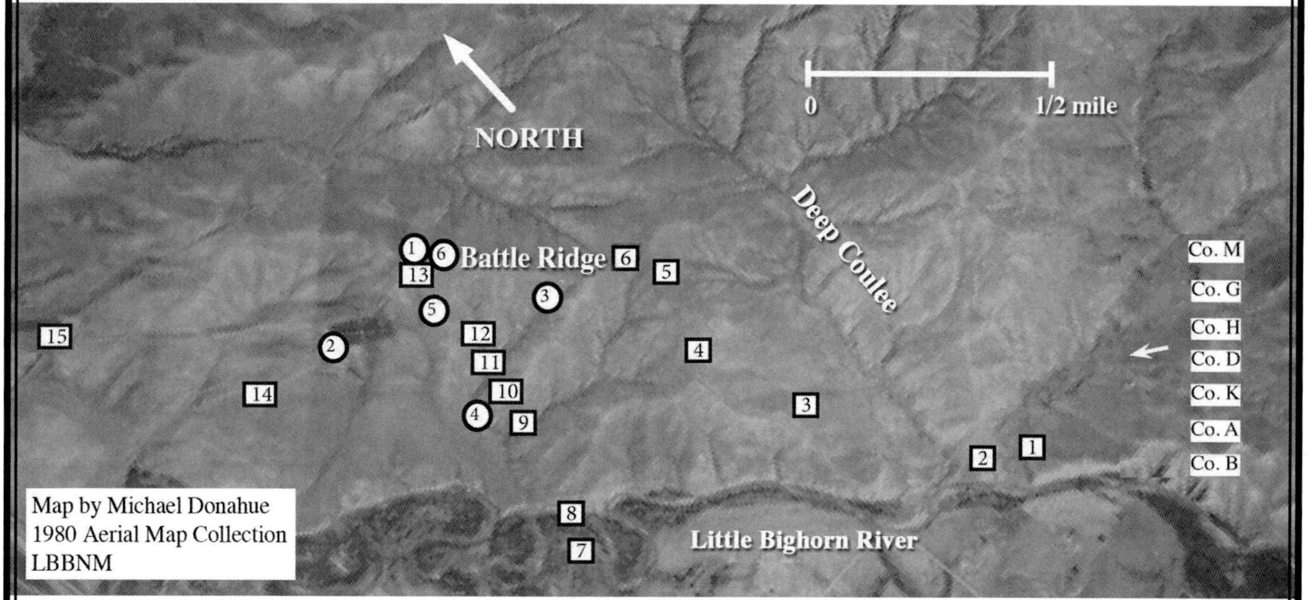

[1] John Foley	John Patton	[10] John Briody	W.V. Reily	Warrior Death Sites
[2] James Butler	Willis Wright	Mitch Buoyer	Henry Voss	(bodies removed)
[3] John Darris	James Bustard ?	Weston Harrington	John Vickory	
[4] Jeremiah Finley	[7] William Brown	[11] Autie Reed	William Teeman	(1) Limber Bones
August Finckle	[8] John Ogden	[12] Boston Custer	Gustav Klein	Closed Hand
[5] James Calhoun	[9] George Brown	[13] George Custer	Charles McCarthy	(2) Open Belly
John Crittenden	Andy Knecht	Tom Custer	Thomas Tweed	(3) Lame White Man
Frederick Walsh	William Huber	W.W. Cooke	[14] Mark Kellogg	Bear With Horns
[6] Myles Keogh	Richard Farrell	Algernon Smith	[15] William Sharrow	(4) Noisy Walking
John Wild	Albert Meyer	George Lord		(Died Later)
Charles Graham	William Rees	George Yates		(5) Black White Man
Edwin Bobo	Robert Hughes	John Parker		(Died Later)
Frank Varden	Frederick Hohmeyer	Edward Driscoll		(6) Unknown Lakota in
	Timothy Donnelly	Francis Hughes		headdress

Fig. 84. Little Bighorn Battlefield Indian and Cavalry Casualty sites.

Other officers were buried in a line with military precision to the left of Tom—Yates, Reily, Smith and Cooke. In army tradition, the surviving officers took the embarrassing, unorganized chaotic positions of the dead and substituted linear order.

One sergeant cut tipi poles into lengths about 5-feet long and laid them near the bodies. A small fire was built on top of the hill, and someone offered up a metal ramrod, likely from an infantryman's rifle. The metal was heated until it was red hot, and then very carefully applied to the top of the pole to create Roman numerals. The poles were then driven into the ground at the head of each officer and Henry Nowlan mapped the burial locations.[93] General Terry and other officers gathered near the officers' graves. When Terry first saw Custer's body, he cried and proclaimed, "The flower of the army is gone." The lawyer officer pulled out a small prayer book, read some passages, and spoke a few words over their lost comrades. They could do no more. The burials were finished, or so they thought.[94]

The exhausted troops slowly rode back to Reno Hill in order to evacuate their wounded and prepare to move.[95] Reno's troops crossed the river and rode through the abandoned village site for the first time. George Herendeen stated that four heads were found in the village with no trunks; the solemn soldiers buried them.[96]

While Reno's soldiers had been burying their dead friends, the 7th Infantry and the remainder of the 7th Cavalry had been busy. The task of burying the rest of Reno's dead in the valley was accomplished. The items left behind by the villagers were destroyed, and the slow, painful task of moving the wounded down the steep ravines and across the river commenced.[97] The 7th Infantry and officer G. C. Doane had been at work making 25 crude hand litters for transporting Reno's wounded. They planned to backtrack toward the Little Bighorn River that evening to meet the *Far West*.[98]

In the valley, Maguire's aide Sergeant Becker mapped the terrain near the area of the Reno timber fight. In a cleared square in the timber where he discovered the bodies of three soldiers, evidence of several chiefs' lodges existed. This ambiguous discovery by Becker brings up some intriguing and possibly horrific evidence of the 'scalp dances' and torture admitted by Two Moons and Little Knife.[99]

At around 6:30 p.m. the men left the 'pestilence-laden atmosphere' and began to move the wounded down the valley. The infantry and cavalry had great difficulty transporting the wounded by hand.[100] The troops were soon stumbling in the darkness through the former Indian camp until midnight—they had traversed only four and a half miles. The camp was situated next to a bend in the Little Bighorn River in the proximity of a dead 'white,' bearded warrior eerily positioned in the tree branches above them. They were still 15 miles from Grant Marsh's boat and relief.[101]

As the sun came up on June 29, the officers deliberated about how to get the 52 wounded to the *Far West* most comfortably. Twenty-one men could not even ride and had to be carried. The hand-carried litters had not worked well, so new mule litters were created by Doane and substituted. While this was being accomplished, the 7th Cavalry soldiers destroyed nearby abandoned village contents.[102]

Due to his ill health and being busy with other tasks, Gibbon had not seen the field of battle. He decided to take advantage of the delay and "No Hip" rode across the river. He stumbled upon the fully clothed body of a white man lying on his back in the grass in an advanced state of decomposition. The man had been scalped and one ear was removed. His clothing was not that of a soldier, and the body wore unusual boots with straps and buckles to tighten the insteps. Gibbon had an orderly cut off the boot, and it was later recognized as belonging to correspondent Kellogg.[103] Mathey buried the newspaperman a half a mile from Custer Hill.[104]

As he rode toward the mounded hilltop covered with dead horses, Gibbon found another unburied soldier. It was clear that the exhausted burial party had done little to proceed past Custer Hill and evidently missed some casualties. Making a wide circle, he forded the river through Deep Ravine and came upon the men destroying articles in the village. The two burial tipis were nothing but a blackened mass. Gibbon saw the paradox of the soldier's actions, recording:

> Riding across the valley toward the bluffs, we passed the site of the two teepies filled with dead Indians, now a mass of charred remains, and approached a clump of small trees, in and near which the Indians had buried a number of their dead, the ponies slaughtered in their honor lying about the remains of their dead masters, now tumbled upon the ground from the destruction of the scaffolding by those human ghouls whose existence seems to be inseparable from a fighting force, after the fighting is over, and whose vandal acts painfully impress one with the conviction that in war barbarism stands upon a level only a little lower than our boasted modern civilization.[105]

The column moved out at 6:00 p.m. on their 15-mile journey.[106] A quick storm appeared and rained down more misery on the troops wandering in the blackness. After midnight, the *Far West* was located and fires were built along the shore to allow the men to see how to board. Eighteen inches of grass overlaid with canvas was prepared on the lower deck for the comfort of the wounded, and Comanche was led to the stern between the rudders.[107]

Fig. 85. Comanche was ridden by Myles Keogh on the day of the battle and was severely wounded, but he survived. Later, the horse became the mascot of the 7th Cavalry. After his death, he was stuffed, and today resides in a museum in Kansas.

National Park Service, Little Bighorn Battlefield National Monument LIBI_00011_07119

On June 30th, the *Far West's* progress was aided by the current of the Bighorn River emptying into the wider Yellowstone, but it was slowed to wait for the troops to catch up so they could be ferried across to the north side of the Yellowstone River. The men were to be refitted and resupplied in order to continue the dogged pursuit of the Lakota and Cheyenne for the rest of the summer. The old news of Crook's fight along the Rosebud finally reached them as they followed the river homeward. It was too little too late.

Grant Marsh guided his boat in record time to get 'the most precious cargo a boat ever carried,' as Terry labeled them, back to Bismarck and Fort Lincoln. Marsh would make history with his winding, speedy transport of the wounded—a trip of 710 miles. After the delays, he would make the trip in 54 hours. The boat's whistle blew in the darkness as it docked in Bismarck at 11:00 p.m. on July 5 across the river from Fort Lincoln. News of the military loss sped through the city, while newspaper editor Clement Lounsberry quickly set forth a story for the telegraph the next morning, sending the first account to the *New York Herald* with the headline: "MASSACRED."

Early the next morning, there was a knock on the door of the Custer home. A doctor accompanied by two officers entered the home. Emma Reed, Maggie Calhoun, and Libbie Custer entered the parlor and received the terrible news. They wept loudly and Mrs. Calhoun could not believe the news; she had lost her husband, three brothers, and a nephew. Libbie wrapped herself in a shawl and went with an officer to break the sad news to the other widows. She was no longer in the army.[108]

In bleak irony, the last book Custer was reading before leaving on the expedition still lay on his desk. It was a novel entitled *Her Dearest Foe*. Custer had underlined the following lines in the book:

> I have sort of faith in my own fortune. I think it shall conquer in the end.[109]

In the eastern United States, the country marveled at its future potential, national vision, and technological advances at the Centennial Exhibition in Philadelphia. The newspaper headlines were filled with stories concerning the fall presidential election between Republican Rutherford B. Hayes and Democratic nominee Samuel J. Tilden, both having refused to make the Sioux War an agenda item in the election. Major issues were the continued reconstruction duty in the southern states, even though the war had been over for over a decade.

The news of the death of Custer and his troops was quickly displaced in the headlines by a massacre that occurred in Hamburg, South Carolina, on July 8, 1876. Racially-motivated whites arrested some black militiamen for not turning over their arms after a nonviolent altercation. The eventual results included the sacking of the town, as well as the murder and mutilation of several of the militiamen. The timing of both events made many Americans realize that there were pressing national social issues that were threatening the stability of the entire nation as it celebrated its power and might on its 100th birthday.[110]

Biased newspapers painted the victors as bloodthirsty savages and glorified a heroic Custer, focusing on his 'Last Stand.' The different newspapers also blamed everyone—Custer, Sitting Bull, Indian agents, Grant, Sherman, Sheridan, Terry, Reno, Congress and others for the military debacle. Little defense was given to the native population fighting for the way of life they had always known. Custer's casualty rate of 262 men was almost one percent of the standing army of the time. Translated into today's figures, it would be the equivalent of losing over 5,000 men in one day of combat.

Congress had been divided on military funding, but such contentions disappeared and an increased budget quickly followed. Reactionary patriots signed up for military service as 'Custer's Avengers' ready to seek revenge on the tribes. Understaffed companies doubled in numbers to allow for more aggressive maneuvers. It was a wake-up call that challenged the hubris of the American people, Congress, and the military.

Grant, Sherman, Sheridan, Terry, Reno, and Benteen quickly instigated the blame game. Custer became the scapegoat and responsibility for the military debacle is still being debated.

The wrath of the nation supported military rule at the reservations. Embarrassed by evidence that some agency natives had participated in the battle and were fed, clothed, and armed by the U.S. Government, a crackdown was initiated. The native population was unhorsed, disarmed, and their ability to wander freely, restricted. The government pressured some of the chiefs to sign away the Black Hills, seizing the territory right after a payment was made to the tribes. True ownership of this area remains in political limbo to this day, since the money was never formally accepted.

Terry and the 7th Cavalry remained in the field for the summer searching for the Cheyenne and Lakota with little success. Sheridan was able to get two new forts funded to act as supply bases. Fort Custer, the 'million-dollar fort,' was built near the battlefield, while Fort Keogh was erected near another Custer fight, the Yellowstone fight of August 4, 1873. In early September of 1876, Crook's forces engaged warriors under American Horse and Roman Nose at Slim Buttes. In an ironic numbers reversal, the troops were able to rout a village of 200 by about 2,000 soldiers. While casualty numbers were low on both sides, the destruction of the camp was a blow to the Lakota, and guidons, saddles, and personal equipment from the 7th Cavalry were discovered in the village.[111]

General Crook, who had a poor showing on June 17, was more successful when subordinate Colonel Ranald Mackenzie found a Cheyenne village camped on the Red Fork of the Powder River. On November 25, 1876, a little over 1,000 bluecoats attacked 183 lodges in the village of Little Wolf and Dull Knife. In a repeat of Crook's first Powder River fight, the Cheyenne tipis, food, and clothing were destroyed. Crazy Horse once again rescued the destitute Cheyenne. However, the dogged pursuit and repetitious destruction of their livelihood had a devastating effect on their psyche.[112] Colonel Nelson Miles, an aggressive admirer and friend of Custer, relentlessly pursued the tribes until, by spring of 1877, hundreds of starving warriors and their families surrendered at the agencies with Crazy Horse among them.

Sitting Bull, Gall, Crow King, and about 400 Hunkpapa Lakota followers decided on a different course of action and fled across the border into Canada. The buffalo in the region began to disappear and pressure was turned up by the Canadian government to force Sitting Bull back across the border. General Miles patrolled the northern border to keep them from hunting in Montana Territory. Malnourished and weakened, small parties realized it was better to submit to the dreaded reservation life than freeze and starve to death.

Sitting Bull finally surrendered at Fort Buford in July 1881. The surrender unofficially ended the bloodshed for nine years until more Lakota and 7th Cavalry blood was tragically spilled along Wounded Knee Creek on a freezing December day in 1890. Large numbers of women and children were killed, and frozen warrior bodies were placed in a mass grave with a small military-type headstone placed over their bodies in a strange image similar to one on Custer Hill. This time the newspaper headlines back east did not read: "MASSACRE."[113]

Chapter 17

Nail the Box Up

Shortly after the battle, various Cheyenne and Lakota participants wandered near the battlefield. A Cheyenne named Lost Leg searched for missing horses lost in the area. As they drew closer, his party could smell death from a great distance, and the rank odor was so potent that they wandered up Reno Creek instead.[1] In the fall of 1876, another visitor to the killing field roamed through the chilly air searching for unfired soldier cartridges. Wooden Leg stumbled upon a soldier's folding pocketknife and picked it up as a souvenir. The young man's mind flashed back to the hot afternoon when he had fought the bluecoats on this bloody ground. He and a small group noted where their dead and wounded had been found and remembered their brave deeds. He described the scene:

> Dirt and sagebrush mounds now were at the places where had been the dead soldiers. In a few places we could see some parts of their bodies exposed. But mostly the graves were good, except they had no stones piled over them. At one end of many different ones of the graves was a straight board stuck into the ground, to stand there. They were straight boards, not crosses. Dead horses were lying in decay here and there among the graves. Wolves had been eating on the horses. I did not notice any place where it appeared wolves had been at the graves.[2]

It was a well-known fact that warriors sometimes dug up graves to pilfer items or scalp the dead. Warriors had returned to open up the graves of the dead soldiers after the Rosebud battle. Custer, in 1873, had feared that the bodies of Honsinger, Baliran, and Tuttle would suffer the same fate and had carefully concealed their graves. The 7th Cavalry commander's friends and enemies had not shared the same concern over his body. The basket and stone-covered grave was a curious, distinctive mausoleum that begged for exploration by passing Indians and non-Indians. By the time a recovery team arrived in 1877, the stones and pegged-down basket protection had disappeared.

A little over a year after the battle, another party started toward the battlefield with a two-fold purpose: to return to civilization the remains of the fallen officers, and to re-inter the bodies of the enlisted men.[3] General Philip Sheridan's brother, Captain Michael V. Sheridan, was in charge. While the friend and mentor of George Custer would insist on overseeing the burial of Elliott, on this occasion, he would defer this responsibility to his brother. Henry Nowlan, who had carefully mapped the burials of

Fig. 86. Curly, a Crow Scout. A rare photograph taken shortly after 1881 near the soldier mass grave and marker placed on Custer Hill. He may have been in mourning, as noted by his cropped hair. Curly lived within sight of the monument in the Little Bighorn Valley below.

Courtesy of Putt Thompson

the officers, joined 88 members of Company I of the 7th Cavalry. George Herendeen and guide, John Baronette, also joined the recovery party. Michael Sheridan refused to allow reporters to accompany the party to avoid publicity about what he feared might be discovered.[4]

Stopping at the confluence of the Little Bighorn and Bighorn rivers at Post Number 2 [later Fort Custer], Michael Sheridan got off the steamer, *Fletcher*, and witnessed men busily constructing small pine coffins. Arriving on July 2, 1877, he stated, "The command was put into camp, and Capt. Nowlan and myself, with a small detachment and some Crow Indians, proceeded to make an examination of the battlefield."[5]

Included in the contingency were Curly, White Man Runs Him, and Half Yellow Face. Michael Sheridan asked Curly what he had witnessed and was convinced fairly early that "he had run away before the fight began, and the greater portion of his tale was untrustworthy."[6] Half Yellow Face, who had only been in the valley and hilltop fights, could give Sheridan little information about the Custer fight.

Camping at the Reno entrenchments, the men formed a skirmish line of 50 men and crossed the entire battlefield searching for graves. Willow sticks were placed at each grave to help mark the dead. That evening, the men were put to the task of cutting 3-foot long cedar stakes that would be used as headboards and driven 1-foot deep into the ground.

On July 3rd, a day before the country celebrated its 101st birthday, the party rode to Calhoun Hill. They removed the remains of Calhoun and placed them in the small box brought for that purpose. Next to him was the body of Crittenden, whose family had requested that he be left on the field where he fell. His body was placed in a coffin and covered in three feet of soil. A wooden cross with his name was eventually erected near his head. Next, Keogh's remains were taken from his grave and the men proceeded to Custer Hill. Sheridan recorded, "There were large and small trenches [graves]. Some contained but few remains."[7] Such an observation reveals that some of the graves were for men, while other heaps of soil covered the severed parts of the soldiers. Michael Sheridan wrote:

> Only the naked bones remained in the trenches. There were no traces of flesh or corruption, and no odor, except that which was wafted from the shoals of wild flowers blooming in the valley below. In a few hours, the thin layer of dirt had been removed from the bones of over 200 soldiers, and the remains re-interred in the same trenches, but rather more

decently that before. Three feet of earth, tastefully heaped and packed with spades and mallets, was put upon each set of remains, and the head marked by a cedar stake.[8]

No questions were asked as to how a small party of men could reinter over 200 corpses 'in a few hours.' It was a falsehood. According to Sheridan, the easily identified remains of William V. Reily, James Calhoun, Algernon Smith, George Yates, W. W. Cooke, Tom, and George Custer were removed and placed in the small wooden boxes hauled in four ox-wagons. Donald McIntosh's body was removed from the valley, and Dr. James DeWolf was taken from his grave above the river crossing. Grass from the valley was cut and used as packing in the small boxes. In some cases, shreds of clothing still existed and tufts of hair were attached to the human skulls. The horse bones that had been bleached by the sun were still in a circular position on the hilltop. Bits of hair and mane also still clung to some of the horse skulls.

At some point, Benny Hodgson's body was removed, as well. Even though his body had been carefully wrapped in two army blankets, along with a piece of tent cloth, and buried deeper than most of the dead, it was noted that "the body was found literally torn from the grave by wolves, and the bones really worse disturbed than those which received but shallow burial."[9] Aboard the steamer *Fletcher*, a beautiful casket for the lieutenant was filled with his remains.

Fig. 87. This photo taken in 1879 indicates the general state of Custer Hill before the soldier clean-up. The Indians did not care for the hard-soled boots and note that the bottom of a boot is located on Custer's grave stake.

Courtesy of John Doerner

Sheridan recorded that the "unmistakable identified" bodies of seven officers were transported back east on the steamer *Fletcher*. This is a revealing statement, since they had recovered nine bodies. He had painted a rosy, yet clearly whitewashed, picture of certain identification while surrounded by the wonderful aroma of wildflowers wafting from the valley a mile away. He could sign his official report, and there was no press there to challenge his version of events. Yet, he was unable to silence the many eyewitnesses to the reburial and recovery effort.

Only two days after the officers' bodies were removed, adventurer Philetus W. Norris, who would later become the first superintendent of Yellowstone National Park, arrived on the battlefield. He had come to remove the remains of his friend Charley Reynolds and borrowed guide Baronette who, no doubt, gave him an account of the actual circumstances. Norris was also a correspondent for the *New York Herald*. He rode over the battlefield and made one of the most accurate maps to date. In addition to recovering Reynold's remains, Baronette described the scene on Custer Hill a few days earlier. Norris wrote:

> Some feet down the western slope were found, partially unearthed, all that remains of the gallant Custer. Beside and below him lay his brother, Captain Thomas Custer, and other officers and friends, surrounded by the partially exposed and bleaching bones of the soldiers and their horses. The true duty of the correspondent being to furnish only necessary, or at least proper information, and not needless recital of horrors, much is omitted that might truthfully be written. Although the remains of General Custer and most of his friends there, as elsewhere, had evidently been disfigured by the coyotes or savages, and probably both, and many, if not most of the skulls there and throughout all the fields were smashed to fragments, mangled, or missing, still what is decided to be and probably were the main portions of the bones of General Custer and his two brothers were secured. Neither [Harry A.] Reed nor the officers who fell with him or Reno were found . . . Though the bodies of Lieutenant McIntosh and Dr. DeWolf were discovered and buried last year, little if any portions of their remains or of Lieutenant Hodgson's could now be found...some were found this year that were not lost . . . Of course, where the remains were partially uncovered, an indescribable odor arose.[10]

Nearby, photographer John H. Fouch set up his camera on the very top of Custer Hill looking toward the river below. Standing in the middle of the dead horse ring, he took the first known stereoscopic photograph of Custer Hill. Spent cartridges from the soldier guns lay among the sun-bleached bones of the horses. At least six cedar stakes that had marked the officer graves can be seen in the photo. Due to the sudden drop in elevation, the rest of the hill is, unfortunately, hidden from the camera lens.[11] Soon, other members of the burial party began to leak a story that differed greatly from Sheridan's sanitized report. A July 13, 1877 *Bismarck Tribune* report by an unidentified individual noted:

> The remains of Yates were in good condition and found as they were left—those buried deep had been dug up by the wolves and the bones scattered to some extent, but those buried near the surface were disturbed but little, Yates not at all.[12]

George Herendeen, who was at the excavations, stated, "Out of the grave where Custer was buried, not more than a double handful of small bones were picked up. The body had been dragged out and torn to pieces by coyotes and the bones scattered about."[13] Battle survivor Theodore Goldin later queried soldiers who participated in the recovery about their discovery and collection of the bones of the officers. One soldier responded:

> No, we took some pains to gather with each of the bodies enough to make a complete skeleton, but as most of the bodies had been exposed by foxes and coyotes, and in several instances portions of bodies had been dragged away, but we were not at all certain as to the identity of any of the bodies, save that we gathered them near the point where the marker indicated they had been originally buried.[14]

Interpreter Thomas LeForge had sat on the ground only 10 feet away watching the disinterment of the remains, and stated:

> The soldiers respectfully go through the motions of disinterring and transferring to a box the body of General Custer . . . But they gathered up nothing substantial except one thigh bone and the skull attached to some portion of the skeleton trunk. Beside these, the quantity of cohering and transferable bodily substance was not enough to fill my hat.[15]

LeForge also recalled, "The body supposed to be that of Captain Custer was less cut up than that of any other officer."[16] One newspaper reporter noted, "The whole matter of collecting the remains was bunglingly and imperfectly done, and [in the] instances [of] the case of Capt. Tom Custer, whose identity, was guessed at by the size and shape of his head, and General Custer's by his hair."[17] These are puzzling statements since Tom's skull was smashed beyond recognition and George's hair was short and clipped by several officers. Sergeant Michael Caddle helped in the exhumations as well as the reburials, and he stated in a letter to friend:

> When they came to the body marked Number One on the list [Nowlan's and Maguire's map] and on the stake at its head, and supposed to be that of General Custer, it was placed in the coffin, and then on the ground was found a blouse on which it had been lying. An examination of the blouse revealed the name of the wearer in an inside pocket, it was that of a corporal [William Teeman]. It was a disconcerting discovery to find that even the General could not be satisfactorily identified, but the Sergeant goes on to state that they found another body and placed it in the coffin. I think we got the right body the second time.[18]

James McNally, a member of the retrieval party, wrote:

> There was a dispute about the remains of Custer as to whether the bones were Custer's. The first two sets of bones were taken from the box and the third put in. Michael Sheridan, a brother of Gen. Sheridan, who was detailed to go with Capt. Nowlan, made the remark, 'Nail the box up; it is alright as long as the people think so.'[19]

A frustrated Michael Sheridan had inherited a mess on Custer Hill and believed all was well since he had not brought any photographers or correspondents along. Instead, he helped create a distorted cover-up story falsely describing easy identifications contrary to other eyewitnesses. Phil Sheridan and other officers decided to make a summer tour of the forts under his jurisdiction. The soldiers stopped at the battlefield only weeks later on July 21, 1877.[20]

Unknown to Phil Sheridan's entourage, a correspondent from the *Chicago Tribune* was racing to join the expedition. He beat the officers by arriving on July 20 and spent the night alone on the battlefield.

Even though only weeks had passed since the re-interment and removal party had been there, the correspondent was shocked by what he witnessed. He wrote:

> So scandalous is it that I will not write the details, but only say that a party of humane prospectors are proposing an expedition to the field to perform this work of humanity which the superior officers have failed to do. From conversation with the rank-and-file, I know the demoralizing tendency of this conduct, and with a demoralized, disheartened, and apprehensive rank-and-file, better no army at all.[21]

Phil Sheridan's party was shocked by the sight of the killing field. Since his brother's departure, only weeks earlier, the burials had again been exposed to the elements, wolves, and human ghouls. Lakota scouts and about 70 men were ordered to rebury exposed bodies. The 5th Cavalry soldiers were divided into three detachments, each commanded by a commissioned officer, and they spent four hours searching for and reburying the bodies. According to one officer, wolves had dug up only a couple of bodies. Seventeen bodies were reburied. Some bodies buried near ravines had been uncovered by hail and a rainstorm only days earlier.

Brigadier General George "Sandy" Forsyth made a revealing appraisal of the battlefield, noting, "I am half inclined to think, strange as it may appear, that nearly all the desecration of graves at the Custer Battlefield has been done by curiosity hunters in the shape of human coyotes."[22] Private Chris Madsen was present and recorded, "To our surprise, we found in remote ravines and depressions and sagebrush thickets a number of skeletons in faded uniforms that had been mauled and dragged about by wolves. General [Phil] Sheridan ordered that all the bones be buried."[23]

John Bourke inserted that they reburied 17 bodies, but he added that 10 had never been buried. He was critical, as well, stating:

> It was hard to go ten yards in any direction without stepping on portions of human anatomy and skeletons of horses, singly or mixed together . . . Pieces of clothing, soldiers' hats, cavalry coats, boots with the leather legs cut off, but with the human feet and bones still sticking in them, strewed the hill . . . Sticking out from the ground in the ravine was the body of a man still clad in the rough garb of a scout, boots and bullet-ridden hat still by him; there was nothing to give the slightest idea as to who he might have been.[24]

A reporter for the *New York Sun* provided a description of the field that was dated July 25, 1877. It disclosed similar observations from a non-military perspective that were accusatory, revealing:

> Their bones, divested of clothing by brutal savages, and their flesh by famished wolves and howling coyotes, are fingered by every curiosity hunter, or kicked by every hostile who may chance over the field . . . Today we had another terrible storm, and at the present moment, there is not a particle of earth on a single bone of the entire 300 skeletons, nor is

> it probable that there ever will be unless placed there by the elements or through the sympathetic efforts of some party of miners.[25]

While possibly exaggerating the numbers and the state of the burials, it was brutal in its attack of the recent military burial parties. It seems the bones of horses and men were mingled, piled and poorly buried in heaps. Numerous newspaper reports were published about the officer misidentifications. A defensive Michael Sheridan shot back:

> No person connected with that paper [*New York Herald*] was with me, and that any statement it may have made was manufactured by the *Herald* office and was untrue. Nowlan and myself properly identified by the stakes driven into the ground at the head of each grave the body of every officer I brought in without the slightest difficulty.[26]

The officer by omission stretched the truth in his defense. The stakes were obviously there, but the bodies that had been resting under them had been scattered.

Corporal Samuel Staples had died with Keogh's company, and by the fall of 1877, his father read about the removal of officers' bodies and the sad state of the burials of the enlisted soldiers. The elder Staples demanded he be allowed to remove his son's remains. He noted, "It is hard for me to understand how the remains of the officers could be in condition for removal and recognition while those of privates and noncommissioned officers had become food for wolves."[27]

Sheridan thought the battlefield area should be declared a National Cemetery since there were so many military dead located at the site.[28]

Colonel Nelson Miles, a defender of Custer's memory, eventually visited the site in 1878 with Mrs. Samuel Sturgis, the wife of the 7th Cavalry commander. She was appalled by the state of the burials and communicated the condition to her uncle, General W. T. Sherman. He ordered Department Commander Terry to take care of the problem once and for all.[29]

While the Sheridan brothers had tried to rebury the dead during July 1877, other bones were on a journey eastward. A little over a year after departing from Fort Lincoln, the supposed remnants of George Custer reemerged. Libbie was not there waiting for him, and the strains of the band joyfully playing *Garryowen* were absent. The officers' skeletons were placed in nicer caskets and taken across the river to Bismarck, where Northern Pacific train cars transported the remains to their families.[30] On August 3, 1877, the bodies of Algernon Smith, Tom Custer, George Yates, James Calhoun, and Donald McIntosh were interred at Fort Leavenworth, Kansas. Later, William Reily was buried in Washington, D. C., W. W. Cooke in Hamilton, Canada, Benny Hodgson in Philadelphia, and Myles Keogh in Auburn, New York.[31]

The very same railroad that Custer had helped advance westward now returned him to the east. Libbie wrote to General John Schofield asking for help with her husband's burial. She wrote:

Col. Sheridan writes me that he had recovered the remains of General Custer, and that they now lie at Fort Lincoln, Dakota, awaiting my decision when the removal shall take place. General Custer asked me to lay him at West Point and I would like to carry out his request. Will you advise me what to do? Can you tell me when to fix the date for his burial? My dear Gen'l. Schofield, will you and Mrs. Schofield select the spot where he will lie - I regret that I must bring myself before you as I do, but I am entirely alone in New York except for an aunt and cousin as helpless as myself and I will trust that you will give me your council [sic] and suggestion as to what I am to do.[32]

Meanwhile, the U.S. Government had decided to focus its attention on another group of natives they believed should no longer roam free. In early October 1877, newspaper headlines emphasized the surrender of Chief Joseph of the Nez Perce, chased by some of the same soldiers who had pursued Sitting Bull only a year before. At about the same time, on October 10, under huge red and orange-leafed trees overlooking the magnificent Hudson River on the grounds of West Point Military Academy, human remains thought to be those of Custer were laid to rest far from that barren and treeless Montana hillside.

A tall, stone obelisk was erected over the grave with a bronze relief sculpture of a buffalo head protruding from one face of the memorial.[33] The very symbol of the Indian way of life was chosen to memorialize Custer. As a soldier and citizen, whose beliefs about destiny and western settlement were akin to many Americans, he had influenced the near demise of the buffalo. The people whose very existence depended upon this animal were no longer able to roam freely and hunt the disappearing beast. It was another paradox that exemplified the life and death of George Armstrong Custer. Most of Custer still remained on the Montana hilltop amid the scores of sun-bleached horse bones. Four years later, a stumpy, gray stone obelisk would be placed only six feet from where his body was found and his name would be carved on that piece of granite, as well.

While wolves and curiosity seekers scavenged the battlefield in January of 1879, critics were picking apart the reputation of Marcus Reno in Chicago. Frederick Whittaker, Custer's first biographer, with the support of Libbie, had convinced the press that Reno's actions during the fight were tainted by cowardice.

Tired of the continued criticism by the press and Custer defenders, Reno requested a Court of Inquiry to clear his name and reputation. Seventh Cavalry officers dressed in their finest uniforms were called one after another to the witness stand. Two of the civilian packers, who could not care less about the formality of the occasion or the reputation of the 7th Cavalry, condemned Reno as a drunkard. The rest of the 7th officers pulled together to protect Reno and the honor of the military regiment—much of the testimony was a cover-up.

As Benteen later stated, "As to queries before the Court of Inquiry, these I would answer now as I did then, and shield Reno quite as much as I did then."[34]

While Reno was not found guilty of any wrongdoing, his actions were deemed less that meritorious. In his report after the battle, Reno made special note of Benteen's bravery, and according to the captain, such deserved recognition caused a great deal of consternation with the other officers of the regiment.[35]

In the spring of 1879, Terry ordered Captain George Sanderson to clean up the battlefield. The soldiers built a mound on top of Custer Hill, only feet from where Custer was found. Digging a deep hole in the mound, he buried the remains of at least five bodies. He built a cordwood obelisk 11-feet high by 10-feet square. Inside, were placed all the horse bones discovered on the field, and Sanderson re-mounded all the individual graves. Fort Custer post photographer, Stanley J. Morrow, captured numerous images of the scattered bones as the work was in progress, and they are some of the most well-known early photos.[36]

By August of 1879, it was decided that the Custer Battlefield would become a fourth-class National Cemetery. Auburn Marble and Granite Works in Cambridge, Massachusetts, was awarded a contract for a permanent memorial marker. The stubby, granite obelisk was created in three sections, and when stacked, it was 11-feet tall. It was 6-feet wide at the base and 3-feet wide at the top. With engraved names, the 38,547-pound monument cost approximately $1,000. The truncated stones were shipped across the Great Lakes to Duluth, Minnesota, and transported via the Northern Pacific Railroad. Arriving at Bismarck, the monument was then loaded on a steamer that conveyed them to Fort Custer where it was stored.

In the summer of 1880, an 8-foot square concrete foundation was poured atop iron bars where the cordwood monument had stood. The horse bones were relocated about 60 feet to the north. It would take almost a month for Lieutenant Charles F. Roe to stack the three stones and gather all the skeletal remains from the field for a fourth and final reburial, which was completed in 1881. Today, most of the soldier and associated military casualties are located in a trench around the granite memorial in a mass grave. The only individual recognition provided is the long list of scout, civilian, and soldier names carved into the four faces of the stone above them.[37]

The inscription reads: In memory of Officers and Soldiers who Fell near this Place, fighting with the 7th United States Cavalry against Sioux Indians, on the 25th and 26th of June, A.D., 1876.

Fig. 88. Last Stand Hill Monument. An early photograph of the mass grave and granite soldier marker after 1881. Note the chips along the monument seams taken by souvenir hunters. Eventually, soldier names were recarved and a fence erected to protect the memorial.

Courtesy of Christopher Kortlander & the Custer Battlefield Museum, Garryowen, Montana.

No mention was made of the Cheyenne, who, as a percentage, took more casualties than the different Lakota bands. The Crow tribe whose reservation surrounds the monument and had participated in the fight is also ignored. Such a lack of recognition spawned the native quote:

The Sioux got the credit,
the Cheyenne did the fighting,
and the Crow got the land.[38]

More visitors began to arrive on the hill, and many wanted a souvenir of their visit. They began to chisel off pieces of the granite from the seams and corners of the monument. A stone engraver had to be summoned to hide the damage by beveling the corners and re-chiseling some of the damaged soldiers' names, as well. By 1883, a 10-foot iron fence was placed around the mass grave and memorial to protect it.[39]

In 1882, Major W. W. Sanders recommended that iron posts be supplied to replace the wooden stakes that had marked the original graves, as prairie fires and ranging cattle were destroying them.

No action was taken, however, until April 1890, when Captain Owen Sweet and a detail of Negro soldiers of the 25th Infantry positioned 217 marble markers at the original gravesites. The task of erecting permanent markers for the dead was an extraordinarily difficult job, since the bodies had been removed nine years earlier. The Reno/Benteen Battlefield was not supplied with marble markers. The solitary and grouped markers made the Custer Battlefield National Monument the only battlefield in the world with individual headstones at the supposed death sites.[40]

Fig. 89. Marble Markers on Last Stand Hill. This rarely seen photograph, taken after Owen Sweet placed individual marble markers shows the damage to stones by souvenir hunters, To protect the resource from the public, an iron fence was erected to keep individuals away from the stones.

Courtesy of the author

As the years passed, more markers were placed over the battlefield for the civilian casualties. The spurious-marker problem was compounded in the early 1930s when some of the markers for the Reno dead were interspersed with those on the Custer field. Today, there are more markers [252] on the Custer field than the 210 men who died there. Archeology has confirmed that many of the paired markers that dot the landscape at one time contained the remains of only one individual.[41]

Visitors and researchers today must be wary of the impact of individual marble markers, because they do not always reflect accurate burial and death sites. The skirmish line on Calhoun Hill that was mapped and recorded in oral testimony is not reflected in terms of markers. While approximately 40 men died on Custer Hill, there are today 52 markers. The area between Custer Hill and Deep Ravine, where approximately 10 men were buried, has 51 markers; there are none in the Deep Ravine, where at least 28 bodies were spotted. Even Custer's individual marker on the hill's slope is incorrect as it is inscribed that he "fell here," while in reality, his body was discovered on the crest of the hill.

These early individual marble markers had to be protected from the public. Since a 10-foot fence protected the large monument, vandals began to chip away at the smaller headstones, destroying many of them. Over the years, others have been stolen from the battlefield. To protect the markers on Custer Hill, a 4-foot iron fence was erected in 1930.[42] Ironically, while differing groups, including 7th Cavalry supporters and American Indian Movement members, have debated 'ownership' of the story and the site, the desecration of the sacred ground forced the army and the National Park Service to limit access. They have not only had to protect the site for the American public, but also from it.

With so many soldiers buried on the hallowed ground, Congress decided to bury more war veterans on the battlefield. The Custer National Cemetery was established in 1879, and 23 fort cemeteries were reinterred on the level area below Custer Hill. Casualties of some of the most famous western Indian fights, including the Fetterman fight, the Wagon Box Fight, and the Hayfield Fight are represented.

Today, neat rows of white markers denote the burial sites of nearly 5,000 veterans of the Indian Wars, the Spanish American War, World Wars I and II, the Korean War, and the Viet Nam War. Never realizing or caring that the burial excavations were disturbing actual 1876 areas of conflict, the army and the National Park Service continued to infringe on the battlefield until the National Cemetery was officially closed in January of 1978.[43]

Like many of the markers on the battlefield, the marble stone that reads 'Custer fell here' and the West Point grave of his final resting place create a history that is tainted with half-truths and romantic illusion. Such distortion alters the story of the fight as well. The Battle of the Little Bighorn was not some strange anomaly where a maniacal leader, anxious to become President, threw his men's lives away carelessly. Some of Custer's motives were no doubt self-serving. He had hoped for the resurrection of his career, a move up the ladder of promotion, and an effort to alleviate financial debts. However, it is clear that his tactical decisions and actions were based on solid past experience.

Custer had learned at the Washita that taking hostages could neutralize an attack on his outnumbered troops, and he had planned the same at Little Bighorn. He also understood that hostage-taking and camping among the valued warriors' families could neutralize their aggressiveness and force them to the reservation. He had learned in the Black Hills that hostage-taking could secure his safety and direct his troops through unknown regions. The threat of hanging a warrior chief got results each and every time he attempted to manipulate the tribes in that manner. Hanging reached deep into the psychology of the warrior as a dreaded and demeaning manner of death, not so different from the way torture and mutilation affected the white soldiers.

At the Washita, and twice along the banks of the Yellowstone, Custer had been outnumbered by his foe, but his tactically aggressive approach had paid off on all three occasions. He felt confident that a body of two companies (a little over 80 men), if effectively handled, could defend itself against a large number of warriors. It was not by mere happenstance that he died alongside two companies. He had learned the value of using the river and terrain as obstacles for the advances of the enemy and had used that in his previous three fights. Doubling up on horses held by their holders worked well and allowed more men to fight when pressed, regardless of the fact that it went against common tactical wisdom. Shortly after the Little Bighorn Battle, the *Wathena Reporter* accurately and perceptively stated, "Perhaps his [Custer's] previous successes as an Indian fighter might have rendered him liable to underrate the foe at whose hands he finally met his death."[44]

In his first three Indian fights, Custer knew that he had support nearby, and officers like Bell and Stanley would not have let him down. Such would not be the case with the reins of his salvation in the hands of an insubordinate Benteen and a demoralized Reno. They were just over the hill, but they would not make even a half-hearted attempt at rescue until it was too late. Instead, the ghosts of the Washita raised their ugly heads, as the results were eerily similar, coming full circle—Custer's aggressiveness had turned him into Elliott, while his subordinate officers rested and nursed their wounds nearby.

That he had planned on taking hostages at the Little Bighorn cannot be debated. John Martin was the only man who heard the strategy and lived to tell about it. The 7th Cavalry movements, along with Indian survivor testimony, make his motives clear. The move might have won Custer another victory, except that he was unable to cross the river and capture a trapped party of noncombatants. The warriors fought with an intensity that Custer had never witnessed except in the early moments at the Washita. On that frigid morning, however, overwhelming numbers had overrun the poorly armed and dismounted Cheyenne. The two attacks by the warriors along the Yellowstone were simply holding actions, since their families were not in immediate peril. Even with single-shot carbines, Custer had performed well against Lakota who had a large number of repeating weapons. His use of terrain near the Yellowstone was superb; however, the undulating landscape along the Little Bighorn would become a liability against the best guerilla fighters of the plains. Custer was able to capture and kill the warriors' horses, leaving them dismounted, early in the Washita fight. This ploy failed miserably at Little Bighorn due to the timidity of his Arikara scouts.

For everything he learned, there were valuable lessons that the 7th commander continued to ignore. While he relied heavily on his allied Indian scouts to locate the villagers, he hesitated when it came to trusting their observations. Instead, he wanted to see the situation with his own eyes. He had, on all four occasions, separated his troops from his ammunition supply train and support forces. It was a disaster simply waiting to happen.

Perhaps his biggest mistaken belief was that his subordinates had the same stamina, aggressiveness, and motivation that he had. To believe that Benteen, who had contested and argued with Custer for several days before the battle, would suddenly fall into line when ordered to do so was an ignorant assumption. Based on Boston's ability to reach his brother, there was ample time for Benteen to join and reinforce his commander.

Another factor unknown to Custer at the time was that the warriors had a dress rehearsal a week earlier at the Rosebud on how to stretch out cavalry companies and cut them off from supporting units. The strategy had come very close to working there; it was perfected along the Greasy Grass. While the warriors' actions had been remarkable in their initial defense of the village, it was their surrounding of the five companies along the Little Bighorn that became a trap Custer could not escape. The warriors' tactic of forcing the cavalry away from flat ground where they could take advantage of their long-range carbines and into the broken, rolling hills above the river, proved to be genius.

The myths surrounding Custer and the Battle of Little Bighorn only escalated in number as time passed. The tragic story appealed to the ideals of heroism against overwhelming odds in a Romantic Victorian mindset and became a battle cry for the inevitability of manifest destiny and the spread of white civilization westward. It consolidated the anger of the American public and a united Congress to relentlessly pursue the tribes fighting to preserve their way of life. While in most instances in recorded history the story is told by the winners, such was not the case at the Greasy Grass. Only a few wanted to listen to the warriors' side of the story or understand their perspectives.

The story of the Indian Wars and Little Bighorn has been rather one-sided in its interpretation from the beginning. In 1925, Mrs. Thomas Beaverheart, the daughter of slain warrior Lame White Man, asked the War Department to place a permanent marker at his death site. Up to that time, his death spot had been marked by a small rock cairn. The War Department that had jurisdiction over the site did not respond to her request. It would not be until 1956 before the National Park Service placed a small wooden marker to honor him. Since the military cared for the battlefield for almost 70 years, the story was heavily steeped in the military perspective. Edward S. Luce, the first and longest serving National Park Service superintendent, and a former 7th Cavalry officer, reinforced such views throughout the 1940s until the 1950s.[45]

It seemed destined that equal importance to the Indian perspectives would be considered sooner or later by their inclusion in the Custer National Cemetery for war veterans. The reservation system and forced assimilation required Lakota, Cheyenne, and Crow natives to fight under the stars and stripes in numerous wars. Many also joined, fought, and died by choice in a long, historic, warrior tradition. Today, visitors wander between the linear rows of white marble markers to see Indian and non-Indian veterans buried side by side.

Change is a slow process, and a protective myth of Custer and the Little Bighorn remained immovable for decades. The premise that the Lakota and Cheyenne were the enemy persisted for over two generations. Both Indian and non-Indians often perceived the Crows and Arikara as traitors to the native nations by misunderstanding the long history of intertribal warfare. The Arapaho, who had a minimal presence at the battle as captives of the Lakota, today proudly claim their role against what they see as an imperialistic U.S. Government.

In lieu of the shifting perceptions in film, literature, and the American psyche, changes in interpretation occurred at the Little Bighorn. By the 1960s, National Park interpretive rangers were telling the multiple perspectives of the sides that fought on the sacred ground. However, in name and physical presence, the Custer Battlefield seemed to memorialize only Custer and the 7th Cavalry. Many Lakota and Cheyenne did not feel welcome at the site, even though it represented one of their greatest victories.

There were several valid reasons for such feelings. First, the area was named Custer Battlefield National Monument, which was not only named after the loser, but it seemed to memorialize Custer with little regard to the Indians. Second, was the lack of physical markers and memorials for the Lakota, Arapaho, and Cheyenne who fought and died there. Another problem was that since the battlefield was within the boundaries of the Crow reservation, the long-time rivalry between the Crow people and their traditional enemies, the Lakota and Cheyenne, played a resistant role as well. Not only was it necessary to re-evaluate non-Indian/Indian relationships, but also Indian interpretive issues as well.

On June 25, 1988, Indians, once again, disturbed Custer's grave. In another strangely disrespectful act, not unlike that which occurred 112 years earlier, Russell Means and members of the American Indian Movement, dug into the soldier mass grave. Equipped with shovels, bags of ready-mix concrete and a large metal plaque, Means gave an inflammatory speech, demanding that an Indian memorial be placed on the site. He threatened to destroy that which existed for the soldiers unless one for the Lakota, Cheyenne, and Arapaho was constructed. It was a political threat meant to challenge the lack of Indian recognition by the National Park Service and the United States Government. The welded words on the steel plaque were equally confrontational:

> In Honor of our Indian Patriots who fought and defeated the U.S. Calvary [sic]. In order to save our women and children from mass-murder. In doing so, preserving rights to our homelands, Treaties, and Sovereignty. 6/25/88 G. Magpie.[46]

This memorial plaque was eventually removed, but it had initiated the much-needed effect of challenging the status quo within the National Park Service. With the hiring of the first Indian superintendent and legislation from the 102st Congress, the name of the battlefield was changed to Little Bighorn Battlefield National Monument, and an Indian Memorial was approved in 1991. It would take more than 11 years to fund and construct the Indian Memorial that was dedicated on June 25, 2003. Lame White Man and many other warriors received red granite markers to designate their general death sites. Included in the new Indian memorial was recognition of all participating tribes who fought in the battle and tributes to the Arikara scouts with white, marble markers, similar to those of the soldiers.

Today, a large circular Indian Memorial with symbolic pictograph warriors made of bronze stands only yards from the soldier memorial. Its design includes a *spirit window* that looks toward the soldier mass grave and monument, allowing the spirits of the soldiers and warriors to commune. 'Peace Through Unity' is the theme of the memorial, and it was proposed to be a living tribute that recognized the sacrifice and perspectives of all the native people who fought and died there. It also encourages Americans and all visitors to be less divisive and confrontational.

Lakota Enos Poor Bear stated eloquently, "If this memorial is to serve its purpose, it must not only be a tribute to the dead; it must contain a message for the living."[47]

Much of the former anger by Indians and non-Indians has disappeared with this more balanced and objective approach to memorials. As a park ranger and historical interpreter on the battlefield for over two decades, I have seen dramatic change for the better. Memorials and interpretation objectively represent multiple perspectives in an attempt to portray an unbiased history. However, the recent changes do not alter the sadness of this special place.

It is easy to wander late into the evening along the hilltop to see the Indian Memorial, the soldier memorial, as well as the lonely red and white markers that stand immobile through the swaying green buffalo grass of late June. With a little imagination, one can hear the eagle bone whistles, trumpets, or the wind blowing through the moving landscape. Visitors can visualize the chaos, smoke, and screams of Irishmen and Germans in their accents, or the distant tremolo of Indian women encouraging their warriors to be brave. As visitors walk down from Custer Hill, they pass the museum with Black Elk's powerful admonition in English and Lakota in large metal letters: "Know the power that is peace."

Over 200,000 visitors annually stop at Little Bighorn. Few stroll through the graves of the approximately 5,000 buried veterans in the national cemetery that speak of sacrifice and service to this country. A large American flag flutters above the cemetery. The beautifully maintained green grass with rows of white marble markers looks more like a finely manicured park. Most first-time visitors mistakenly think that the markers represent the 7th Cavalry soldiers who were killed at the site.

The National Cemetery does not reveal the layers of tragic death that occurred at other far away battlefields or on the blood-soaked ground made red by soldiers and warriors in 1876. Instead, there is a tree-lined concrete walkway extending under non-indigenous spruce trees that the U. S. Army planted decades ago, in order to create shade above the inhospitable rolling ground.

Trees and grasses, not native to the region, struggle to survive in the hot summer sun of the Great Plains. Nearby, native plants flourish, living as they always have. However, each year they seem to grapple against the invasive species that are brought in unintentionally by new visitors from faraway places. Tons of concrete and asphalt, along with buildings, make the battlefield accessible, but it has dramatically modified the way the place appeared in 1876. Man-made memorials and signage continue to sprout along the banks of the Greasy Grass. Interstate 90 runs through the area where the village once stood, while abandoned cars, run-down trailers, and prosperous wheat fields cover the valley. Neon lights flash from the Crow casino that can be seen from the battlefield.

Each summer morning, while I am in residence as a park ranger, a walk from on-site housing takes me to the battlefield museum. Surrounded by symbols of death and violence, the journey through the landscape is still beautiful and serene. The shade provided by the invasive trees is welcoming on hot days. Each June, the red robins build nests in the tree branches, representing new life, fresh beginnings, and a revitalized season. Yet, even in these new beginnings, there is death present as well. These baby birds are knocked from their nests to die on the sacred ground below, and their desperate mothers' chirps seem like mourning songs. Big black and white magpies fly into the nests that were already occupied by the robins and knock out the former occupants. They are aggressive, larger, louder, and more powerful. It seems that even nature repeats the tragedy on the ground overlooking the Little Bighorn. It is a paradox that George Armstrong Custer might have understood.

Epilogue

Albert Barnitz was severely wounded in a fight with a warrior at the Washita. He survived the battle; however, it effectively ended his military career. After his disability discharge, he worked primarily as a newspaper columnist. He kept extensive diaries and letters about his military service that were published after his death. He died in Asbury Park, New Jersey, in 1912, and is buried in Arlington National Cemetery.[1]

James M. Bell came to the rescue of Custer with additional ammunition during the Washita fight. He was on leave during the Little Bighorn campaign and received a captaincy due to the death of George Yates. He fought against the Nez Perce at the Battle of Canyon Creek and continued to fight throughout the west for 20 years. He was a brigadier general of volunteers during the Spanish American War and fought in the Philippines as well. Promoted to brigadier general in the regular army in 1901, he then retired. He lived in Pasadena, California, until his death on September 17, 1919.[2]

Frederick Benteen's military service led him to participate in the 1877 Nez Perce campaign during which he received rank of a brevet colonel. He became a major of the 9th Cavalry in 1882 and eventually wrote an accusatory letter about George Crook that was similar to one penned after the Washita fight. This spirit of insubordination that clung to the officer, combined with his alcoholism, helped end his career. After being court-martialed for drunkenness on duty, using obscene language, and indecent exposure, his punishment of dismissal was altered to one year's suspension by President Grover Cleveland. Suffering from severe back problems, Benteen moved to Atlanta and asked for a disability discharge from service after his suspension. In 1890, he received a brevet of brigadier general for his bravery in the Little Bighorn and Nez Perce campaigns. Letters written late in life to various friends never lessened in their intense hatred of the dead Custer. Benteen died on June 22, 1898, after suffering a stroke. He originally was buried in Atlanta, Georgia, and later reinterred in Arlington National Cemetery in 1902.[3]

Black Elk's life story is more exciting than any Hollywood movie script. After the Little Bighorn, he traveled with Buffalo Bill's Wild West Show in 1887 and met Queen Victoria. He was left behind in Paris, but he eventually found his way back to America—in time to be a participant in the Wounded Knee disaster in 1890. He fought the 7th Cavalry again and rescued a child whose mother was killed. He eventually became a Catholic who combined some of his native beliefs with those of his newfound religion. He died at age 87 on August 19, 1950, on the Pine Ridge reservation. He prophesied events throughout his life; before his death, he predicted the night sky would be filled with lights when he died. Pine Ridge had one of the most spectacular displays of the northern lights ever seen on the night of his wake.[4]

Charles Braden survived the thigh wound he suffered on August 11, 1873 during the Yellowstone expedition. He went on permanent sick leave and was promoted to first lieutenant in 1875, before retiring in 1878 because of his injury. Braden taught candidates entering West Point and was secretary of the Association of Graduates of West Point for 31 years. He was awarded the Distinguished Service Cross after his death on January 15, 1919.[5]

James Bradley was the leader of the Crow scouts during the 1876 campaign, and the first solider to see the Custer casualties. He kept extensive diaries about the campaign that were published after his death. He died in the Battle of the Big Hole, leading a mounted detachment of 7th Infantry against the Nez Perce on August 9, 1877. He is buried at Big Hole Battlefield Cemetery.[6]

Charles Brewster, a supporter of Custer during his court-martial, was caught up in the 'Benzine boards' which attempted to eliminate unworthy officers from the regular army. In 1870, Brewster requested a discharge rather than face the board. Having second thoughts, he unsuccessfully tried for the next 30 years of his life to be reinstated into military service. Many of the officers, including Benteen, Mathey, Moylan, Godfrey, and Bell, signed letters against Brewster in 1878. Brewster died on July 20, 1904 in Nebraska.[7]

John Burkman always regretted that he had not died with his beloved commander. He fought the Nez Perce in 1877, and after falling off a horse in 1879, was discharged from the army due to his injury. A biography called *Old Neutriment* was penned while he lived in the National Military Home in Los Angeles. He moved close to the battlefield at Billings, Montana, where he resided for 30 years. He greeted Libbie in 1910 when she attended the dedication of the Custer Memorial in Monroe, Michigan. He was rumored to carry two coins in his pockets dated 1839 and 1876—the birth and death years of the hero he worshipped. On November 6, 1925, he committed suicide and was found with a bag of candy in one hand and a revolver in the other. He is buried in the Custer National Cemetery only a few hundred yards from where Custer's body was found.[8]

Ben Clark continued to serve as a mediator for Custer's 1868-69 campaigns on the southern plains until the 7th commander dismissed him. He later became post interpreter at Camp Supply and Fort Reno and acted as a scout for General Nelson Miles during the Red River War. Clark served with Crook during the Sioux campaign of 1876 and participated in the Dull Knife fight. Despondent after the death of his third Cheyenne wife, and suffering from paralysis himself, Clark committed suicide by a gunshot to the temple on July 24, 1914. He is buried at Fort Reno, Oklahoma.[9]

John Charles 'Chip' Creighton [Chesterwood]survived the hilltop fight with Reno and was discharged from the army in 1877. Two weeks later, he re-enlisted with Company E of the 7th Cavalry, where he became a sergeant until another discharge in 1882. He eventually became a house painter. He attended the 50th anniversary of the battle and died on January 30, 1935, in Tacoma, Washington.[10]

Crazy Horse stayed with his people and fought Nelson Miles at the Battle of Wolf Mountain on January 8, 1877. With his community starving and cold, he eventually surrendered at Camp Robinson on May 5, 1877. He was promised his own reservation by George Crook, but ended up at the Red Cloud Agency when the promise was unfulfilled. Influential to young men, disliked by elders, and feared by the military, Crazy Horse was led to the stockade at Camp Robinson to be incarcerated. Seeing the bars and understanding what was in store, he resisted and was bayonetted by guard William Gentiles. He died later that day, September 5, 1877, and was eventually buried in a secret gravesite by his family.[11]

George Crook, after his less than stellar performance at the Rosebud, continued to chase the Lakota though the summer of 1876. He finally struck a blow at Slim Buttes and the Dull Knife fight, ending his campaign in December of the same year. In 1882, he spent four years chasing Geronimo and the Chiricahua Apache as Department Commander of Arizona. After Geronimo escaped, he was replaced by Nelson Miles as Department Commander and took up the same role for the Department of the Platte. He became a major general in 1888 and commander of the Department of the Missouri. Realizing late in life that the U.S. Government and the army had mistreated and betrayed the promises made to Indian tribes, he spent the rest of his career defending Indian rights. He died suddenly at age 61 on March 21, 1890, and is buried at Arlington National Cemetery.[12]

Crow King lost two brothers in the Little Bighorn battle. A member of the Silent Eaters Lodge society, he surrendered at Camp Poplar, Montana, in 1881. He became a policeman at Standing Rock and served in that capacity until he died in 1884.[13]

Curly, the Crow scout who, on occasion, denied he had been in the fight, became a sought-after witness by the media. Sometime called the 'only survivor of Custer's Last Stand,' his story continued to change throughout his life. He became a rancher and policeman, building a cabin below the battlefield at Crow Agency. He attended the 10th anniversary reunion where he was challenged by Gall about his escape story. Curly eventually fell out of favor with the other three Crow scouts, who also challenged his story. Curly was granted a pension late in life and died of pneumonia on May 21, 1923. He is buried in the Custer National Cemetery.[14]

Elizabeth Bacon Custer never remarried and dedicated the remainder of her life to defending her husband's reputation against detractors, eventually outliving most of them. She moved into a Park Avenue apartment in New York City, far from the hardships of the plains she had come to know. She traveled the world and wrote three best-selling books about her experiences with her husband: *Tenting on the Plains, Following the Guidon*, and *Boots and Saddles*. Dying of a heart attack four days before her 91st birthday in 1933, she was interred at West Point by the 'supposed' remains of her husband.[15]

Charles De Rudio survived the battle by hiding in the timber and added another incredible story to his repertoire. He became a captain in 1882, later serving at Fort Meade, Fort Sam Houston, and Fort Bayard. He retired from military service in 1896. He came out of retirement to become a major, only to retire again in 1909. He died a year later on November 1, 1910, and his ashes were buried in the San Francisco National Cemetery.[16]

Winfield Scott Edgerly was promoted to first lieutenant effective June 25, 1876, and later was a recruiter for the cavalry in 1883. As a captain, he fought with the 7th Cavalry at Wounded Knee and later moved to Texas for the years 1892-1895. In 1898, he earned the rank of major of the 6th Cavalry and ended up back with the 7th Cavalry in Cuba. Rising in rank to colonel, he served in the Philippines and eventually became a brigadier general in 1905. He served in Germany and became commander of the Department of the Dakota, as well as serving as an instructor at the Fort Riley Cavalry and Artillery School. He retired in 1909 and died on September 10, 1927. Arlington National Cemetery now holds his remains.[17]

Thomas French was remembered for his brave actions at Little Bighorn, but like Weir, he was haunted by the deaths there and took solace in the bottle. French fought in the Nez Perce campaign and was slightly wounded at the Canyon Creek fight. He was highly critical of Reno after the battle but was unable to testify at the Reno Court of Inquiry due to an impending court-martial and confinement. His drinking continued to affect his behavior, and he was sentenced to dismissal. Crook and President Hayes intervened, reducing his rank and pay for one year. French eventually resigned in 1880, and within two weeks, his home burned to the ground under suspicious circumstances. He died of apoplexy at age 39 on March 27, 1882. He was buried at Fort Leavenworth until his remains were eventually moved to Holy Rood Cemetery in Georgetown, Washington, D. C.[18]

Gall fled with Sitting Bull into Canada after the Little Bighorn. He began to quarrel with the Lakota medicine man about how long they could remain there, as people were suffering. Gall, with about 300 followers, surrendered in 1881. He became an agency leader on behalf of his people at the Standing Rock Agency, later becoming a judge. He attended the 10th anniversary of the battle, telling his account of the fight to the press and 7th Cavalry officers. In 1889, he traveled to Washington as part of a Lakota delegation to deal with problems on the reservation. At the end of his life, Gall weighed over 300 pounds and suffered from a series of maladies, including the bayonet wound he received from the soldiers, years earlier. According to his friend, photographer D. F. Barry, he died from an overdose of weight loss medication and his death date is debated. He was buried in December 1894, in St. Elizabeth's Cemetery, South Dakota.[19]

Frederic Gerard survived the Little Bighorn by hiding in the timber. Two years later he was named Morton County assessor in Dakota Territory. He was critical of Reno in the 1879 Court of Inquiry. He was employed as post interpreter at Fort Stevenson and later opened a store in Mandan. In 1890, he became a salesman for Pillsbury Mills, and in 1908, he was paralyzed while lifting a trunk. He died on January 30, 1913, and was buried behind the College of St. Benedict in St. Cloud, Minnesota.[20]

John Gibbon, a year after the Little Bighorn, was able to sneak up on a village at Big Hole, Montana. Chief Joseph and his Nez Perce were caught off guard but quickly recovered and fought to a standoff, severely wounding Gibbon in the thigh. He commanded the Department of the Dakota in 1878-79 and the Department of the Platte in 1884. He was promoted to brigadier general a year later and commanded the Department of the Columbia until he retired. He died in Baltimore on February 6, 1896, and is buried in Arlington National Cemetery.[21]

Moses Flying Hawk, a nephew of Sitting Bull who fought at Little Bighorn, eventually surrendered and lived out his life at the Pine Ridge Agency. His brother Kicking Bear was one of the main leaders of the Ghost Dance uprisings. Flying Hawk died on Christmas Eve, 1931.[22]

Francis Gibson, Benteen's second in command of Company H, had temporary command of the company after Little Bighorn. He participated in the Nez Perce campaign and in 1880 was appointed the rank of captain. In 1891, he resigned due to a disability caused by chronic gastroenteritis. He died in 1919 and is buried in Arlington National Cemetery.[23]

Edward Godfrey, the workhorse of the Washita and the savior of Weir's company, served in the Nez Perce campaign the following year. At the Snake Creek fight, he was severely wounded in the hip. In 1894, he received the Medal of Honor and was breveted the rank of major for his bravery in that action. Godfrey testified at the Reno Court of Inquiry in 1879. He taught cavalry tactics at West Point for four years, until 1883. In 1890, Godfrey commanded Company D at the Wounded Knee tragedy and while returning from the campaign, was injured in a train wreck from which he never fully recovered. He later served with the 1st Cavalry, the 7th Cavalry again in Arizona and New Mexico, and the 12th Cavalry in Cuba. He was eventually commissioned as a colonel in the 9th Cavalry, serving in the Philippines. In 1907, he was promoted to brigadier general. Godfrey attended the 50th anniversary of the battle as one of the few officers still alive. In a ceremony in which Godfrey 'buried the hatchet' with warrior survivor White Bull, they shook hands over the grave of an unknown 7th Cavalry soldier at Garryowen, Montana. He died at age 88, on April 1, 1932, and is buried in Arlington National Cemetery.[24]

Goes Ahead left Reno Hill with Crow scouts White Man Runs Him and Hairy Moccasin and returned to Crow Agency. After the battle, various writers and researchers interviewed him. He died on May 31, 1919, at Crow Agency and is buried in the Custer National Cemetery.[25]

Hairy Moccasin left Reno Hill with White Man Runs Him and Goes Ahead, and they were the first to tell Colonel Gibbon's forces of Custer's defeat. He eventually became a farmer and occasionally was interviewed about his experiences in the fight. He died on October 9, 1922, and his burial site is an unmarked grave near St. Ann's Catholic Church, Lodge Grass, Montana. The church no longer stands on the site, but evidence of the graves still exists on private property.[26]

Half Yellow Face, the leader of the Crow scouts, survived the valley fight and helped rescue White Swan. He died shorty after the battle in 1879 and little is known about his life after the battle. He was supposedly buried in an unmarked tree burial near Wyola, Montana.[27]

Winfield Scott Hancock, hero of Gettysburg and the Mexican-American War, found himself in charge of the Fifth Military District, which included Texas and Louisiana. After his poor showing in the 1867 campaign, he continued to serve as a departmental commander, but in the Department of the Dakota until late 1872. He also became commander of the Departments of the Atlantic and the East until 1886. He ran for president as a Democratic nominee in 1880, but he lost the election to James Garfield. He died on February 9, 1886, and is buried in Montgomery Cemetery in Norristown, Pennsylvania.[28]

Luther Hare was appointed first lieutenant on June 25, 1876, commanding Company L for the summer. He was in the Canyon Creek fight during the Nez Perce campaign as a regimental engineer. He was sick and missed the Wounded Knee tragedy, but he was involved in the Apache campaign in 1896. He became a lieutenant colonel for the 1st Texas Cavalry and, a month later, was serving as a colonel in Cuba. He fought with the 33rd Infantry in the Spanish-American War, where he won a Silver Star for gallantry at the battles of San Jacinto and Northern Luzon in 1899. He was appointed brigadier general of volunteers in 1900 and was a major of the 12th Cavalry a year later. He retired due to a disability but came back to service as an instructor of militia in Austin, Texas. He later became a professor of military science and tactics at the University of Texas in 1911 before retiring. Hare became a colonel and retired again in 1916, only to serve as a commander in charge of training army students at Simmons College in Kansas. He died of cancer of the throat on December 22, 1929, and is buried at Arlington National Cemetery.[29]

He Dog, a shirt wearer in the Oglala tribe and warrior at Little Bighorn, continued fighting at the battles of Slim Buttes and Wolf Mountain. He surrendered at Fort Robinson, Nebraska, on May 6,

1877. He later fled to Canada to join Sitting Bull, when the Red Cloud Agency was relocated. He gave up at Fort Keogh in 1880. He was transferred to the Standing Rock Agency in 1881, and later, he was sent to Pine Ridge. He Dog was sent to Washington, D. C., as a delegate to represent his tribe. He became a prominent peacetime leader and eventually became a judge for the Court of Indian Offenses in the 1890s. He Dog died in 1936 on the Pine Ridge Reservation.[30]

George Herendeen continued his service as an army scout throughout the Nez Perce campaign in 1877. He lived in various cities in Montana before becoming an employee on the Fort Belknap Indian reservation. Pneumonia took his life on June 17, 1918, and he is buried in Harlem Cemetery, Harlem, Montana.[31]

Daniel Knipe [Kanipe] survived death with Custer by carrying a verbal message to hurry the pack train. That summer, he suffered a controversial, self-inflicted revolver wound to his calf. Knipe claimed it was accidental and later unsuccessfully applied for a disability pension. He became a first sergeant and re-enlisted for another five-year stint on August 7, 1877, remaining in the 7th until his discharge in 1882. Returning to Fort Lincoln, he courted and married his former first sergeant's wife, Missouri Ann Wykoff Bobo, and they had nine children. He moved to Marion, North Carolina, where he was employed by the U. S. Revenue Department and chased moonshiners. He became a guide and informant for researcher Walter Camp in his studies of the Little Bighorn. Their planned book was never published. During World War I he was appointed by the governor as a captain in the 19th North Carolina Militia. He died on July 18, 1926, and is buried in Oak Grove Cemetery in Marion, North Carolina.[32]

Little Raven was persuaded to go to the reservation in January 1869 by Custer, along with his southern Arapaho band. Always recognized for his oratory skills, he toured Washington, D. C., and New York City in 1871 as a peace advocate. During the Red River Wars of 1874-75, he was able to keep most of his warriors from abandoning their peace promises. Residing at Cantonment, Oklahoma, he died in 1889.[33]

Little Robe surrendered at Custer's request at Fort Cobb after the Washita and opposed the Dog Soldier leaders who wished to keep fighting. He continued to promote peace and was part of a Cheyenne delegation sent to Washington in 1871. In 1873, he met with President Grant, and, during the Red River War of 1874-75, he remained on the reservation when many of his people left. He died in 1875.[34]

Little Wolf was almost killed immediately after the Little Bighorn battle due to his late arrival. He was in Dull Knife's village when it was attacked in November 1876. He was shot seven times but somehow managed to survive. In May 1877, he surrendered and was placed on the reservation at Fort Reno in present day Oklahoma. Considered an exile from his home in the Tongue River area, he, along with Dull Knife and 300 Cheyenne, headed north without permission. The 1,500-mile chase included several engagements with soldiers. The bands split up, and Dull Knife surrendered at Camp Robinson, while Little Wolf made it to Montana. After promising to scout for General Nelson Miles, he was allowed to remain at the Red Cloud Agency. While under the influence of alcohol, he killed an adversary and was banished from the reservation and lost his prominent position with the tribe. He died along the Rosebud River in 1904.[35]

Lone Wolf, the Kiowa leader, who was almost hanged with Satanta by Custer, surrendered and was exiled to Fort Marion in St. Augustine, Florida, instead of being tried before a military court. Remaining in Florida for four years, he was eventually allowed to return to his people, where he soon died of malaria.[36]

Low Dog lost an older brother at Little Bighorn and fled to Canada with Sitting Bull. He surrendered at Camp Poplar, Montana, in 1881. Settling with the Miniconjou tribe at Cheyenne River Agency, he died in 1894.[37]

Magpie survived the battle of the Washita and was present when Custer smoked the pipe with the Cheyenne arrow keeper along Sweetwater Creek. He surrendered with Medicine Arrow's band in 1869. In the 1870s, he traveled north to join the Northern Cheyenne and eventually united with Sitting Bull's Lakota. Magpie was wounded twice at the battle of the Rosebud and was present in the Little Bighorn camp on June 25, 1876. He returned to the reservation and enlisted as an Indian scout. He was elected to the Cheyenne Chiefs' Council and later became a Christian. He died around 1936.[38]

John Martin, Custer's last surviving messenger, continued to serve for the next three years as a trumpeter and participated in the Nez Perce campaign of 1877. Martin testified in the Reno Court of Inquiry and later shared his accounts with researcher Walter Camp. In 1879, he re-enlisted for five years in the 3rd Artillery Regiment stationed in New York City. Later, he was transferred to the 4th Artillery in 1893 and eventually signed up for three more years with the unit. In 1900, he enlisted with the 7th Cavalry, serving in Cuba and received a promotion to master sergeant, before his mandatory retirement in 1904. After an unsuccessful candy business and equally unsuccessful marriage, he became a 'ticket chopper' for the New York subway system. He eventually became a mechanic's assistant at the Brooklyn Naval Yard. Having escaped the dangers of Little Bighorn, he discovered the true dangers of the streets of New York City when his luck ran out. Already suffering from poor health, he was struck by a taxi as he attempted to cross a street. He died a few days later on Christmas Eve, 1922, and he is buried at Cypress Hills National Cemetery in Brooklyn, New York.[39]

Edward Mathey survived his mule train escort duty to become a captain during the Nez Perce campaign a year later. In 1878, he was an escort of the movement of the Cheyenne tribe in Nebraska. He was on detached service during the Wounded Knee episode and retired with a disability in 1896. He acted as professor of military science at Baylor University from 1901-1903. Mathey died in Denver in 1915, following a prostate operation and is buried in Arlington National Cemetery.[40]

Thomas McDougall helped escort Chief Joseph and his Nez Perce after their surrender at Fort Lincoln. He retired due to a disability in 1890. He later became a major, retiring again in 1904. He died in Vermont on July 3, 1909, and is buried in Arlington National Cemetery.[41]

Monahsetah was returned to the Cheyenne in June 1869, and her daughter, Bird Girl, died probably before her second birthday. Some Cheyenne oral stories suggest that she gave birth to a boy named Yellow Swallow by Custer, soon afterward, but inconsistencies in the location of his burial site, attribution to Monahsetah's mother as his parent, and the probability that Custer was probably sterile from mercury treatments from an earlier bout with gonorrhea, makes the story suspect. She reunited with her former husband, Little Eagle, whom she shot before her army captivity. Marital bliss was to be elusive to

Monahsetah, as she was married and divorced to five different husbands and had at least five more children; several died before her own death. Known later as 'Morning Alfrey,' she died on January 25, 1922, and was buried on her land allotment near Kingfisher, Oklahoma.[42]

Anna Morgan, who was rescued from the Cheyenne village along Sweetwater Creek in the Texas panhandle, returned to her husband. She gave birth several months later to a half-breed child, who was the product of her repeated rapes. The child, who was loved by his mother, died at the age of three. Her husband and brother were unable to accept what had happened to her, and she was plagued by guilt. Rejected by her family and distraught over the loss of her child, she died in Kansas at the Topeka Insane Asylum on June 11, 1902, at the age of 58.[43]

Myles Moylan, who survived the valley and hilltop fights, was fighting the Nez Perce a year later. He suffered a severe wound in the Battle of Bear Paw Mountain before Chief Joseph's surrender. He received the Medal of Honor and the brevet rank of major for his distinguished gallantry in 1890. In the same year, he commanded a company that participated in the action at Wounded Knee. He was eventually promoted to major of African-American troops in the 10th Cavalry, in 1892. Retiring a year later, he settled in San Diego, California, where he died of stomach cancer on December 11, 1909.[44]

Henry Nowlan, the officer who mapped the graves of Custer and his officers, was promoted to fill Keogh's rank of captain with Company I effective from June 25, 1876. He fought the Nez Perce at the Canyon Creek battle, where he was breveted the rank of major. He commanded Company I at the Wounded Knee fight. In 1895, he earned a promotion to major and died on November 10, 1898, in Hot Springs, Arkansas. He is buried in Little Rock National Cemetery.[45]

One Bull remained with Sitting Bull in Canada and was one of the last, along with his leader, to surrender in 1881. He placed his children in the agency's Congregational school and became a policeman. As the Ghost Dance spread fear among the military, they asked him to remove Kicking Bird from the reservation. According to Sitting Bull's relatives, One Bull turned on the spiritual leader and spied on him for Agent McLaughlin. He remained active in tribal affairs after Sitting Bull's death. He died on June 23, 1947, at the age of 94 and was buried in the mission cemetery near Little Eagle, South Dakota.[46]

Pawnee Killer successfully tormented Custer during the 1867 Hancock campaign. The Oglala Lakota war leader did much of the raiding in Kansas during that period and helped wipe out the party under Lieutenant Lyman Kidder, who was a messenger to Custer. He likely fought at 1868 Beecher Island and left before the Battle of Summit Springs in 1869. He joined his nemesis Custer on a buffalo hunt for the Russian Grand Duke Alexis in Nebraska in 1872. He finally ended up at the Red Cloud Agency in Nebraska, but he left not long afterward. The Oglala then slipped into the mists of an unknown and unrecorded history.[47]

Dr. Henry Porter saved many lives on Reno Hill as the only surviving doctor from the Little Bighorn Fight. He returned to private practice in Bismarck before signing another contract with the army at Camp Hancock, serving in that capacity until 1877. He served as Burleigh County physician and was president of the Medical Society of North Dakota. He later became a world traveler, visiting Europe, Mexico, South America, Cuba, and Puerto Rico. He died of a heart attack on May 3, 1903, while in Agra, India, and is buried there.[48]

Rain-in-the-Face's actions at the Little Bighorn are still questioned today, and his vow to cut out and eat the heart of Tom Custer grew to a mythic legend. Henry Wadsworth Longfellow's poem, *The Revenge of Rain-in-the-Face*, created the myth that he killed George Custer himself. He fled to Canada with Sitting Bull, then surrendered in 1880 at Fort Keogh. He was at Standing Rock agency when he was accidently shot in the leg, which incapacitated him for the remainder of his life. His fame grew as a participant of the Little Bighorn battle when he toured with Buffalo Bill's Wild West Show and was an attraction at the 1893 World's Fair. His stories became self-serving and exaggerated. He died on September 14, 1905, and is buried near Aberdeen, South Dakota.[49]

Red Feather, Oglala warrior and close friend of Crazy Horse, surrendered in 1877. He became an advocate for reservation life and helped the non-Indians during the Ghost Dance uprisings. He became a devout Catholic and an Indian policeman on the Pine Ridge reservation. After a leg amputation, he was unable to fulfill those responsibilities and details of his death are scarce.[50]

Red Star, also known as 'Strikes the Bear' and 'White Calf' survived the battle. Varnum listed this Arikara scout as missing in action on June 26. He became a policeman from 1898 until 1901. He died on June 7, 1929, and is buried in the Indian Scout Cemetery in North Dakota.[51]

Marcus Reno never recovered from the effects of alcohol and the 1876 battle. In 1877, he was court-martialed for flirting with another officer's wife and was to be dismissed from service. President Rutherford Hayes changed the punishment to a two-year suspension. Reno's heavy drinking and the striking of a junior officer also caused disdain from fellow officers. Once again, charges were eventually dropped, and Reno remained in the 7th Cavalry. Looking for a scapegoat for her husband's death, Libbie and author Frederick Whittaker blamed Reno for the debacle at Little Bighorn. He requested a Court of Inquiry in 1879, in Chicago, to defend his reputation and was partially successful. However, the dark cloud that seemed to always be over Reno reappeared at Fort Meade when he was caught window-peeping at the daughter of Colonel Sturgis. He was dishonorably dismissed from service in 1880. Falling on hard times, he died of tongue cancer on March 30, 1889, and was buried in a pauper's grave in Washington, D.C. In the mid-1960s, the family of Reno asked that his discharge be reviewed and Reno was pardoned. His remains were reinterred in the Custer National Cemetery on September 9, 1967, on the famous ground that had haunted him for much of his career.[52]

Alfred Terry quickly attempted to avoid the responsibility for the failure of the 1876 campaign. He earned the rank of major general in 1886 and became the commander of the Department of Missouri. In 1877, he pleaded with Sitting Bull to return to the U. S. and receive amnesty, but he was unsuccessful. He retired due to poor health and died on the day after Sitting Bull's death in 1890. Terry is buried in the Grove Street Cemetery in New Haven, Connecticut.[53]

By 1877, **Two Moons** had been convinced to give up his fight and was hired as a Cheyenne scout to fight the Nez Perce. He was sent to Washington as part of a delegation to represent the Northern Cheyenne nation. He later gave several interviews about his memory of the Little Bighorn fight, and he was actually one of the models for the Indian on the buffalo nickel in 1914. He died in 1917, and is buried at Busby, Montana.[54]

Thomas Rosser, Custer's nemesis during the Civil War and friend during the Yellowstone campaign, showed his admiration of the 7th commander by condemning Reno's actions after the Little Bighorn. He and Reno had a short literary duel in the newspapers, with Rosser defending Custer. By 1881, Rosser became chief engineer for the Canadian Pacific Railroad until he retired to farming. With the coming of the Spanish-American War, his desire for military service was rekindled, and he was appointed brigadier general but did not see action because of illness. He later became postmaster of Charlottesville, Virginia, until his death on March 29, 1910, and is buried in Riverview Cemetery in Charlottesville.[55]

John Ryan, the man who fired the first and last shots of the Battle of the Little Bighorn and fought at the Washita, was discharged from the 7th Cavalry just six months after the fight. He moved to West Newton, Massachusetts, where he lived for over 50 years. In 1878, he joined the Newton Police Department, and in 1903, he was promoted to serve as captain until his retirement in 1913. In later life, he spoke about his experiences in the Civil and Plains Indian wars, until he suffered a stroke in 1924. He died on October 14, 1926, and is buried in Calvary Cemetery, Waltham, Massachusetts.[56]

Satanta, the influential Kiowa leader and warrior who had escaped being hung by Custer, continued his fight against the intrusion of whites into his southern plains homeland. He was involved in raids along the Red River in Texas, and he was arrested and placed in prison there. Eventually paroled, his association with other raiders and his constant challenge of white authority caused him to be placed in prison again where his spirit dwindled. He noted, "I cannot wither and die like a dog in chains." After an unsuccessful attempt at suicide by slashing his arteries, he leaped to his death from the second floor landing of the prison hospital in Huntsville, Texas, on October 11, 1878. He was buried in a pine coffin in a potter's field called Perkerwood Hill until later reinterred at Ft. Sill, Oklahoma.[57]

Stanislas Roy survived the hilltop fight and was awarded a Medal of Honor for bringing water to the wounded there. He fought at the Snake Creek fight against the Nez Perce the following year. He spent 32 years in Company A of the 7th Cavalry and fought in the Wounded Knee fight. He received a pension and attended the unveiling of the Custer Memorial in Monroe, Michigan, on June 4, 1910. He was buried in Greenlawn Cemetery, Columbus, Ohio, after his death on February 10, 1913.[58]

Spotted Tail was appointed Red Cloud Agency chief in 1876, but his appeal to the tribes went unheeded and many joined Sitting Bull. He encouraged the wrath and jealousy of other leaders that was intensified when members of his family entered the Carlisle Indian school. They converted to Christianity and became farmers. Red Cloud and Crow Dog were especially incensed and attempted to have him removed from his position. Crow Dog and Spotted Tail had a confrontation that may have dealt with a woman, and the former threatened to shoot Spotted Tail. Though he backed down at this time, Crow Dog later assassinated him with a rifle on August 5, 1881. Spotted Tail was buried near Rosebud Agency.[59]

David Stanley contended successfully with Custer and his problems with alcohol throughout the Yellowstone expedition. He was transferred to Texas in 1879, with the 22nd Infantry Regiment and promoted to brigadier general in 1884, commanding the Department of Texas until his retirement in 1892. He ran the soldiers' home in Washington, D.C., where he was buried after his death on March 13, 1902.[60]

Samuel Sturgis always blamed the death of his son Jack on Custer, and he became a vocal critic of the officer. He took field command of the 7th and fought against the Nez Perce in 1877. He was appointed the governor of the soldiers' home in Washington, D.C., from 1881-1885. He was commander of Fort Meade until his retirement in 1886. He is buried in Arlington Cemetery following his death from diabetes on September 28, 1889.[61]

Michael V. Sheridan had acted as an aide to his brother throughout the Civil War and attained brevets up to the rank of lieutenant colonel. He received a staff appointment with his brother after the war and became a lieutenant colonel in 1870. After his retrieval of the Little Bighorn officer bodies, he continued to serve his brother until Philip's death in 1888. He served in the Spanish-American War as a brigadier general of volunteers, retiring as a regular brigadier in 1902. He died on February 21, 1918, in Washington, D.C., and is buried in Arlington National Cemetery.[62]

Philip H. Sheridan remained commander of the Department of the Missouri with most of the final wars against the Indians occurring during his tenure. He climbed into the rank of General-in-Chief after Sherman resigned the position. He was instrumental in the development of Yellowstone National Park. He died of a heart attack while holding that office on August 5, 1888. He is buried in Arlington National Cemetery.[63]

William T. Sherman, who helped mastermind the Sioux campaign of 1876, continued to act as General-in-Chief. In 1875, he published his memoirs in two volumes and tried to avoid being drawn into the political arena. It was an unhappy time for the famous Civil War general, and he retired from his Washington post in 1883. Sherman remained a national hero for the Union until he died on February 14, 1891, in New York City. His body was transported to Calvary Cemetery in St. Louis, Missouri, where he was buried.[64]

Sitting Bull, the great resister, stayed in Canada until pressure from the U. S. and Canadian Governments, along with the decimation of buffalo there, forced his dwindling band to Fort Buford. Proud until the very end, on July 20, 1881, he handed his rifle to his son to relinquish to authorities, "I wish to be remembered that I was the last man of my tribe to surrender my rifle. This boy has given it to you, and he now wants to know how he is going to make a living." Forced to the Standing Rock agency under agent James McLaughlin, the two strong-minded men continually clashed. William "Buffalo Bill" Cody was able to pry the great spiritual leader from the reservation for four months in 1885 and featured the 'slayer of Custer' in his Wild West Show. This was a lucrative outlet for Sitting Bull, as he held rights to his photos and autographs. He had a great fondness for the ladies and became close friends with Annie Oakley, who toured with him. In the late 1880s, some Lakota were inspired to create special shirts that would supposedly make them bullet-proof. They believed that a ritual dance would bring back the buffalo and the white men would disappear. The "Ghost Dance" worried the military as a possible uprising and even though Sitting Bull was not a strong supporter, the paranoid McLaughlin decided to arrest Sitting Bull. On December 15, 1890, native agency police arrived at his cabin at daybreak to arrest him. As he was being removed, followers and family members encouraged the passive Lakota to resist. Chided by his son and others, one of his supporters fired into the police escort. The police officer immediately shot and killed Sitting Bull. The melee caused the deaths of seven followers and family members of the great chief.

There were also four dead policemen and three wounded, two of which died later. Like Custer, Sitting Bull's remains were dug up from their original grave and possibly scattered elsewhere. The final resting place of Sitting Bull's bones is still in question, like those of Custer.[65]

William O. Taylor survived the Little Bighorn Battle and was discharged on January 17, 1877, with the reputation of having been a poor soldier. He became a metal polisher in a sewing machine factory after his service in the military. He died of heart disease on February 19, 1923, and is buried in Orange, Massachusetts.[66]

Peter Thompson was awarded a Medal of Honor on October 5, 1878, for 'voluntarily bringing water to the wounded in which effort he was shot through the right hand and forearm.' He participated in the Northern Cheyenne campaign during 1878-79 and was discharged in 1880. He worked as an amalgamator for the Homestake Mining Company for 18 years before becoming a rancher in Alzada, Montana. He died on December 3, 1928 from heart problems and is buried in Lead, South Dakota.[67]

Charles Varnum, chief of Custer's scouts, was promoted to first lieutenant with Company C, on June 15, 1876. While he was involved with the Nez Perce campaign, he never engaged their warriors. He eventually became the regimental quartermaster in 1879 and also testified at the Reno Court of Inquiry the same year. Varnum was in the Wounded Knee fight, and the next day, he bravely covered another company at White Clay Creek. For that action, he received the Medal of Honor in 1897. He became a professor of military service at the University of Wyoming in 1895 and was involved in the Spanish-American War, ending up in Cuba. Varnum contracted typhoid fever and was sent home where he climbed the ranks from major to lieutenant colonel, until he retired from a disability in 1907. He continued to teach military tactics at the University of Maine and for the Idaho National Guard. He retired a second time as a colonel in 1918. Varnum attended the 50th anniversary of the battle with Godfrey. He was the last officer survivor from the Little Bighorn. He died on February 26, 1936, and is buried in the San Francisco National Cemetery.[68]

George Wallace, who helped select Custer's burial spot, was promoted effective from June 25, 1876, to the rank of first lieutenant before replacing Cooke as regimental adjutant in August of the same year. Wallace was in the Canyon Creek fight with the Nez Perce a year later and did service with recruiting duties at Jefferson Barracks in Missouri. He also taught at the cavalry school at Fort Leavenworth, Kansas. In 1885, he became a captain. At Wounded Knee, he was commanding Company K as they attempted to disarm the Lakota and was killed by a shot in the head and abdomen on December 29, 1890. Today he lies in his family plot at Rose Hill Cemetery, in Yorkville, South Carolina.[69]

Thomas Weir, a devoted supporter of Custer, never fully recovered from the experience and severe guilt of Little Bighorn, and he was dead within six months. Writing Mrs. Custer after the battle, he had promised to provide incriminating information about fellow officers' actions, but he died from the effects of alcoholism on December 9, 1876, while on recruiting duty in New York City. He was buried on Governor's Island, New York, but later reinterred in Cypress Hills National Cemetery, Brooklyn.[70]

Robert West, the man who had been involved in the court-martial of Custer, continued to battle the bottle. Barnitz, who had a changing view of Custer, had a reappraisal of West. Recalling West as "a thorough soldier and a polished gentleman—of the reliable type," he later called him "crafty, scheming, envious, selfish, malicious." Knowing Custer would come after him for his role in the court-martial, West pleaded with Sheridan for the sutlership at the newly established Fort Sill. Resigning his commission, other officers chipped in to help West with his initial investment and stock, but the plan never developed, likely due to the officer's drunkenness. He died at Fort Arbuckle, Florida, on September 3, 1869.[71]

John Weston had come to Custer's rescue on the Yellowstone expedition but missed the Little Bighorn due to a transfer to the Commissary Department, where he received a promotion to captain. He remained there for most of his career until becoming a brigadier general and serving in the Spanish-American War. He commanded the Department of Luzon in the Philippines in 1906, until his retirement in 1909. He died on August 3, 1917, and is buried in Arlington National Cemetery.[72]

Sarah White rejoined her family after being held in captivity by the Cheyenne and freed by Custer. Gossip about her life with the Indians made life difficult for her as Victorian attitudes of the time defined her as 'damaged goods'. She eventually became a teacher and married Erastus Brooks in 1869. They had six children. She would talk about her experiences on occasion, but it became clear it was something she wished would be forgotten. She died on May 11, 1939, at the age of 89, only a mile from the spot where the warriors had abducted her.[73]

White Bull, who had counted many coups at Little Bighorn, decided not to follow his uncle Sitting Bull into Canada and surrendered his small band at the end of 1876. He later became a judge and fought for the Indian claims of the Black Hills. In 1926, he represented the Lakota at the 50th anniversary where he shook the hand of Edward Godfrey as a sign of peace. He called it the "happiest day of his life" and Godfrey gave him a much-treasured American flag. He died in South Dakota, in 1947.[74]

White Man Runs Him returned to Crow Agency after leaving Reno Hill. He attended the 50th anniversary ceremonies and was interviewed many times about his participation in the Battle. His stories, like that of the other Crow scouts, changed over the years. He briefly appeared in the movie, *Red Raiders*, in 1927, and died on his ranch near Lodge Grass, Montana, on June 2, 1929. He is buried in the Custer National Cemetery.[75]

White Shield, the southern Cheyenne warrior who fought at Little Bighorn eventually surrendered and was placed on the reservation. There, he refused to cooperate with the assimilation process. However, his son Harvey White Shield graduated from the Haskell Indian College in Kansas. White Shield died on May 2, 1918.[76]

White Swan, who was severely wounded in the valley fight, never fully recovered from his injuries. He attended the 10th anniversary of the Battle and died on August 12, 1904, at age 53. He is buried in the Custer National Cemetery. The old, stone superintendent's house holds the White Swan Library, named after this brave Crow scout.[77]

Charles Windolph was awarded the Medal of Honor on October 5, 1878, for helping secure water for the wounded on Reno Hill. He was discharged from service in 1883, and for 49 years he worked for the Homestake Mining Company. He lived in Lead, South Dakota, and attended the 50th anniversary ceremonies on the battlefield. He died at age 98 as the oldest living soldier survivor of the conflict on March 11, 1950. He is buried in the Black Hills National Cemetery.[78]

Wooden Leg, the young teenager who had fought bravely at the Rosebud and Little Bighorn fights, was chased by the military until surrendering with his Cheyenne brethren at Fort Keogh, a year later. Wooden Leg eventually ended up on the Pine Ridge Indian reservation, and in 1889, he enlisted as a scout at Fort Keogh and guided the soldiers to Wounded Knee. He lived out the reminder of his life on the Northern Cheyenne reservation, only 30 miles from the Little Bighorn Battlefield. He joined other warriors at the 1906 reunion held on the Battlefield, and in 1908, he was baptized into Christianity. Dr. Thomas Marquis recorded his story in *Wooden Leg—A Warrior Who Fought Custer.* Fearful of government reprisals, he claimed that most of the soldiers committed suicide, a story he later recanted. At about the same time, in 1930, he was sent to Washington as part of a Cheyenne delegation as a representative. He became a judge on the reservation and died in 1940.[79]

Yellow Nose, the captured Ute turned Cheyenne, fled with the Cheyenne after the fight at Little Bighorn. Ranald Mackenzie attacked his village along Crazy Woman Fork in November 1876, and he fought the soldiers, allowing the noncombatants to escape. After his surrender, he became a noted shaman after he settled in Oklahoma Territory. He later did a series of drawings on his life exploits and died in 1910.[80]

Appendix 1

Field and Map Research for the August 4, 1873 Yellowstone Fight

To date, no one had discovered the exact location of Custer's ambush by warriors near the Yellowstone River, on August 4, 1873. Since Custer's 80 men carried 100 rounds each for their Sharps carbines and expended most of it, there should be close to 7,000-8,000 casings and as many bullets. In 2010, I interviewed three knowledgeable local authorities on the fight. They lived in the vicinity of Miles City and had metal-detected on Fort Keogh property and on adjacent privately-owned land. Two of them shared maps and allowed me to photograph their collections. None of them had located the site. Sanctioned surveys by Vance Haynes, Jason Pitsch, and the author did not find the site either.

Fort Keogh relics discovered near the Yellowstone River (dated in the late 1870s) seem to indicate that flooding had not washed the relics from this period away. Maps made by the military and the railroad survey party indicate the battle and campsite were below what is called the 'big hill,' west of Miles City. The route of the wagon train notes that it descended the bluffs next to the Yellowstone River, and according to Custer, he rode to a small cluster of cottonwoods two miles ahead where the fight began.

One theory that has gained popularity was that Custer was attacked under the 'big hill' by warriors hiding in a nearby river loop that is now dry and would offer concealment. An 1877 map created by Edward Maguire shows that the river flowed through the loop, making it impossible for warriors to hide there. The crude 1873 military map, made during the expedition, also indicates water in this loop.[1] The fight site is immediately to the west of this river loop and noted as a set of crossed sabers on the map. [See Fig. 33.]

An overlay of the military map with a current Google aerial map shows that the area where Custer concealed his horses and fought is now in the Yellowstone River and that the water has moved northward, washing away the soldier fight site. This would account for no one finding the numerous casings ejected by Custer's men. Bullets may remain in the terrain some distance from the river, but unfortunately, this area had been heavily altered by farming and agriculture for decades. The author has been unable to metal detect in the field due to crops in the area. A ravine that could hide several hundred warriors exists to the west of this area.

Other evidence of a site farther west of the big hill area is noted in a logbook by steamboat Captain A. C. LaBarge, who describes the river in the area of the fight.[2] First Lieutenant Lafayette E. Campbell of the 22nd Infantry, who was on the 1873 campaign, noted in his diary, "This morning we leave camp early and pass over the scene of Genl. Custer's battle with the Indians." The August 4 campsite is clearly mapped both by the military and railroad surveyors below the 'big hill.' This means the fight site was west of the camp and along the route noted in the surveyor map.[3]

Some evidence in terms of bullets should be present in the disturbed farmland, and further research will hopefully help pinpoint the site at a future date.

Appendix 2

Field Research and Archeology for the August 11, 1873 Fight

The second fight in which the Lakota warriors attacked Custer on August 11, 1873 is certain. The camp was located in privately owned land in Pease Bottom. One long-time landowner showed me the only probable crossing areas for the warriors due to the terrain and high bluffs in the area. This contradicts those presented in other published maps. Archeological work by Vance Haynes and others helps to reconstruct the battle and where fighting took place. The artifacts were presented to the Little Bighorn Battlefield, where the author had a chance to study them along with maps made of the surveys. Vance Haynes has promised a long-awaited book on this episode, and the author does not want to infringe on that publication. I have presented only general information here to reinforce battle movements.[1]

One of the most important finds was the fuse casings from one of Stanley's artillery pieces found on a prominent hilltop overlooking the camp. Private Hills noted that the gun was fired from the area below the Braden fight, that, in turn, gives us the area where his detachment fought and where the officer was wounded. Hill's testimony also designates this knoll as Custer's headquarters location before the cavalry charge toward the warriors.

It is also clear that warriors took positions along the bluffs that overlooked the soldier camp and fired into it. Written testimony shows that Yates' Company F was ordered by Custer to clear the area of warriors and to offer support near Braden as well. Yates' company was one of two companies using the experimental bolt-action Ward-Burton carbines. Discovery of this type of cartridge casing shows where this action occurred and matches the oral testimony of Yates' in these areas.

Soldier and warrior casings were found near the river, stretching from the bluffs where Braden fought. This indicates that many of the troops who charged the warriors rode along the bluffs overlooking the river. Indian cartridges in a ravine near the river indicate a crossing used by many of the warriors who attacked from the west. Maps included here [Fig. 35-38] are based on personal field research and artifact discoveries in combination with the historical narratives.

Appendix 3

Field Study and the Little Bighorn Battle Timeline

There is great debate and no consensus on when the Little Bighorn battle began and exactly how long it lasted. Several researchers such as Greg Michno, John Gray, and Frederic Wagner have attempted to do a minute-by-minute assessment of the battle. The time/motion studies by Gray attempt to match locations of various individuals as they interacted, and his time estimates are based on horse-gait speeds; even though he never rode across the field on horseback. Unfortunately, interpreting the battle sequences using this concept, with its lack of flexibility as well as inaccuracy, quickly back these researchers into a corner with major errors occurring. I believe in using *general relative time* estimates, since no one can prove without a doubt what happened during every minute, and to do so is an effort in futility. *General time* is an approximate time estimate and not an attempt to prove something minute-by-minute. *Relative time* takes into account that actions are happening simultaneously and comparing these various events is imperative in establishing chronology as well as timing. By cross-referencing actions/sightings on the battlefield and actually riding a horse across the terrain, one can estimate the timing of various events.

Few researchers have had the opportunity to ride horses across the battlefield and most either walk over the ground or drive along the modern winding battlefield road. Walking or driving across the battlefield gives the researcher an unrealistic sense of time, space and distance—a *slowing down* of troop and warrior movements. I was shocked at the rapidity of movement on the battlefield when riding on horseback for the first time. This author soon realized how quickly the troops and warriors could move from one sector to another, and how time estimates could be even faster when riding at a fast trot or a gallop. The author presented this new timeline in a lecture for the 18th Annual Symposium for the Custer Battlefield & Historical Association on June 25, 2004.[1]

In 1877, General Nelson Miles, in an attempt to condemn the lack of movement by the rest of the 7th Cavalry for not coming to the aid of Custer, timed a soldier riding from Reno Hill to Calhoun Hill. His estimate was 15 minutes. My 2003 video-timed ride on horseback proved he was accurate. This surprises many researchers who have not done proper field research. Variables, including the state of the horses, the changing gaits, and differing terrain make it impossible to give absolutes, still there can be general time estimates created by modern horse rides.

Frederic Wagner did an admirable job of determining that some soldier watches were on St. Paul time (there were no time zones in 1876), and other soldiers may have been using what he called 'sun time' or a guess based on the position of the sun in the sky. These differences seem apparent when looking at the contradictory times given by various soldier participants, with most relying on the watch of George Wallace—the official time recorder.

Wallace has Custer's column moving over the Reno Creek divide around noon and attacking around 3:00 p.m. This is in direct contrast to the warriors, who suggested the battle commenced at the time for mid-day watering of horses at noon or shortly after. Both Martin and Reno testified that the troops crossed the divide earlier, and I believe they are correct. This makes the warrior time estimate of a noon battle more valid. Even after dancing all night, it seems implausible that the warriors were still asleep in their lodges at 3:00 p.m. Wallace recorded this time as the beginning of the battle. Even more revealing is the quick speed reported by soldiers as they advanced down Reno Creek. Since a horse walks at 4 miles an hour and the soldiers had to advance 12 miles from the divide to the Little Bighorn River crossing, it makes no sense that it took Custer three hours to reach the village, or that he slowly walked his horses when he was in 'hurry mode.'

Most researchers believe that the battle lasted approximately two hours, and I agree, but their estimates are based on soldier watches and movements. They have avoided Indian estimations of time passage, because they are normally anecdotal, such as "the battle was over in the time it took a hungry man to eat his lunch" or, "it was over by the time the sun moved from one tipi pole to the next." I have used John Stands in Timber time estimates instead, because they are based on warrior testimony. John Stands in Timber was a tribal historian who spoke English as well as Cheyenne, fluently. He walked the battlefield with his father Wolf Tooth and others who fought in the battle. With his ability to live in both worlds and not to rely on an interpreter, it can be said there could be few opportunities for errors in translation that plagued some interviews. Stands in Timber was able to translate some approximate times in terms of the white man's watch giving us a different approach in determining the length of the battle.

Stands in Timber relayed that the early fight along Nye/Cartwright Ridge was about 30 minutes in length, the wait by troopers west of Custer Hill near the river as around 20 minutes, and the last stand as less than 30 minutes.[2] The author, using soldier testimony, believes that Custer's fight near the river in Medicine Tail Coulee lasted about 20 minutes. Timed rides conducted by the author discovered it would take approximately 20 minutes for Custer to ride from Ford B (Medicine Tail Coulee) to near the fleeing noncombatants and back to the National Cemetery area, as described by Stands in Timber. The accumulative timing is close to two hours.

The commencement of the fight along Nye/Cartwright Ridge took place after Reno and Benteen's command joined forces. Private Jacob Hetler on Sharpshooter Ridge saw one of Custer's battalions riding along the ridge before it began fighting. It was the same movement described by Wolf Tooth, who soon attacked the soldiers. Since Reno's valley fight and retreat probably lasted about 30 minutes, and the fact that Custer's fight near the river began about 10 minutes later than Reno's initial attack, Custer's fight near the river lasted about 20 minutes before the action along Nye/Cartwright commenced. The volleys of gunfire were heard shortly after soldiers arrived on Reno Hill, verifying the fighting along Nye/Cartwright and this timing sequence. The fighting in the vicinity of Medicine Tail Coulee and Nye/Cartwright Ridge took around 45-50 minutes, using John Stands in Timber's information and testimony based on Martin, Hetler, and Wolf Tooth. This would leave almost an hour for the demise of Custer's troops along battle ridge and supports the warrior testimony that the military collapse was quick and deadly.

Appendix 4

Field Research on the Trail of George Custer and John Martin

The trail of George Custer's five companies along the bluffs and to the battlefield where they died, has been the subject of much debate. Accepted by many historians, and at one time by this author as well, was that the troops rode down a cedar-filled ravine, known today as Cedar Coulee, and away from the village. They supposedly turned left into Medicine Tail Coulee and rode to the edge of the river where they attempted a crossing. Such a move would have hidden Custer's troops but taken them some distance away from the village that was, at the same moment, being attacked by Major Reno. If speed was of the essence, then this route makes no sense.

Researching the origins of this accepted route, I found that researcher Walter Camp was responsible for the concept. He toured the battlefield and created several maps with help from Peter Thompson, Curly and Daniel Knipe in 1908-09. Unfortunately, the man who could have easily confirmed Custer's route, John Martin, was never able to go to the battlefield with Camp. Deeper research suggests than Curly was the main proponent of the route, but both he and Knipe left the area before Custer advanced into Medicine Tail Coulee.

Camp, in his letters to Martin, could never convince the messenger to admit that he turned left and descended down Medicine Tail Coulee. Martin refused to accept Camp's route simply because it did not match his memory or route. Martin later showed Benteen where he had received Custer's message—600 yards from the river.

There is a ravine that empties into Medicine Tail Coulee there, but it is not Cedar Coulee. Martin, indeed, never turned into Medicine Tail Coulee. Instead, it is the ravine that is situated immediately behind the line of bluffs that overlooks the river. It is the shortest viable route to the village. The three remaining Crow scouts witnessed Custer's route and shared it with photographer E. S. Curtis who mapped it.[1] [See Fig. 65.] Edward Maguire, who mapped the field immediately after the battle, also draws the same trail.[2] [See Fig. 64.] Benteen, who followed Custer's trail on June 27, also mapped it.[3] The trail matches Martin's verbal testimony. All these maps agree that Custer rode down the first ravine behind the bluffs.

Such a route takes the column over Weir Point, which is the highest ground in the area. Martin, in a rarely published recollection, stated that he saw the entire Indian village about halfway to the river. As the troops came over Weir Point, they would have seen the entire village for the first time, matching Martin's observation. Such a view would not have been possible if he had ridden down Cedar Coulee. Thompson last saw the command going "'out of sight over the hill." In an unpublished letter from Captain Thomas French, he noted that the command went 'over the saddle' to their deaths beyond. There is a small saddle behind Weir Point that matches this description. John Gibbon and Theodore Goldin, who both saw Custer's trail, stated it went over the sharp, high peak. Martin also testified that Custer went to a high hill with his nephew and brother, where the Crow scouts looked down. From his remarks, it seems that Martin was left below in the saddle of Weir Point and that he is not talking about Sharpshooter Ridge.

In 2003, historian John Doerner and I videotaped two re-enactors riding the route from the end of Sharpshooter Ridge down Cedar Coulee, then down to the river via Medicine Tail Coulee. Riding at a fast trot, they reached the river in 11 minutes. Custer had ordered Martin to take the same trail back. If he did, it would have taken 22 minutes total, or a little longer for the return uphill ascent. Reno's men would have been in the timber and not falling back at this point, as witnessed by Martin.

In 2013, I repeated the exercise with friend George Kush present as a witness, but instead, had the re-enactor ride over Weir Point following the trail noted in the maps made by the Crow scouts. The trip from Sharpshooter Ridge to Medicine Tail Coulee and returning to Weir Point took 13 minutes.

This is important due to what Martin saw as he reached the high ground near Weir Point—Reno's skirmish line was falling back toward the timber and Custer was already engaged behind him. Most witnesses agreed that Reno's men fought in skirmish line formation for only 15 minutes. The 13-minute timed ride matches the Weir Point route rather than that of Cedar Coulee.

Another interesting discovery during the field research was the small hill or 'knoll' that Martin described watching the Custer fight for a brief few seconds. Halfway to Weir Point is a small hill with a single tree. Martin described looking into the village and seeing warriors waving blankets in what he believed was an attempt to frighten the soldier's horses. On the return route, it is impossible to see the river crossing or the village. In a 2013 field test, ranger Jerry Jasmer stood on the hill and looked into the valley. I was situated in the village site, took a blanket, and waved it at some nearby horses stampeding them. He could see me clearly. However, I was not near the ford. There are warrior accounts of attempts to herd horses in the village, and I believe that this is what Martin actually witnessed. This hill is the only location where one can see into the village, until reaching Weir Point. Walter Camp must have also been puzzled by this in his field research and designates this hill on one of his maps. I believe he had surmised this was Martin's return route.[4]

Such important field study helps us to prove Custer's route and establish the timing of the two attacks by Custer and Reno. Custer's secondary attack occurred only minutes after Reno's initial assault—probably within 10 minutes of one another. Martin also saw Custer's troops falling back from the river, simultaneously. The premise that if Reno had continued his attack, then Custer could have successfully crossed the river to attack, is false due to the sequences of the events. It also destroys the myth that Custer waited some 30-45 minutes in Medicine Tail Coulee before his attack.

It is clear from the trails noted on the Benteen and Maguire maps that the five companies divided near the river and that one battalion went toward the battlefield, while another rode to Nye/Cartwright Ridge. This matches the warrior accounts and contradicts some theories that Custer divided his forces as he rode down Medicine Tail Coulee—another story created by Curly who was not present.

Appendix 5

Field Research and Sight Tests for Custer Waving His Hat

Upon reaching the heights near Sharpshooter Ridge, Custer saw the village for the first time before Reno had initiated his charge across the valley. He made a brief speech to his nearby troops, where he took off his hat and waved it to them as they sat on their horses on nearby low ground. Curly, White Man Runs Him, Martin, and Thompson witnessed the incident.[1] Some researchers have suggested that Custer was waving his hat to inspire Reno as he charged across the valley floor, but none of the eyewitnesses with Custer mention that. Two soldiers who were with Reno claim to have seen Custer waving his hat from a great distance in the valley floor. Charles De Rudio claimed to have recognized both Custer and Cooke from his position near the river timber—a distance of over one mile. De Rudio claimed that he recognized the officers from their dress and Cooke from his long whiskers—such is an absurdity.[2] Henry Petring claimed that as he galloped across the plain, he and others around him saw Custer waving his hat. This was supposedly about 15 minutes before the De Rudio sighting.[3]

In the summer of 2006, I decided to do sight tests. Rangers Steve Littleton and Tom Smith were situated on the bluffs. One stood on Weir Point and the other along the top of Sharpshooter Ridge. I stood near the timber area where De Rudio supposedly saw Custer wave his hat and recognized them by their dress. My vision is 20/20 and I have excellent long-range sight. Not only could I barely see that humans were standing at the two hills, I certainly could not make out which ranger was which. Equipped with radios, I asked them to wave their hats. It was impossible to see someone wave a hat or recognize the individual from that distance.

In the summer of 2011, Carolyn Bennett helped me with another sight test. We stood near a trailer in the valley that would have likely been the closest route Reno would have taken in his attack and had ranger Tom Smith standing at the very edge of the bluffs. Both positions were much closer than the previous study. Neither Carolyn nor I could see the ranger waving his hat, due to distance.

Another aspect that helps to disprove the hat-waving incident is timing. Martin noted that he and Custer did not see Reno charging across the valley floor when they reached the bluffs. Charles Varnum, who was at the head of Reno's charge, noticed the gray horse company was moving at a trot. Fifteen minutes later De Rudio, who was near the timber retreat. stated that he saw Custer waving his hat on the bluffs. It is clear from Martin's testimony that Custer was already moving and engaged near the river by that time.

It also seems clear that the surviving soldiers, including Martin and Thompson, told their story to other soldiers on the Reno hilltop later, and the soldiers creatively incorporated the hat-waving incident into their own personal testimony. However, timing and site distance tests prove De Rudio and Petring were fabricating their sightings.

Appendix 6

Field and Map Research on the Men Killed in Deep Ravine

Historical records and maps note that at least 28 soldiers were killed in the Deep Ravine. There is no written record that they were removed in the mass grave reburials in 1881. In 1984, archeologists metal detected and did geomorphology work in the ravine but found little.[1] After the study, I found an original map made by Walter Camp who surveyed the battlefield with Daniel Knipe, in 1908. Knipe saw the bodies lying in the ravine and the two men mapped the location using Camp's surveying equipment. They noted the '28 men found here' as being in a straight line around 2,200 feet from the mass grave monument.[2]

The Deep Ravine was described as having walls that were straight and perpendicular to the surrounding higher ground, and during the original burials, soldiers kicked dirt from above in an attempt to cover the bodies. Today, the walls of the Deep Ravine have slumped into the bottom of the formation and would have deeply buried the original floor. A 1937 photograph taken by Charles Kuhlman notes that the walls were indeed more precipitous than today, which seems to agree with the earlier description in 1876.[3] It is clear that the Deep Ravine looks quite different today.

In 1994, the National Park decided to do ground penetrating research that probed the entire ravine and found a strange anomaly located and deeply buried in one location—it is the area noted on the Camp map. A concentration of six warrior casings from a Henry or Winchester (or both) was found directly above this same area, near the bend in Deep Ravine, in 1984. The firing pin impressions on the back of the .44 casings prove they were fired in two separate weapons.[4] It is my belief that there are still some soldiers 'missing in action' and deeply buried at this location.

Appendix 7

Field Research and the Death Site of Mark Kellogg

In 1988, historical interpreter Michael Moore and I began to look for the death site of newspaper correspondent Mark Kellogg. This site had remained uncertain since the battle, due to contradictory accounts of the body location. Colonel John Gibbon discovered Kellogg's corpse on June 29 and provided a detailed written description of the location. Using a Maguire map that noted Gibbon's camp on that day, we were able to find an approximate match of the route he took. Our findings were published in the *Greasy Grass*, Vol. 7, May 1991. The site was northwest of the National Park battlefield boundary.
In the early 1990s, battlefield superintendent Neil Mangum asked me to search for missing markers on the battlefield, using maps and photos, since we suspected that some had been stolen. I located at least two missing stones—one on Finckle/Finley Ridge and another in Horse Holders' Ravine. While searching the photographs of Ken Roahen of Girl Scouts coming up the new tour road in 1958, I spotted what looked like a soldier marker in the ravine to the west of the new road.[1]

I decided to create a mock marker of the same size as well as proportions and place it in the exact location. I photographed the scene and overlaid it on the Roahen image. The images matched perfectly, leading me to believe that another marker was missing in that location. Darrell Linthicum, who played echo taps in this area for memorial services held at the old (now removed) amphitheater before 1958, recalled seeing such a marker in the area. This was confirmed by some of his other relatives. With this new evidence, it seems highly likely that the burial site in the ravine was that of Kellogg. It was a short distance from the site proposed in the first *Greasy Grass* article.

It carefully matched the oral account by Gibbon. A survey of the area by historian John Doerner and myself failed to find any original bricks that were used to steady early markers or artifacts at the exact site. Battle-related artifacts were found nearby, consisting of two army bullets, a sutler's token (likely post battle), and a post-dated 1886 blank casing. The later casing may have been fired by Lt. Baldwin's infantry that re-enacted Custer's skirmish lines on the 10th anniversary, in 1886. This revised information was published in the *Greasy Grass*, Vol. 19, May 2003.

Endnotes

Prologue

1 Justus, *The Saga of Clara H. Blinn at the Battle of the Washita*, 11-13.
2 Broome, *Dog Soldier Justice*, 62; Keim, *Sheridan's Troopers*, 150. Keim believed that Satanta and the Kiowa were part of the group of abductors.
3 Hardorff, *Washita Memories*, 41-43.
4 Justus, *The Saga of Clara H. Blinn at the Battle of the Washita*, 11-20. William Griffinstein was the Fort Cobb trader who attempted to secure the release of Clara Blinn.
5 *Leavenworth Bulletin*, December 22, 1868.
6 Custer, *My Life on the Plains*, (Quaife version), 211; Davis, *A Summer on the Plains*, 307.

Chapter 1: Under the Magnifiying Glass

1 Dustin, *The Custer Tragedy*, 202-203.
2 Kidd, *Recollections of a Cavalryman*, 316.
3 Frost, *The Custer Album*, 17-52.
4 Custer, *My Life on the Plains*, (Quaife), 25-260.
5 Frost, *The Court-Martial of General George Armstrong Custer*, 151.
6 Frost, *General Custer's Libbie*, 169; Custer, Elizabeth B., *Tenting on the Plains*, or *General Custer in Kansas and Texas*, 403.
7 Swanson, G. A. *Custer-His Life and Times*, 35; Utley, *Cavalier in Buckskin: George Armstrong Custer and the Western Military Frontier*, 13-35. This description of George Custer is based on a tintype taken in February 1869, after his Washita campaign, which is likely the way he looked during this campaign.
8 Custer, My Life, (Quaife), 216.
9 Gibson, *The Washita*, 15.
10 *New York Sun*, May 14, 1899.
11 Carroll, *The Benteen-Goldin Letters on Custer and his Last Battle*, 99. Theodore Goldin who survived the Battle of the Little Bighorn made this statement.
12 Custer, *My Life*, (Quaife), 216.
13 One Custer sympathizer was a former New Yorker lawyer with unruly hair and a goatee named Charles Brewster who served with him in the Civil War and had forged a strong loyalty that extended to his current service. (Klokner, *The Officer Corps of Custer's Seventh Cavalry-1866-1876*, 48); A personal correspondence from Brewster to Custer before the court-martial gave him warning of the current standing with regimental officers. Brewster wrote: "Lieutenant [David] Wallingford has expressed to me that if subpoenaed, he should give everything the best coloring for you, that he could. That is also my resolution. There is not another officer here whom you could count on. Neither [Joel] Elliott, [Edward] Meyers [sic], [Samuel] Robins [sic], [Albert] Barnitz, [Frank] Commagere, [Owen] Hale, [Henry] Jackson nor [James] Levey [sic]. Nor [Dr. I.T.] Coates either. [Louis] Hamilton I think is friendly toward you. [Myles] Keogh

is not. [Robert] West is very bitter it seems; and you may expect the worst from him." (Quoted in Frost, *The Court-Martial*, 91).

14 Carroll, *The Benteen-Goldin Letters*, 259.

15 Benteen to D.F. Barry, September 28, 1895, LBBNM collection. Benteen wrote that Custer during their first meeting criticized one of his former commanders Major General J. H. Wilson.

16 Klokner, *The Officer Corps*, 107-108; Frost, *The Court-Martial*, 89. During the Civil War, Captain Robert West served in various units in the artillery, infantry, and cavalry and concluded the war with the highest brevet rank of his career as a colonel for gallantry at the Battle of Five Forks.

17 Custer, *My Life on the Plains* or, *Personal Experiences with Indians*, (Stewart version) 206-210; Klokner, *The Officer Corps*, 60; Broome, *Custer into the West with the Journal and Maps of Lieutenant Henry Jackson*, 39-40; Court-martial records of Privates Robert Hamilton, John Fisher and Charles Mooney, Record Group 153, Box 2363, National Archives. Elliott backed and endorsed several soldiers he thought were unjustly punished by his Custer. Benteen wrote, "Elliott and West, the latter being a Div. [Division] Comdr. [Commander] . . . was a distinguished man, but given at times to hellish periodical [drinking] sprees. Those two had, in underhand ways been 'peppering' Custer, thinking he was not aware of it, but he was." (Carroll, *The Benteen-Goldin Letters*, 252-253).

18 Klokner, *The Officer Corps*, 68.

19 Klokner, *The Officer Corps*, 56.

20 *New York Sun*, May 14, 1899.

21 Klokner, *The Officer Corps*, 50-51.

22 Utley, *Cavalier in Buckskin*, 60-62.

23 Lieber, *Instructions for the Government of Armies of the United States in the Field*, 7-8.

24 Athearn, *Sherman and the Settlement of the West*, 130-131.

25 Lieber, *Instructions for the Government*, 17-18.

26 Keim, *Sheridan's Troopers on the Borders: A Winter Campaign on the Plains*, 103.

27 Utley, *Cavalier in Buckskin*, 61.

28 Mann, *America, Found and Lost*, 32-55.

29 Donovan, *A Terrible Glory: Custer and the Little Bighorn*, 22-23.

30 Hardorff, Washita *Memories*, 54-55. At his death in 1868, Black Kettle was either 56 or 67 years old, depending on which of his friends one believes—broad latitude at best. Known to his people as *Moketaveto*, he was born in the Black Hills around 1800.

31 Hyde, *Life of George Bent*, 156-157. *Homyike*, also known as George Bent, was the mixed-race son of a Cheyenne woman and white trader. He was present in the village and survived the carnage of Sand Creek.

32 Ibid., 162.

33 Liberty, *Cheyenne Memories*, 169-170.

34 Monnett, *Reimagining Transitional Kansas Landscapes: Environment and Violence*, 261-266.

35 Gwynne, *Empire of the Summer Moon*, 224-225.

36 Utley, *Cavalier in Buckskin*, 58-59; Gwynne, *Empire of the Summer Moon*, 230.

37 Barnett, *Touched by Fire*, 154. Edmund Guerrier was born of French and Cheyenne parents and was living with the tribe at the time of the brutal revenge raids in Kansas. He recalled the war parties consisted of warriors from the bands of Little Rock, Black Kettle, Medicine Arrow and Bull Bear with "nearly all the different bands of Cheyenne has some of their young men in the war party which committed the out rages and murders along the Solomon and Saline [Rivers]. Red Nose, and The-man-who-breaks-the marrow bones, *Hoehamoahoe*, were the two leaders in this massacre: the former belonging to the Dog Soldiers, and the later in Black Kettle's band." (Cordry, *True Heart-The Story of Edmund Guerrier*, 14-21).

38 Hardorff, *Washita Memories*, 32. The white settlements fell under the jurisdiction of the Department of the Missouri and depredations were recorded for a four-month period: 78 acts of Indian aggression, including the deaths of 131

civilians and 12 soldiers, the rape of 15 women, the capture of several children, the burning of 12 farms, 2 stage buildings, a ranch, and 3 wagon trains, and the capture of 735 horses, as well as 931 head of stock.

39 Greene, *Washita*, 102; Brill, *Custer, Black Kettle, and the Fight on the Washita*, 128.

40 Keim, *Sheridan's Troopers*, 159-160.

41 Hoig, *The Battle of the Washita*, 93.

42 Harrison, *Monahsetah-The Life of a Custer Captive*, 34. Some writers such, as Peter Harrison, suggested that the separation of Black Kettle's camp was due to the fact that he was a moderate and unwelcome in the larger camp. Black Kettle's recollection of the Sand Creek massacre must have weighed heavily on his mind as he considered camping with the other tribes. Learning just hours earlier that troops were looking for them, it seems more likely that Black Kettle had second thoughts about his separated camp's own security as he considered his wife's admonition to move the next day. Others in Black Kettle's camp planned to move the next day proving they were not unwelcome. Their delay would prove fatal regardless of the reason for their isolation.

43 George Kush Collection, W. W. Cooke to Dr. A. H. Cooke, December 5, 1868.

44 Hardorff, *Washita Memories*, 14-15, 427-431, Letter to the author from Kevin Mohr, December 15, 2016. There is debate among scholars as to how many women Black Kettle married. Hardorff notes four wives: Arapaho Woman, Little Sage Woman, Medicine Woman Later, and Sioux Woman. Others note that he had only one wife.

45 Harrison, *Monahsetah*, 4-42; Hardorff, *Washita Memories*, 44. Little Rock's tipi was described as decorated by many painted figures and trophies of warpath and chase. It was so beautiful that Custer did not burn it, but instead gifted it to Phil Sheridan.

46 Harrison, *Monahsetah*, 21-22. Custer later mistakenly believed that *Mahwissa* was Black Kettle's sister, but George Bent who lived with the Cheyenne stated that he had no sister.

47 Hoig, *The Battle of the Washita*, 93.

48 Hardorff, *Washita Memories*, 311. Bad Bank was also known as High Bank.

49 Powell, *People of the Sacred Mountain*, Vol. 2, 598.

50 Puckett, *History of Oklahoma and Indian Territory and Homeseeker's Guide*, 98.

Chapter 2, Washita: From the Ashes the Phoenix Rises

1 Utley, *Cavalier in Buckskin*, 63-64.

2 Gibson, *The Washita*, 22.

3 Hardorff, *Washita Memories*, 202-203. Clark served with the 6th U. S. Cavalry during the Civil War.

4 Hoig, *The Battle of the Washita*, 77; Keim, *Sheridan's Troopers*, 111. During the 1867 Hancock campaign Milner was demoted when he was found in a drunken stupor leading the troops. Custer enjoyed describing the picturesque scout in his book *My Life on the Plains*, since such personalities made his book more interesting and appealing to Eastern readers.

5 Custer, *Following the Guidon*, 22-23.

6 Ryan, *Ten Years, The Newton (Mass.) Circuit*, May 28, 1909.

7 Custer, *My Life*, (Quaife), 284-286.

8 Klokner, *The Officer Corps*, 65.

9 Klokner, *The Officer Corps*, 67-68.

10 Gibson, *The Washita*, 26.

11 Utley, *Life in Custer's Cavalry*, 215-218; W. W. Cooke letter to Dr. A. H. Cooke, December 5, 1868, George Kush collection. Unbeknownst to the pursuing soldiers, the trail was made by warriors returning to their camps after raiding and killing settlers in the Smoky Hill country, according to Cooke. This is contradicted by some of the villagers, who stated they were looking for war with the Utes. Evidence from the settlements found in the captured village was likely the source for Cooke's conclusions.

12 Cozzens, *Eyewitnesses to the Indian Wars*, 356.

13 Custer, *My Life*, (Quaife), 306.

14 Greene, *Washita*, 100-102.

15 Klokner, *The Officer Corps*, 78. Mathey emigrated from France and participated in numerous engagements as an infantryman during the Civil War.

16 Mills, Charles K., *Harvest of Barren Regrets*, 214.

17 Carroll, John M., *Camp Talk: The Very Private Letters of Frederick W. Benteen of the 7th U.S. Cavalry to his Wife, 1871-1888*, 76-77.

18 Brewster, *Battle of the Washita*, National Tribune, May 18, 1899.

19 Steves, *Benteen from the Tome*, 28. Later a few officers were asked to leave their sabers behind when near the village, making it clear that a few were retained. A mistaken Frederick Benteen later commented on the saber issue by writing, "Not a sabre in the command" suggesting most were left behind. Dogs also accompanied the advance either by habit or accident.

20 Klokner, *The Officer Corps*, 43; Camp IU Notes, *Interview with James Bell*, Folder 1, Box 2. Edward Godfrey recalled that seven wagons loaded with ammunition and forage were cut loose from the main train—one for the headquarters staff and the rest for the remaining companies. Bell, who was in charge of the new train, recalled two wagons and four ambulances. Frederick Benteen was ordered to act as escort for Bell's newly compacted wagon assemblage.

21 Custer, *My Life*, (Quaife), 312-314.

22 Barnard, *Ten Years with Custer-A 7th Cavalryman's Memoirs*, 279.

23 Brewster, *Battle of the Washita, National Tribune*, May 18, 1899.

24 Custer, *My Life*, (Quaife), 318-320.

25 George Kush Collection, W. W. Cooke to Dr. A. H. Cooke, December 5, 1868.

26 Camp BYU Notes, *William C. Stair Interview*, MSS 57, Box 3. Before advancing, some officers, who evidently ignored orders, had to leave their sabers stacked in the snow to avoid the clanking of metal. Custer first muzzled one of his own canines with a cook's apron; other pets were killed to silence them. Finally, Tom and George took a lariat and choked one of their greyhounds to death to silence the animal.

27 Godfrey, *Extract of Narrative Account. The Cavalry Journal*, October 1928.

28 Klokner, *The Officer Corps*. 101,110.

29 Camp IU Notes, *Interview with Ben Clark*, Box 2, *Battle of the Washita*, Box 3.

30 Klokner, *The Officer Corps*, 74. Company E under Myers and I under Charles Brewster attacked from the southwest of the village. Myers, suffering from snow blindness, allowed his company to be commanded by Lieutenant J. M. Johnson. The young lieutenant had been commended for gallantry at Vicksburg and had left the service to enroll at West Point where he graduated shortly before the Washita campaign.

31 Klokner, *The Officer Corps*, 48-49.

32 Klokner, *The Officer Corps*, 45-46. Custer advanced with the largest of the contingencies--Companies A [Hamilton], C [Berry], D [Weir], K [West], the Osage and white scouts, as well as 40 sharpshooters led by Cooke.

33 Greene, *Washita*, 112-113.

34 Custer, *My Life*, (Quaife), 336.

35 Barnard, *Ten Years with Custer*, 95.

36 Custer, *My Life*, (Quaife), 325-329. California Joe predicted an overwhelming victory if they could maintain the element of surprise, but also expressed concerns about running into more warriors than they could handle, fearing the army could be whipped.

37 George Kush Collection, W. W. Cooke to Dr. A. H. Cooke, December 5, 1868.

38 Custer, *My Life*, (Quaife), 330-331; Francis Gibson noted, "The morning star still shone in majestic splendor like a beacon light, as if to warn the silent village with its sleeping 'braves' of approaching danger." (Gibson, *The Washita*, 37).

39 Camp Notes, *Interview with Edward S. Godfrey,* Box 2, BYU; New York Sun, May 14, 1899.
40 Camp Notes, *Interview with Dennis Lynch,* Box 3, BYU.
41 Hardorff, *Washita Memories,* 307.
42 Ediger and Hoffman, *Some Reminiscences of the Battle of the Washita,* 137-138.
43 Ibid.,138-139.
44 Hoig, *The Battle of the Washita,* 128.
45 Reedstrom, *Garryowen-Custer's Seventh Cavalry Regimental Fighting Song,* 2-6.
46 Charles Bates collection, Edward Godfrey to Richard Henry Pratt, May 11, 1921.
47 Liddic and Harbaugh *Camp on Custer,* 37. This observation was made by Sergeant John Henley.
48 Garryowen is today a suburb of Limerick, Ireland, dominated by inexpensive housing complexes and a racetrack. Translated "Owen's Garden," the song lyrics relish in drinking and fighting against the British authorities. "To pay the reck'ning on the nail" refers to a large nail mounted on a stone on which notice of tax payment was impaled. Such a nail still exists in the Limerick City Museum. *Yankee Doodle* was one of Custer's favorite songs during the Civil War and he realized the power of music to inspire men in combat; it "excited the enthusiasm of the entire command to the highest pitch, and made each individual member feel as if he were a host in himself." (Urwin, Custer Victorious, 94).
49 *Kansas City Star,* December 4, 1904.
50 Custer, *My Life,* (Quaife), 335.
51 Langley, *With Custer at the Washita, 1868.* Trooper Henry Langley of Company C made this statement.
52 Puckett, *History of Oklahoma and Indian Territory and Homeseeker's Guide,* 1906.
53 George Kush Collection W. W. Cooke to Dr. A. H. Cooke, December 5, 1868; Hardorff, *Washita Memories,* 160. These accounts of soldiers unable to speak Cheyenne and to hear this over the sounds of chaos and gunfire, make such testimony questionable.
54 Brill, *Custer, Black Kettle, and the Fight on the Washita,* 155-156; *New York Sun,* May 14, 1899; Hardorff, *Washita Memories,* 307; Ediger and Hoffman, *Some Reminiscences,* 140. The youth, Magpie, claims that Black Kettle was attempting to ride away with his wife on horseback. Magpie, in his 1930 interview, did not mention personally witnessing the chief's death.
55 Hyde, *Life of George Bent,* 319; Harrison, *Monahsetah,* 38-39.
56 Graham, *The Custer Myth,* 202.
57 Hyde, *Life of George Bent,* 317.
58 Steves, *Benteen from the Tome,* 29.
59 Hoig, *Battle of the Washita,* 128; *New York Sun,* May 14, 1899.
60 Merington, *The Custer Story-The Life and Intimate Letters of General Custer and his Wife Elizabeth,* 222; Hoig, *Battle of the Washita,* 132; Hyde, *Life of George Bent,* 317.
61 *Eone-Ah-Pah,* the only Kiowa Indian Who Participated in the Battle of the Washita, on November 27, 1869 [sic], Folder I, Box 38, *William Cary Brown collection.*
62 Oklahoma Historical Society, Interview with Mrs. B. K. Young Bird, June 22, 1937.
63 Hardorff, *Washita Memories,* 338.
64 Ediger and Hoffman, *Some Reminiscences of the Battle of the Washita,*139.
65 *Watonga Republican,* June 18, 1997.
66 *Eone-Ah-Pah,* the only Kiowa Indian Who Participated in the Battle of the Washita, on November 27, 1869 [sic].
67 Ediger and Hoffman, *Some Reminiscences of the Battle of the Washita,*139.
68 Brill, Custer, *Black Kettle and the Fight on the Washita,* 161-162.
69 *New York Sun,* May 14, 1899.

70 Utley, *Life in Custer's Cavalry: Diaries and Letters of Albert and Jennie Barnitz*, 1867-1868, 226-227.
71 Camp Notes, Interview with Godfrey, Box 2, BYU.
72 Hammer, *Men With Custer: Biographies of the 7th Cavalry*, 304-305; Barnard, *Ten Years with Custer*, 77-78.
73 Custer, *Following the Guidon*, 42.
74 Keim, *Sheridan's Troopers*, 117.
75 *New York Sun*, May 14, 1899.
76 Ibid.
77 George Kush Collection, W. W. Cooke to Dr. A. H. Cooke, December 5, 1868.
78 *New York Sun*, May 14, 1899.
79 Camp Notes, Interview with Godfrey, Box 3, BYU.
80 Ediger and Hoffman, *Some Reminiscences*, 139-141.
81 *Kansas City Star*, December 4, 1904.
82 Camp Notes, Interview with Dennis Lynch, *Box 2, BYU*.
83 George Kush Collection, W. W. Cooke to Dr. A. H. Cooke, December 5, 1868.
84 Godfrey, *Battle of the Washita, November 27, 1868*, Collection S-1314, Ser. 3, Bk. 16, Folder 237, *Charles F. Bates collection*, Beinecke Library, New Haven, Conn.
85 Hyde, *Life of George Bent*, 320; Cozzens, *Eyewitnesses to the Indian Wars*, 398-399; Harrison, *Monahsetah*, 39-41. A warrior named Bob-Tail-Bear rode up and tomahawked the Sergeant Major Walter Kennedy. Little Chief is the name given in Harrison's account.
86 Keim, *Sheridan's Troopers*, 146.
87 Hyde, *Life of George Bent* pg. 320.
88 Godfrey, *The Cavalry Journal*, October 1928. The camp circles below contained a young girl about 10 years old named Wolf Belly Woman who explained the arrangement of the village observed by Godfrey, "There was an Arapaho camp [nearest to that of Black Kettle], then another Cheyenne camp, then my father's [Cheyenne Whirlwind], then a Kiowa camp." (Ediger and Hoffman, *Some Reminiscences*, 141).
89 Godfrey, *The Cavalry Journal*, October 1928.
90 Custer, *My Life*, (Quaife), 344.
91 Hardorff, *Washita Memories*, 23, 320. James Bell arrived in the village around 8 a.m.
92 Custer, *My Life*, (Quaife), 347.
93 *New York Sun*, May 14, 1899.
94 *Kansas City Star*, December 4, 1904.
95 Carroll, *General Custer and the Battle of the Washita*, 38; Custer, *My Life on the Plains*, (Quaife), 348-349. Custer's official report found here in Carroll's notation lists among the items captured: 241 saddles, 523 buffalo robes, 210 axes, 140 hatchets, 35 revolvers, 47 rifles, 535 pounds of powder, 1,500 pounds of lead, 4,000 arrows, 90 bullet molds, 35 bows and quivers, 12 shields, 300 pounds of bullets, 775 lariats, 940 buckskin saddle bags, 470 blankets, 93 coats, 700 pounds of tobacco, as well as dried buffalo meat, meal, flour, and other provisions.
96 Carroll, *General Custer and the Battle of the Washita*, 60.
97 Barnard, *Ten Years With Custer*, 80.
98 George Kush Collection, W. W. Cooke to Dr. A. H. Cooke, December 5, 1868.
99 Custer, *My Life*, (Quaife), 356-362; Harrison, *Monahsetah*, 48-50. Some have suggested that the young girl offered to Custer was Monahsetah, but that is unlikely since there was a slight age difference and he did not tie her to the incident. The rationale for such a ceremony would have been to secure safety from a family member in the minds of the Cheyenne captives believing such a marriage would protect them. It is possible the young girl was 16-year-old Mrs. Lone Wolf, who wrote of being part of the interrogation process. (Puckett, *History of Oklahoma and Indian Territory and Homeseeker's Guide*, 1906).

100 Puckett, *History of Oklahoma and Indian Territory and Homeseeker's Guide*, 1906.
101 Camp Notes, Interview with James Bell, Box 2, I.U.
102 Barnard, *Ten Years with Custer*, 79.
103 Godfrey, *The Cavalry Journal*, October 1928. Hardorff, *Washita Memories*, 25.
104 Steves, *Benteen from the Tome*, 29-30.
105 *Kansas City Star*, October 22, 1910.
106 Gwynne, *Empire of the Summer Moon*, 240.
107 Statement by Sarah C. Brooks, July 27, 1935, Joseph G. Master collection.
108 Ediger and Hoffman, *Some Reminiscences*, 139.
109 Puckett, *History of Oklahoma and Indian Territory and Homeseeker's Guide*, 1906.
110 Godfrey, *The Cavalry Journal*, October 1928.
111 Custer, *My Life*, (Quaife), 351.
112 Ibid., 352.
113 Ibid., 372.
114 *New York Sun*, May 14, 1899.
115 Camp Notes, Interview with Godfrey, March 3, 1917, BYU.
116 Hardorff, *Washita Memories*, 308-309.
117 Ediger and Hoffman, *Some Reminiscences*, 140.
118 Custer, *My Life*, (Quaife), 372-373.
119 Keim, *Sheridan's Troopers*, 119-120.
120 Ediger and Hoffman, *Some Reminiscences*, 140-141.
121 Custer, *My Life*, (Quaife), 374-375.
122 Custer to Phil Sheridan, Field letter on the Washita River, November 28, 1868, LBHNM-00019-02622.
123 Ibid., 392-393.
124 Hardorff, *Washita Memories*, 393-395.
125 Langley, *With Custer at the Washita, 1868, Winners of the West*, December 15, 1925.
126 George Kush Collection, W.W. Cooke to Dr. A. H. Cooke, December 5, 1868.
127 Camp Notes, Interviews with Dennis Lynch, October 1908, February 8, 1909, Box 2, BYU.
128 Camp Notes, Interview with Edward Godfrey, March 3, 1917, Box 3, BYU.
129 *New York Sun*, May 14, 1899.
130 Greene, *Washita*, 215-216.
131 Benteen, Annotations in his personal copy of W. L. Holloway, *Wild Life on the Plains and Horrors of Indian Warfare*, no date, as quoted in Hardorff, *Washita Memories*, 78-79. George Bent, who gathered information from the Cheyenne afterward, also believed the number of casualties was inflated. See Appendix C in Greene's *Washita*, 212-214 for his contradictory casualty numbers and the sources for this information. His list includes 40 men, 12 women, and 6 children. Two mixed-blood scouts, Jack Fitzpatrick and John Poisal with the 7th, stated that Custer's casualty tally was "very much exaggerated" and believed not over 20 warriors were killed. Quoted in Hardorff, *Washita Memories*, footnote, 78-79.
132 Custer's official report, November 28, 1868 as quoted in Carroll, *General Custer and the Battle of the Washita*, 38.
133 Burkey, *Custer, Come at Once! The Fort Hays Years of George and Elizabeth Custer 1867-1870*, 68-73. The young captives eventually ended up at the wooden stockade of Fort Hays. In April of the following year, General Nelson Miles gives tantalizing clues as to the disposition of these captives who may have been in the post hospital. He asked that the larger of the two Indian boys in his care be returned to the stockade, while the smaller one be placed in the care of soldier. Miles described the smaller fair-skinned boy as being about two years of age. He stated, "Said child is, in my opinion, the son of white parents. The color of his hair is light, his features and complexion indicate strongly that

he belongs to the white race. The Indians assert that he is quarter-breed, but from their cruel treatment of him, I judge he must have been one of their captives or the child of some settler. His health is much impaired owing to his improper treatment. His history is and must be a mystery. While he remained with the Indians he was placed in the most exposed part of their quarters and his food and clothing taken from him and thrown away. I would earnestly recommend that in view of these facts he be sent to a government hospital or asylum, where he can receive proper care . . . If he is compelled to remain with the savages he will undoubtedly die from exposure and ill treatment." Ibid., pg. 68. The child was eventually sent to a group of Catholic nuns at Leavenworth, Kansas.

134 Hoig, *Battle of the Washita*, 188-189.

135 Camp notes, Box 5, BYU.

136 House of Representatives Executive Document No. 240, 41st Congress, 2nd Session, in Carroll, *General Custer and the Battle of the Washita*, 77-78.

137 *New York Sun*, May 14, 1899.

138 Graham, *The Custer Myth*, 202, 206; Steves, *Benteen from the Tome*, 30.

139 Hofling, *Custer and the Little Big Horn-A Psychobiographical Inquiry*, 72. In the words of Charles Hofling, "Psychologically speaking, the Washita expedition had two aspects for Custer. The first was that it greatly enhanced his fame and that of his regiment. The second, slower to develop, was that it produced some adverse public criticism and heightened tensions within the regiment." Ibid.

Chapter 3, Hostages, Threats, and Different Rules of Conduct

1 *New York Tribune*, December 3, 1868.

2 Custer to Philip Sheridan, Field letter on the Washita River, November 28, 1868, LBHNM-00019-02622.

3 Custer, *My Life*, (Quaife) 414; Harrison, Monahsetah, 68-71.

4 Custer, *My Life*, (Quaife), 414.

5 Ibid., 415-416.

6 Burkey, Custer, *Come at Once*, 72-73.

7 Burkey, Custer, *Come at Once*, 66-67; Liberty, *A Cheyenne Voice*, 245; Marquis, *Custer on the Little Bighorn*, 35.

8 Langellier, Cox, and Pohanka, *Myles Keogh-The Life and Legend of an "Irish Dragoon" in the Seventh Cavalry*, 114; Camp Notes, Interview with Clark, Box 2, Folder 4, I.U. Equally intriguing is a statement in a letter from Keogh who missed the Washita conflict but was with the officers after the battle—"We have knocked the Indians into a cocked hat . . . We now have about 90 squaws from our last fights. Some of them are very pretty. I have one that is quite intelligent. It is usual for officers to have two or three hanging around."(Langellier, Cox, Pohanka, *Myles Keogh*, 114). Asking researcher Walter Camp not to quote him, Clark revealed that some of the officers had sexual relations with the captives. Interpreter 'Romeo' Romero was in charge of the hostages and supposedly sent women to the officers' tents every night. Clark said that Custer picked out a fine looking one and had her in his tent every night with old Mrs. __________ for a guise. While the name of the 'guise' is omitted in the text, it could only be referring to Mrs. Courtenay, the Irish cook that was the only non-Indian female noted on the campaign. Such commentary from Benteen, Keogh, Clark, Cheyenne oral history, as well as the later reported sadness of Monahsetah learning of Custer's death, seem to point to possible impropriety by some of the officers.

9 *New York Herald*, January 4, 1869.

10 Spotts, *Campaigning with Custer*, 75; Keim, *Sheridan's Troopers*, 143-144.

11 Camp Notes, Interview with Dennis Lynch, October 1908 & February 8, 1909, BYU.

12 *New York Herald*, January 4, 1869.

13 Custer, *My Life*, (Quaife) 425-426.

14 *The New York Herald*, January 4, 1869. The 'philanthropic rifles' were weapons given to warriors as part of treaty negotiations and to help them hunt game.

15 Cozzens, *Eyewitness to the Indian Wars*, 272; Hardorff, *Washita Memories*, 235; Justis, *The Saga of Clara H. Blinn at the Battle of the Washita*, 17. Ben Clark later heard that an Indian woman had killed Willie Blinn, because he got in the way of their escape. Mrs. Blinn was killed when she refused to leave her son's body. According to U. S. Indian agent A. G. Boone in a letter to the Harrington family, he noted that a dead squaw was close to Mrs. Blinn. The Indian woman's death remains a mystery.

16 Justus, *The Saga of Clara H. Blinn at the Battle of the Washita*, 18. There is debate and uncertainty in which camp circle that Clara and Willie were discovered. While Ben Clark believed they were found in the Arapaho camp, others believed their bodies were in the Cheyenne or Kiowa circles. Mrs. Blinn and Willie were moved to Fort Gibson when Fort Arbuckle was closed. Today they are buried beneath markers that bear the sad inscription "Unknown" and "Unknown Woman."

17 Merington, *The Custer Story*, 218.

18 Barnett, *Touched by Fire*, 183.

19 *New York Times*, April 7, 1869.

20 Hadley, *The 19th Kansas Cavalry and the Conquest of the Plains Indians*, 428-456.

21 Charles F. Bates Collection, Godfrey letter, no date, Call number S-1314, Box 7, Folder 85, Beinecke Library, Yale University.

22 D. F. Barry Collection, Benteen to D. F. Barry, May 12, 1897, LBBNM.

23 Mills, *Harvest of Barren Regrets*, 11-18.

24 Ibid., 19.

25 Ibid., 39-45.

26 Carroll, *Camp Talk*, xx-xxi.

27 Mills, *Harvest of Barren Regrets*, 212.

28 Ibid., 35.

29 Ladenheim, *Custer's Thorn-The Life of Frederick W. Benteen*, 31.

30 Mills, *Harvest of Barren Regrets*, 114.

31 Ibid., 62-120.

32 Ibid., 177.

33 Ibid., 166.

34 Donovan, *Brazen Trumpet-Frederick W. Benteen and the Battle of the Little Big Horn*, 25-26.

35 Mills, *Harvest of Barren Regrets*, 178.

36 Graham, *The Custer Myth*, 206.

37 Ibid., 208.

38 Ibid.

39 D. F. Barry Collection, Benteen to Barry, August. 29, 1895, March 29, 1897, LBHNM.

40 Klokner, *The Officer Corps*, 49; Hammer, Men With Custer, 50.

41 Klokner, *The Officer Corps*, 83-84.

42 Hammer, *Men With Custer*, 384.

43 Klokner, *The Officer Corps*, 110; *Hammer, Men With Custer*, 437: *Donovan, A Terrible Glory*, 61.

44 Carroll, *The Benteen-Goldin Letters*, 206.

45 Spotts, *Campaigning with Custer*, 77.

46 Robinson III, *Satanta*, xv. This scene was made famous by artist Charles Schreyvogel's oil painting appropriately titled, *Custer's Demand*.

47 Keim, *Sheridan's Troopers*, 156.
48 Spotts, *Campaigning with Custer*, 81.
49 Robinson III, *Santana*, 1-9.
50 Whittaker, *A Life of Major General George A. Custer*, 457.
51 Custer, *My Life*, (Quaife) 440-442.
52 Custer, *My Life*, (Quaife) 442; Slotkin, *The Fatal Environment-The Myth of the Frontier in the Age of Industrialization 1800-1890*, 392. Hancock had suggested such a tactic along Pawnee Fork when Custer was ordered to sweep down on the village and seize the women and children hostage as a show of good faith. To Custer, this Kiowa escape seemed like act II of the same production.
53 Merington, *The Custer Story*, 225; Barnard, *Ten Years with Custer*, 86; Robinson III, *Satanta*, 93.
54 Barnard, *Ten Years with Custer*, 90.
55 Spotts, *Campaigning with Custer*, 80.
56 Cozzens, *Eyewitnesses to the Indians War*, 274.
57 Custer, *My Life*, (Quaife), 453.
58 Custer, *My Life*, (Quaife), 454-461.: Hutton, *Phil Sheridan and His Army*, 89.
59 Spotts, *Campaigning with Custer*, 95.
60 Custer, *My Life*, (Quaife), 469-476.
61 Spotts, *Campaigning with Custer*, 103; Custer, *My Life on the Plains*, (Quaife), 473-474.
62 Whittaker, *A Life of Major General George A. Custer*, 461.
63 Custer, *My Life*, (Quaife), 497.
64 Keim, *Sheridan's Troopers*, 273.
65 Spotts, *Campaigning with Custer*, 118.
66 Custer, *My Life*, (Quaife), 543.
67 Broome, Jeff, *Dog Soldier Justice*, 52-55.
68 Spotts, *Campaigning with Custer*, 140.
69 *The New York Times*, April 7, 1869; Hadley, *The 19th Kansas Cavalry and the Conquest of the Plains Indians*, 428-456.
70 Custer, *My Life*, (Quaife), 550-551; Hadley, *The 19th Kansas Cavalry and the Conquest of the Plains Indians*, 428-456.
71 Spotts, *Campaigning with Custer*, 152-154; Cozzens, *Eyewitnesses to the Indian War*, 300.
72 Barnard, *Custer's First Sergeant John Ryan*, 124.
73 Custer, *My Life*, (Quaife), 554-555.
74 Cozzens, *Eyewitnesses to the Indian War*, 319; Custer, *My Life on the Plains*, (Quaife), 555.
75 Mills, *Harvest of Barren Regrets*, 185-186.
76 Custer, *My Life*, (Quaife), 556-560; Brill, *Custer, Black Kettle*, 223-226. Custer did not mention the ash incident in his own narrative, probably dismissing or not understanding the warning, which many Cheyenne today considered a fulfilled prophecy.
77 *The New York Times*, April 7, 1869.
78 Ibid.
79 Custer, *My Life*, (Quaife), 567.
80 Barnard, *Custer's First Sergeant*, 125.
81 Hadley, *The 19th Kansas Cavalry and the Conquest of the Plains Indians*, 428-456.
82 Cozzens, *Eyewitnesses to the Indian War*, 276-277.
83 Custer, *My Life*, (Quaife), 568-573.
84 Hyde, *The Life of George Bent*, 326; Barnard, *Ten Years with Custer*, 91; Hadley, *The 19th Kansas Cavalry and the Conquest of the Plains Indians*, 428-456; Custer, *My Life* on the Plains, (Quaife), 578; Spotts, *Campaigning with Custer*,

154. Ryan named the four chiefs as Bull Head, Dull Knife, Cut-Em-Up, and One Stab. David Spotts and Hadley included Medicine Arrow as a hostage. It seems that from the process of deduction, and a photograph of the three chiefs that remained, that Medicine Arrow was the hostage messenger.

85 Hyde, *The Life of George Bent*, 326.

86 Hadley, *The 19th Kansas Cavalry and the Conquest of the Plains Indians*, 428-456; *Ten Years with Custer*, 92-93.

87 Spotts, *Campaigning with Custer*, 158.

88 Cozzens, *Eyewitnesses to the Indian War*, 321; Mills, *Harvest of Barren Regrets*, 187; Spotts, *Campaigning with Custer*, 159; E. B. Custer Collection, Microfilm roll 6, March 24, 1869; Barnard, *Custer's First Sergeant*, 125; Hadley, *The 19th Kansas Cavalry and the Conquest of the Plains Indians*, 428-456.

89 Custer, *My Life*, (Quaife), 590-591; *The New York Times*, April 7, 1869; Spotts, *Campaigning with Custer*, 157-160.

90 Custer, *My Life*, (Quaife), 595-596; *The New York Times*, April 7, 1869.

91 Hadley, *The 19th Kansas Cavalry and the Conquest of the Plains Indians*, 428-456.

92 Burkey, *Custer, Come at Once*, 68-71.

93 Merington, *The Custer Story*, pg. 228.

94 Ibid.

95 Mills, *Harvest of Barren Regrets*, 187-189.

96 Custer, *The Black Hills Expedition: Dispatches and Letters*, Army and Navy Journal 12, no. 2 (August 22, 1874), 26-27; Grafe, & Horsted, *Exploring with Custer*, 54.

97 Grafe & Horsted, *Exploring with Custer*, 55.

98 Liddic, *The Book That Never Was-Lemmon's Developing the West and Mary Crawler's Battle of the Little Big Horn Account, Research Review*, 16.

99 Lieber, *Instructions for the Government of Armies of the United States in the Field*, 16-17. Section III, Rule 54 is quoted here.

100 Encyclopaedia Britannica, Eleventh Edition, 801-802; Camp Notes, Field notes of Hugh Lenox Scott material, 186, I.U. General Hugh Lenox Scott served with the U.S. cavalry and was sympathetic to the Indians. Researcher Camp recalled that Scott regarded Custer's keeping of the hostages (who were held for the bringing in of the white women captives) as an outrage on the part of Custer—a piece of bad faith. However, in a contradictory observation, Scott proclaimed that Custer had the right ideas of Indian fighting and that Indians could not be maneuvered by 'threatening their flank' as in civilized warfare—they had to be grabbed, as at the Washita, and as Nelson Miles afterward did. (Camp Notes, Field notes of Hugh Lenox Scott material, 186, I.U.)

101 Hoig, *The Battle of the Washita*, pg. 86; Barnett, *Touched by Fire*, 172-182.

102 Barnett, *Touched by Fire*, 181.

103 Ibid., 173, 179. Many men of the period thought that such a 'damaged woman' was better off dead, while the victims themselves had an opposing view.

104 Even Hollywood, which has taken almost every conceivable stance on Custer, has avoided the theme. Only two films deal with Custer and hostage taking: *"Stolen Women, Captured Hearts"* (1997) and *"Son of the Morning Star"* (1991). In the first movie, historical accuracy has been shelved in order to create a romance story between the white women and their Cheyenne captors. One of the women is converted to the wonderful life of the plains Indians in a scenario comparable to Lieutenant Dunbar in *"Dances with Wolves."* Both the captured women and Dunbar realize that the life of the 'noble red man' is preferred to the life of the corrupt and uncaring white society. In *"Son of the Morning Star"* Custer is shown as being a treacherous scoundrel and an officer of dubious character, when he refuses to free his captive chiefs upon the release of the captive white women. Both films use the event to twist historical record and villainize whites, while playing the Cheyenne as innocents. The author was an 'extra' in the second miniseries playing one of Custer's trumpeters in several scenes. Like most propaganda pieces, they are simplistic at best, and do not deal with the complexity of the people or the situations necessary to accurately portray the story.

105 Dippie, *Nomad-George A. Custer in Turf, Field, and Farm*, 20.
106 Urwin, *Custer Victorious*, 174-175.

Chapter 4, Decoy along the Yellowstone

1 Utley, *Cavalier in Buckskin*, 109-111.
2 Hammer, *Men With Custer*, 291; Nichols, *In Custer's Shadow*: Major Marcus Reno, 18-21.
3 Fougera Collection, Gibson to Yates, April 28, 1915, LBBNM.
4 Libby, *The Arikara Narrative*, 74.
5 Nichols, *In Custer's Shadow*, 25-38, 56-59.
6 Hammer, *Men With Custer*, 291.
7 Nichols, *Reno Court of Inquiry*, 585.
8 Mills, *Harvest of Barren Regrets*, 194-203.
9 Ibid., 204-205.
10 Frost, Lawrence A., *Some Observations on the Yellowstone Expedition of 1873*, 17.
11 Lubetkin, *Jay Cooke's Gamble*, 3-25.
12 Frost, *Some Observations*,16; Lubetkin, *Custer and the 1873 Yellowstone Survey*, 23-29.
13 Lubetkin, *Custer and the 1873 Yellowstone Survey*, 45.
14 Ibid., 47-48.
15 Slotkin, *The Fatal Environment*, 412.
16 Stanley, *Personal Memoirs of Major-General D. S. Stanley, U.S.A.*, 60-226.
17 Frost, *Custer's 7th Cav & the Campaign of 73'*, 62.
18 Kirshner, *The Class of 1861-Custer, Ames and their Classmates after West Point*, 5, 66-86. Rosser was wounded an incredible nine times and survived the bloody carnage; although, he numbed himself with occasional bouts with the bottle throughout the Civil War.
19 Klokner, *The Officer Corps*, 46-47.
20 Burdick, *The Army Life of Charles "Chip" Creighton*, 5-12.
21 Hills, *With General George A. Custer on the Northern Pacific Surveying Expedition in 1873*, 143-160.
22 Stanley, *Personal Memoirs*, 244-245; Charles Larned Collection *Yellowstone Expedition*, September 21, 1873, West Point Military Academy.
23 Mills, *Harvest of Barren Regrets*, 209; Frost, *Custer's 7th Cav & the Campaign of 73'*, 23. The Englishmen were the Honorable Louis Clifford, son of Lord Clifford, and Louis Molesworth, son of Sir Paul Molesworth, as well as a London University colleague, Mr. R. G. Frost of St. Louis.
24 *New York Tribune*, September 9, 1873.
25 *New York Tribune*, July 25, 1873.
26 *Sioux City Journal*, July 18, 1873.
27 Frost, *Some Observations*, 45;Lubetkin, *Cooke's Gamble*, 192; Merington, *The Custer Story*, 253.
28 Custer, *Boots and Saddles*, 226.
29 Klokner, *The Officer Corps*; 70; Larned, *Expedition to the Yellowstone in 1873: Letters of a Young Cavalry Officer*, 191; *The New York Tribune*, Sept. 6, 1873; Johnson and Taunton, *More Sidelights of the Sioux Wars*, 70. Hart's wing consisted of his own Company C along with Myles Moylan [Company A], Tom Custer [Company B], Donald McIntosh [Company G] and Thomas French [Company M]. The second wing was commanded by Benteen and included his Company H as well as George Yates [Company F], Owen Hale [Company K], John Weston [Company L], and Thomas McDougall [Company E]. Companies D and I were absent from this duty, escorting a surveying party to the

far north. Major Marcus Reno was charged with the protection of the Neschim Boundary Commission as they at tempted to survey the exact location of the U.S./Canadian boundary.

30 Larned diary, July 28, 1873.

31 Braden, *The Yellowstone Expedition*, 218-241.

32 Stanley, *Personal Memoirs*, 239.

33 Merington, *The Custer Story*, 251-252.

34 Frost, *Some Observations*, 25, 48.

35 Stanley, *Personal Memoirs*, 240.

36 Barnard, *Ten Years*, 195; Captain William Ludlow, who accompanied the expedition, later noted, "He [Custer] liked to act independently of others and take all the risk and all the glory to himself. He frequently got himself in trouble by assuming more authority than really belonged to his rank. It was on the Yellowstone expedition, where he came into collision with Stanley . . . and perhaps the quarrel would have never occurred if the two generals had been left alone to themselves without the intervention of camp gossips, who sought to foster the traditional jealousy between the infantry and cavalry." (Dustin, *The Custer Tragedy*, 202).

37 Frost, *Some Observations*, 49.

38 Carroll, *Benteen-Goldin Letters*, 240.

39 Frost, *Some Observations*, 32.

40 Barnett, *Touched by Fire*, 244.

41 Barnett, *Touched by Fire*, 244-245; Mills, *Harvest of Barren Regrets*, 208-209.

42 Frost, *Some Observations*, 33.

43 Barnard, *Ten Years*, 193.

44 Frost, *Custer's 7th Cav & the Campaign of '73*, 57; Merington, *The Custer Story*, 250.

45 Frost, *Custer's 7th Cav.*, 63.

46 Lubetkin, *Custer and the 1873 Yellowstone Survey*, 194-197, 202; Merington, *The Custer Story* ,258. Self-taught engineer Albert O. Eckelson took over his duties as chief engineer. On their return to Fort Buford the small party of eight soldiers including Grant and a correspondent were almost capsized by a sudden storm but made it safely.

47 Barnard, *Ten Years*, 195; Carroll, The *Benteen-Goldin Letters*, 256.

48 Stanley, *Personal Memoirs*, 248.

49 Liddic & Harbaugh, *Camp on Custer*, 35.

50 Graham, *The Custer Myth*, 247.

51 Mills, *Harvest of Barren Regrets*, 210.

52 Carroll, *Camp Talk*, 7.

53 Ibid., xi.

54 Merington, *The Custer Story*, 257.

55 Stanley, *Personal Memoirs*, 241; Custer, *Boots and Saddles*, 232.

56 Barnard, *Ten Years*, 195.

57 Frost, *Some Observations*, 58-59; Liddic & Harbaugh, *Camp on Custer*, 40.

58 Innis, *Bloody Knife: Custer's Favorite Scout*, 19-27, 46-47. Hunkpapa warrior Gall was particularly brutal to him and an intense hatred developed between the two. Bloody Knife helped arrest Gall in the winter of 1865-66 and when the Lakota warrior was bayoneted, Bloody Knife stepped up to finish him off aiming a rifle at Gall's head. Stopped by an army officer, Gall, left for dead, recovered to fight at Little Bighorn.

59 Frost, *Custer's 7th Cav*, xi, 75-79; *New York Tribune*, September 8, 1873.

60 Vestal, *Warpath-The True Story of the Fighting Sioux Told in a Biography of Chief White Bull*, 3-7; Campbell Collection, Interview with White Bull, Box 105, Folder 24, pg. 22, University of Oklahoma Western History Collection.

61 Utley, *The Lance and the Shield*, 3-13. This book is the most authoritative work on the life of the great Lakota leader.

While Sitting Bull's exact year of birth is a mystery, he was approximately in his mid to late 40s at the time of the Little Bighorn fight. Ernie LaPointe, the great grandson of the spiritual leader, was told by family members that Sitting Bull was born on the Yellowstone River. His biography based on family oral history is *Sitting Bull—His Life and Legacy*. Some historians suggest his birthplace was on the Grand River. Sitting Bull stated he was born on the Missouri River.

62 Utley, *Lance and Shield*, 11-26.

63 Klokner, *The Officer Corps*, 104-105.

64 Hutton, *Custer Reader*, Custer, *Battling the Sioux*, 206.

65 *New York Tribune*, September 8, 1873.

66 Frost, *Some Observations*, 60.

67 Barnard, *Ten Years*, 197; Carroll, *Custer's Chief of Scouts*, 44-45.

68 Hutton, *Custer Reader*, 207-208.

69 Ibid., 208.

70 Frost, *Custer's 7th Cav.*, 66; Frost, *Some Observations*, 61.

71 Frost, *Some Observations*, 61.

72 *New York Tribune*, September 8, 1873 ; Custer, *Boots and Saddles*, 238. Custer remembered, "Among the Indians who fought us on this occasion were some of the identical warriors who committed the massacre at Fort Phil Kearney, and they no doubt intended a similar programme when they sent the six warriors to dash up and attempt to decoy us into a pursuit past the timber in which the savages hoped to ambush us. Had we pursued the six warriors half a mile further, instead of halting, the entire band of warriors would have been in our rear, and all the advantage of position and numbers would have been with them." (Custer, *Boots and Saddles*, 240).

73 Carroll, *Custer's Chief of Scouts*, 45.

74 Lubetkin, *Clash on the Yellowstone*, 18. Custer's estimate was larger, stating, "I beheld bursting from their concealment between three and four hundred Sioux warriors mounted and caparisoned with all the flaming adornments of paint and feathers which go to make up the Indian war costume . . . they moved in perfect line, and with as seeming good order and alignment as the best drilled cavalry." (Hutton, *Custer Reader*, 209). Calhoun believed there were over 300 warriors. (Frost, *Some Observations*, 61)

75 Campbell Collection, Interview with White Bull, 22, U.O.; Lubetkin, *Custer and the 1873 Yellowstone Survey*, 311; Hutton, *Custer Reader*, 209-210.

76 Hutton, *Custer Reader*, 210.

77 Carroll, *Custer's Chief of Scouts*, 45. Every fourth trooper stayed mounted while the other three men took the small leather link strap hanging from their horse's headstall and snapped it to that of the horseholder's mount. After handing the reins of the three horses to the single mounted soldier, the other troopers quickly formed a skirmish line.

78 Hutton, *Custer Reader*, Custer, 213; *New York Tribune*, September 8, 1873. As the soldiers retreated, several of their horses along with Private J. R. Crow were wounded in the withdrawal.

79 Carroll, *Custer's Chief of Scouts*, 45-46.

80 Hutton, *Custer Reader*, 213.

81 Custer, *Boots and Saddles*, 239; Liddic & Harbaugh, *Camp on Custer*, 42-43.

82 Hutton, *Custer Reader*, 215. In one account Bloody Knife is given credit for killing the warrior, while Custer notes he helped in a coordinated shot. It seems likely this is the same event.

83 Campbell Collection, Interview with White Bull, 22, U.O.

84 *New York Tribune*, September 6 & 8, 1873.

85 *New York Tribune*, September 6, 1873; Frost, *Some Observations*, 63.

86 Hutton, *Custer Reader*, 215.

87 Braden, *The Yellowstone Expedition*, 218-241.

88 Hills, *With General George A. Custer,* 124.
89 Frost, *Custer's 7th Cav.,* 68.
90 *New York Tribune,* September 8, 1873.
91 Barnard, *Ten Years,* 196.
92 Hills, *With General George A. Custer,* 124.
93 *New York Tribune,* September 8, 1873.
94 Hills, *With General George A. Custer,* 124; Lubetkin, *Cooke's Gamble,* 248-249.
95 Hardorff, *Baliran, Honzinger, and the Custers: The Facts and Fictions of the Rain in the Face Myth,* 8; Frost, *Custer's 7th Cav.,* 77; Liddic & Harbaugh, *Camp on Custer,* 40-42.
96 Hutton, *Custer Reader,* 216-217.
97 Ibid.
98 Campbell Collection, *Interview with White Bull,* 23, U.O.
99 Hutton, *Custer Reader,* 217.
100 Hutton, *Custer Reader,* 218; Carroll, *Custer's Chief of Scouts,* 45; Frost, *Custer's 7th Cav.,* 69.
101 Custer, *Boots and Saddles,* 204.
102 Braden, *The Yellowstone Expedition,* 134.
103 Ibid.
104 Hills, *With General George A. Custer,* 124; Barnard, *Ten Years,* 197.
105 Hutton, *Custer Reader,* 218-219.
106 Stanley, *Personal Memoirs,* 242.
107 Lubetkin, *Cooke's Gamble,* 246-247.
108 Frost, *Custer's 7th Cav,* 213.
109 Lubetkin, *Cooke's Gamble,* 247.
110 Campbell Collection, *Interview with White Bull,* 22-23, U.O.

Chapter 5, Traps near the Bighorn River

1 *New York Tribune,* September 9, 1873; Frost, *Some Observations,* 66.
2 Larned, *Yellowstone Expedition,* 4; Frost, *Some Observations,* 65-66.
3 Frost, *Custer's 7th Cav.,*76-78. Father Valentin Sommereisen was the priest.
4 Frost, *Some Observations,* 66; Frost, *Custer's 7th Cav.,* 77; Larned, *Yellowstone Expedition,* 4; *New York Tribune,* September 9, 1873.
5 Frost, *Custer's 7th Cav.,* 79; *New York Tribune,* September 9, 1873. The abandoned Hunkpapa site was strewn with coffee, hatchets, axes, paint, and lodge poles.
6 Larned, *Yellowstone Expedition,* 195.
7 Campbell Collection, Interview with White Bull, 23, U.O.; Lubetkin, *Cooke's Gamble,* 254; *New York Tribune,* September 9, 1873; *Army and Navy Journal,* August 30, 1873.
8 Stanley, *Personal Memoirs,* 242.
9 Frost, *Some Observations,* 68.
10 *New York Tribune,* September 6 & 9, 1873; Braden, *The Yellowstone Expedition,* 218-241; Frost, *Some Observations,* 72; Larned, *Yellowstone Expedition,* 4. One historian believes that Custer was not serious about catching the warriors since he brought cattle along which would slow him down.
11 Hills, *With General George A. Custer,* 143-160.
12 *Sioux City Journal,* August 14, 1874; Camp Collection, Unclassified Envelope 75, 452, I. U.

13 *New York Tribune*, September 9, 1973.
14 Larned, *Yellowstone Expedition*, 4.
15 Hills, *With General George A. Custer*, 143-160; Frost, *Some Observations*, 70-71; *New York Tribune*, September 6, 1873.
16 Frost, *Some Observations*, 73.
17 Ibid.,71.
18 Vestal, *Warpath*, 163; Campbell Collection, Interview with White Bull, 23.
19 Carroll, *The Yellowstone Expedition of 1873*, 57-58; *New York Tribune*, September 6, 1873.
20 Hills, *With General George A. Custer*, 143-160.
21 Carroll, *Yellowstone Expedition*, 58.
22 *New York Tribune*, September 9, 1873.
23 Frost, *Some Observations*, 73.
24 Hutton, *Custer Reader*, 196.
25 *New York Tribune*, September 6, 1873.
26 Klokner, The Officer Corps, 61-62; Hammer, Men With Custer, 115-116.
27 Barnard, *Ten Years*, 198; Liddic and Harbaugh, *Camp on Custer*, 46.
28 Hutton, *Custer Reader*, 196.
29 *New York Tribune*, September 9, 1873.
30 Hills, *With General George A. Custer*, 143-160.
31 Ibid.
32 Burdick, *Army Life*, 13.
33 *New York Tribune*, September 6, 1873; *New York Tribune*, September 9, 1873; Braden, *The Yellowstone Expedition*, 218-241; Frost, *Some Observations*, 74.
34 Braden, *The Yellowstone Expedition*, 218-241.
35 Burdick, *Army Life*, 13-14.
36 Carroll, *Custer's Chief of Scouts*, 47.
37 Campbell Collection, *Interview with White Bull*, 23-24, U.O. Years later, White Bull could not remember who led the attack on the soldiers.
38 *New York Tribune*, September 9, 1873.
39 *New York Tribune*, September 6, 1873; Liddic and Harbaugh, *Camp on Custer*, 44-45.
40 Taunton, *More Sidelights of the Sioux Wars*, 82-83. The right wing commanded by French included McIntosh's Company G, Tom Custer's Company B, Moylan's Company A, and French's Company M. Colonel Verling Hart commanded the left wing—Yates' Company F, Weston's Company L, Hale's Company K, and McDougall's Company E. Custer stated that each commander had two "squadrons" that consisted of two companies each. Hale and Yates became subordinate squadron leaders with the left wing.
41 Frost, *Some Observations*, 75. It appears from the soldier testimony that French took only two companies [G and M] to watch the eastern crossing.
42 *New York Tribune*, September 8, 1873.
43 *New York Tribune*, September 9, 1873.
44 Braden, *The Yellowstone Expedition*, 218-241.
45 *New York Tribune*, September 9, 1873.
46 Frost, *Custer's 7th Cav*, 84; Braden, *The Yellowstone Expedition*, 218-241.
47 Hills, *With General George A. Custer*, 143-160.
48 *New York Tribune*, September 9, 1873.
49 Braden, *The Yellowstone Expedition*, *Journal of the U.S. Cavalry Association* 15, 289-301.

50 Ibid.
51 *New York Tribune*, September 9, 1873.
52 *New York Tribune*, September 6, 1873. Ward Burton cartridges used by Company F discovered west of Braden's position reinforces this newspaper account of his action in the area. Artifact map from Little Bighorn Battlefield National Monument in the author's collection.
53 *New York Tribune*, September 6, 1873.
54 Hills, *With General George A. Custer*, 143-160.
55 *New York Tribune*, August 25, 1873 & September 9, 1873.
56 *New York Tribune*, September 6 & 9, 1873. Ward Burton expended cases in this area show where Yates' Company F fought on top of the bench land overlooking the soldier camp. Artifact map from Little Bighorn Battlefield National Monument in the author's collection.
57 Liddic & Harbaugh, *Camp on Custer*, 47.
58 Frost, *Custer's 7th Cav.*, 86.
59 Carroll, *Custer's Chief of Scouts*, 48. Varnum recalled, "I climbed this bluff in front of my troop and the Indians pressing in pretty fast, Custer ordered the troop dismounted and put in action to cover our position. As the men climbed the bluff, I forming them, Custer called to me to wheel it to the right across a ravine and drive the Indians out. This I did and was nicely remembered in Custer's report." (Ibid).
60 Frost, *Some Observations*, 77; Frost, *Custer's 7th Cav.*, 86.
61 Lubetkin, *Cooke's Gamble*, 263-264.
62 Frost, *Custer's 7th Cav.*, 84; Taunton, *More Sidelights of the Sioux Wars*, 80.
63 *New York Tribune*, September 9, 1873.
64 Liddic and Harbaugh, *Camp on Custer*, 46.
65 Hills, *With General George A. Custer*, 143-160.
66 Frost, *Some Observations*, 75.
67 Frost, *Custer's 7th Cav.*, 84.
68 *New York Tribune*, September 6, 1873.
69 Liddic & Harbaugh, *Camp on Custer*, 47.
70 Hills, *With General George A. Custer*, 143-160.
71 *New York Tribune*, September 6, 1873.
72 Frost, *Custer's 7th Cav.*, 84.
73 Liddic & Harbaugh, *Camp on Custer*, 48-49; Hutton, *Custer Reader*, 197; Carroll, *Custer's Chief of Scouts*, 48.
74 Frost, *Some Observations*, 76.
75 Carroll, *Custer's Chief of Scouts*, pg. 48; Hills, *With General George A. Custer*, 143-160. Fuse caps from the artillery pieces were found on the hilltop below Braden's bluff that verifies Hill's account of Stanley shelling from the soldier camp. Varnum implies firing from Stanley after arrival in Custer's camp as well. Archeological maps made by Vance Haynes in the author's collection from a copy at LBBNM.
76 Barnard, *Ten Years*, 199.
77 Frost, *Custer's 7th Cav.*, 85.
78 Lubetkin, *Cooke's Gamble*, 264; *New York Tribune*, September 6, 1873; Lubetkin, *Custer and the 1873 Yellowstone Expedition*, 311.
79 *New York Times*, September 10, 1873. Calhoun who was present believed 500 warriors crossed the river.
80 Frost, *Custer's 7th Cav.*, 84-85. John Ryan and other officers estimated the total warrior numbers from 800 to 1,000.
81 Liddic & Harbaugh, *Camp on Custer*, 34.
82 Liddic & Harbaugh, *Camp on Custer*, 50.
83 Braden, *The Yellowstone Expedition*, 218-241. To keep the ends of his shattered femur from rubbing together, an

open-topped wooden box was constructed to protect Braden's leg. Attached to the box was an iron rod that went around Braden's neck. All was attached with plaster to keep the leg from shortening.

84 Frost, *Custer's 7th Cav.*, 87, 96-98; Frost, *Some Observations*, 78; Barnard, *Ten Years*, 200; Pompeys Pillar National Monument Visitor Guide.

85 *New York Times*, October 13, 1873.

86 Frost, *Custer's 7th Cav.*, 98.

87 *New York Times*, October 13, 1873; Frost, *Custer's 7th Cav.*, 98-101.

88 Frost, *Custer's 7th Cav.*, 98-101.

89 Carroll, *Benteen-Goldin Letters*, 270.

90 Frost, *Custer's 7th Cav.*, 101; *New York Times*, October 13, 1873.

91 Dustin, *The Custer Tragedy*, 203.

92 Frost, *Custer's 7th Cav.*, 101-102.

93 Lubetkin, *Cooke's Gamble*, 266.

94 *1873 Official Report of the Yellowstone Expedition*, pg. 13, War Department Library. While some historians have indicted Custer as a gold-hungry entrepreneur, whose purpose in traversing the Black Hills was to discover the gold metal that makes white men go crazy, such was not the case. It is clear that the same occurred a year earlier without the same results or notoriety. There were three real differences in the two expeditions—the Black Hills discovery was on Indian held land, the country was attempting to survive a financial panic, and the press focused on the later discovery.

95 *New York Times*, September 11, 1873; Lubetkin, *Custer and the 1873 Yellowstone Survey*, 314; Frost, *Custer's 7th Cav.*, 101.

96 *The Weekly Rocky Mountain News* (Denver) September 24, 1873.

Chapter 6, *Paha Sapa* and the Road to War

1 Dustin, *The Custer Tragedy*, 203.

2 Elizabeth B. Custer Collection, Custer Battlefield Museum, Garryowen, Montana. One example of George Custer's humor that survived in Libbie's collection of notes may have been meant as funny, or perhaps contained a more serious insinuation. Custer had clipped a divorce notice from a classified newspaper in Bismarck and pasted his wife's name over the divorcee's name. The ad read, "*NOTICE. Whereas, my wife, Mrs. E Custer [cut out & pasted name] having left my bed and board without just cause or provocation, all persons are hereby forbidden harboring or trusting her on my account, as I shall pay no debts of her contracting. Bismarck, D. T., Nov. CUSTER [cut out & pasted name]*. Ibid.

3 Frost, Lawrence, *The Custer Album*, 114-115. In 1873, Custer had the opportunity to attend a troop reunion and reception line that consisted of President Grant, General Sheridan, General Sheridan, and himself. Depicted in an engraving from *Frank Leslie's Illustrated Newspaper*, the four men are seen kissing beautiful women. Each officer had opportunity to kiss little girls in the line, but it quickly turned into a fun game of kissing the older girls. An observant reporter kept a score of the officer kisses to babies, little girls, and ladies: Grant kissed 19 babies, 20 little girls, and 38 ladies, while Sherman kissed 34 toddlers, 27 juveniles, and 28 women. Sheridan's score was 17 babies, 11 little girls, and 63 ladies. Custer held the record for fewest babies and the most ladies—13 babies, 43 little girls, and 67 ladies. While a humorous recollection, it presented some underlying truths. Many women were drawn to the chivalrous and dashing cavalryman over the stumpy, bearded older officers.

4 Mueller, Jim, *Swinging Around the Circle with Custer: George Armstrong Custer's Role in the 1866 Election*, 41.

5 Mueller, Jim, *Swinging Around the Circle with Custer*, 41.
Kazanjian, Howard and Enss, Chris, *None Wounded, None Missing, All Dead-The Story of Elizabeth Bacon Custer*, 103.

6 Leckie, *Elizabeth Bacon Custer*, 130; Merington, *The Custer Story*, 251; Blake, Bill, *Was Weir A Hero?* 4-8.

7 Merington, *The Custer Story*, 251.

8 Leckie, *Elizabeth Bacon Custer*, 126.

9 Leckie, *Elizabeth Bacon Custer*, 125-127; Merington, *The Custer Story*, 250. In December of 1870, Custer promised to "cease so long as I am a married man to play cards or any other game of chance or its equivalent." (Leckie, *Elizabeth Bacon Custer*, 125).

10 Merington, *The Custer Story*, 258.

11 Ibid., 264.

12 Katz, *Custer in Photographs*, 123-129.

13 This image is from the George Kush Collection and published with his courtesy.

14 E. B. Custer Collection, Garryowen. The author has attempted to locate other letters written in the hand of Nellie Wadsworth without success. A sample of her handwriting might help prove the author of the love note.

15 For a comprehensive and detailed look at the Black Hills Expedition, a must read is *Exploring with Custer-The 1874 Black Hills Expedition* by Ernest Grafe and Paul Horsted published in 2002.

16 Olson, James C., *Red Cloud and the Sioux Problem*, 15- 26.

17 Hyde, George E., *Red Cloud's Folk*, 54-55, 64, 88, 317.

18 Hyde, *Red Cloud's Folk*, 77, 80, 85, 88, 119, 138, 150-151, 156, 163; Olson, *Red Cloud and the Sioux Problem*, 29-30.

19 Utley, *Little Bighorn Battlefield-A History and Guide to the Battle of the Little Bighorn, 15-26; Kappler, Indian Treaties 1778-1883*, 998-1003. Red Cloud and his followers promised to live in peace on the huge reservation, as well as live off the promised government annuities and learn to farm. Schools, rations, medical supplies, and buildings were part of the promised services. While the treaty banned further incursion into reservation lands from white settlers, it did allow access for government employees including the army and government agents. Debate about the treaty wording and its implications reverberate even today. As a consolation, the U.S. Army promised to abandon all three forts located along the Bozeman Trail—Fort Reno, Fort Kearny, and Fort Smith, which were burned to the ground by the Lakota as soon as the soldiers exited. The tribes promised to stop raids on all whites and to allow the railroads to pass through their reservation.

20 Horsted, Grafe & Nelson, *Crossing the Plains with Custer*, 12.

21 *The Cheyenne Daily Leader* (Wyoming), October, 9, 1875.

22 Horsted, Grafe & Nelson, *Crossing the Plains with Custer*, 12, 24.

23 Olson, *Red Cloud and the Sioux Problem*, 173.

24 Fougera, *With Custer's Cavalry*, 254-256; Ladenheim, *Custer's Thorn*, 144-146.

25 Donovan, *A Terrible Glory*, 89. Little noted in the history books is the diplomatic Custer who was involved in attempting to ease the appalling situation on the reservations and to broker a peace treaty between the Lakota, their Indian adversaries, as well as the military in March of 1875. Some treaties had been completed in May and June, with praise by the Commissioner of Indian Affairs for Custer's efforts. Custer had written,"As the representative of the Military in this vicinity and as one who earnestly desires to encourage the maintenance of peace not only among Indians themselves but between the Indians and the white men I extend to them a cordial and friendly invitation to come to this point and take part in the proposed council." (Ibid).

26 Olson, *Red Cloud and the Sioux Problem*, 200-201.

27 Ibid.,180-181, 209-212, 214-215. Contrary to the popular myth that none of the Lakota wished to sell the hills, Red Cloud and Spotted Tail agreed to sell. Before the meeting began, Spotted Tail had told the delegation that they would take $6 million dollars for the Black Hills, which is what the commissioners later offered. It seems as the bantering began, the Lakota increased the price, with the amount of $70 million dollars coming to the table. A *New York Herald* reporter who was present, concluded that there was likely an error in translation and that the chiefs were actually requesting $7 million rather than the $70 million figure. A baffled commission offered $400,000 for the mining rights to the hills as long as the lease should run or $6 million dollars to be paid in 15 annual installments for the

outright purchase of the Black Hills. The counter-offer seemed too small to the chiefs who were present, and Red Cloud did not even show up for the final meeting. The two groups departed on friendly terms with no decision having been reached.

28 Olson, *Red Cloud and the Sioux Problem*, 214-215.

29 *The Cheyenne Daily Leader* (Wyoming) October 9, 1875.

30 Harper, *The Fights on the Little Horn-Unveiling the Mysteries of Custer's Last Stand*, 2.

31 Gray, *Centennial Campaign*, 35. A warning shot became apparent to the reservation Lakota on January 18, 1876, when government commissioners ordered all sales of ammunition and arms to Indians to cease at the agencies. When no rations or those of inferior quality were issued, the stage was set for more Lakota to leave the reservation confines as their stomachs groaned due to the inability to hunt with firearms. The 'roamers' or non-reservation Lakota, including Crazy Horse and Sitting Bull, stayed in the massive unceded hunting territory which still held enough game to allow sustenance. A prosperous illegal trade of guns and ammunition was created, with the exasperated Lakota trading ration cards for ammunition, and government cattle hides as well as buffalo skins for Henry and Winchester repeating weapons.

32 Donovan, *A Terrible Glory*, 90; Barnett, *Touched by Fire*, 103, 109.

33 Utley, *Little Bighorn Battlefield-A History and Guide to the Battle of the Little Bighorn*, 21-36.

34 Mangum, *Battle of the Rosebud: Prelude to the Little Bighorn*, 11-21.

35 Gray, *Centennial Campaign*, 54-58. Crook led five companies each of the 2nd and 3rd Cavalry and two companies of the 4th Infantry and proceeded northward from Fort Fetterman. He acted as an observer with a less-than-confident Reynolds in command, but it was clear that the brigadier general was the campaign's actual leader. The bluecoats surprised the camp that was incorrectly identified as Crazy Horse's village.

36 Marquis, *Wooden Leg*, 165-167. The soldiers had four dead and six wounded. Wooden Leg recalled that one Cheyenne was killed and several were wounded.

37 Gray, *Centennial Campaign*, 55-56; Marquis, *Wooden Leg*, 170-171.

38 Evans, *Custer's Last Fight*, 59-61.

39 Donovan, *A Terrible Glory*, 101-115.

40 Utley, *Cavalier in Buckskin*, 162.

41 Ibid.

42 Ibid.,163.

43 Willert, *The Terry Letters*, xv.

44 Mills, *Harvest of Barren Regrets*, 195-199.

45 Carroll, *The Custer Autograph Album*, 75; Utley, *Cavalier in Buckskin*, 169.

46 Dustin, *The Custer Tragedy*, 67; Willert, *To the Edge of Darkness—A Chronicle of the 1876 Indian War*, 31, 47.

47 Nichols, *Men With Custer*, 36; Willert, *Edge of Darkness*, 46.

48 Willert, *Edge of Darkness*, 46. The 'squaw men' were Thomas LeForge and Bernard Prevo.

49 Charles A. Booth papers, Booth letter, April 17, 1876, *Little Bighorn Battlefield National Monument*. Booth's letters are published here for the first time. As a lieutenant with the 7th Infantry, he was stationed at Fort Benton, Montana Territory from 1872-75 then was transferred to Fort Shaw, where he departed with Gibbon.

50 Willert, *Edge of Darkness*, 68-71.

51 Dustin, *The Custer Tragedy*, 77.

52 Booth papers, May 1, 1876; Willert, *Edge of Darkness*, 80-86.

53 Booth papers, May 18, 1876.

Chapter 7, Two Trails to Destiny

1 Utley, *Little Bighorn Battlefield*, 32.
2 Chambers, *Fort Abraham Lincoln: Dakota Territory*, 91.
3 Brady, *Indian Fights and Fighters*, 393.
4 Donovan, *A Terrible Glory*, 91-92.
5 Utley, *Cavalier in Buckskin*, 109-110, 154-155.
6 Utley, *Cavalier in Buckskin*, 154-155; Donovan, *A Terrible Glory*, 95-96.
7 Dustin, *The Custer Tragedy*, 203; Donovan, *A Terrible Glory*, 96.
8 Van de Water, *Glory Hunter: Life of General Custer*, 301.
9 Gray, John S., *Centennial Campaign*, pg. 89. One of the greatest misconceptions of the Little Bighorn campaign is that Custer and Terry were unaware of the possible high numbers of Lakota they might encounter.
10 Utley, *Little Bighorn Battlefield*, 44.
11 Klokner, *The Officer Corps*, 100. Colonel Samuel Sturgis was the senior officer of the 7th Cavalry and should have been at Fort Lincoln, but Custer and the older officer did not get along. Sheridan solved the situation by sending Sturgis to Saint Louis for desk duty, while Custer became the actual field commander.
12 Fougera, *With Custer's Cavalry*, 258-259; Hammer, *Men with Custer*, 338-339.
13 Mueller, *Shooting Arrows & Slinging Mud-Custer, The Press, and The Little Bighorn*, 32-33; Barnard, *I Go With Custer-The Life and Death of Reporter Mark Kellogg*, 99-110; Hammer, *Men With Custer*, 184. James Calhoun and another unnamed correspondent were also sending dispatches to the *New York World* cleverly dodging the title of "newspaper men." *New York Tribune* publisher Whitelaw Reid wanted to send along a reporter with the anticipation that it would be the most important Indian expedition since the Civil War. Rather than send a full-time reporter, young Dick Roberts, a friend of the Custers, was hired as a herder for the cattle brought along for food. His newspaper dispatches would end when he was left behind at the Powder River depot along with the cattle.
14 Hunt, *I Fought with Custer*, 3-4; Hammer, *Men With Custer*, 379-380.
15 Hunt, *I Fought with Custer*, 53; Taylor, William O., *With Custer on the Little Bighorn*, 18.
16 Taylor, *With Custer on the Little Bighorn*, 20.
17 Hammer, *Men With Custer*, 92-93; *Isaiah Dorman and the Custer Expedition*, 351; Pengra, Lilah, *Isaiah Dorman Was Not a Runaway Slave*, 1- 10.
18 Woods, Marvin, *Custer's Arikara Indian Interpreter-Frederic Francis Gerard*, 1-17, 110. Gerard's name is sometimes spelled 'Girard,' which is the Canadian spelling of his name.
19 Hunt, *I Fought with Custer*, 54; James, *Custer's Last Gun*, 62-68; Banfill, *The Story of John Burkman's Experience with General George Custer and the 7th Cavalry*. While some soldiers, including Godfrey and Windolph, recalled that Custer carried two such revolvers. Only one is seen in Custer's photograph of his gun rack at Fort Lincoln. Burkman, who was with him almost constantly, recalled seeing only one and it being .45 caliber. There is great debate about what Custer actually carried as a sidearm. Some historians have also suggested he carried a small-caliber experimental sidearm that was given to him. One of Custer's three skin suits is actually cowhide or leather that is today housed at the Smithsonian Museum.
20 Wagner, *Old Neutriment*, 121-122.
21 Hunt, *I Fought with Custer*, 53; Custer, *Boots and Saddles*, 217-218; Barnard, *Ten Years*, 287.
22 Utley, *Little Bighorn Battlefield*, 32; Custer, *Boots and Saddles*, 218.
23 Nichols, *Men With Custer*, 90; Evans, *Custer's Last Fight*, 118. George Custer had also hired Boston as a forage master for the Black Hills Expedition.
24 Klokner, *The Officer Corps*, 74-75, Langellier, Cox, Pohanka, *Myles Keogh*, 49-63; Nichols, *Men With Custer*, 210-211. During the Papal Wars, *Myles Keogh* was awarded the *Medaglia Pro Peri Sede* which recognized his service in the

conflict. He also had earned the Cross of the Order of St. Gregory the Great, which recognized his superb service for the Pope and for 'distinguishing himself in a sortie.'

25 Solimine, *Custer's Bugler-The Life of John Martin (Giovanni Martino)*, 15-47.

26 Hunt, *I Fought with Custer*, 53.

27 Abrams, *James Calhoun's Forgotten Dispatches in the New York World*, 12.

28 Wagner, *Old Neutriment*, 123.

29 Ibid., 124.

30 Evans, *Custer's Last Fight*, 116. Orders generated on May 12 broke down the 12 companies into four battalions of three companies each. The first battalion consisted of Companies B, C and I under Keogh, and Yates commanded the second with Companies E, F and L. The third battalion included Companies A, D and H under Weir, while Captain Thomas French controlled Companies G, K and M.

31 Abrams, Marc, *James Calhoun's Forgotten Dispatches*, 5- 18; Hunt, *I Fought with Custer*, 57, 63.

32 *The Washington Times*, November 4, 1906; Pohanka, Brian C., *A Study of Company Discipline: General Courts-Martial of Enlisted Personnel, 7th Cavalry, 1875-1876*, 35-36.

33 Willert, *Little Bighorn Diary*, 67-69.

34 Evans, *Custer's Last Fight*, 88, 91.

35 Willert, *Little Bighorn Diary*, 112-115, 137.

36 Willert, *To the Edge of Darkness*, 113-120.

37 Philbrick, *The Last Stand-Custer, Sitting Bull, and the Battle of the Little Bighorn*, 1-3.

38 Nichols, *Men With Custer*, 174.

39 Evans, *Custer's Last Fight*, 124; Hammer, *Custer in '76*, 149.

40 Willert, *To the Edge of Darkness*, 137-139, 154-161.

41 Graham, *The Custer Myth*, 236.

42 Evans, *Custer's Last Fight*, 123; Willert, *Terry Letters*, 19. Many writers have surmised that there was some sinister motive by Terry to punish Custer for insubordination or to prove who was the expedition's chief officer. Like a chess player, Terry already had a secondary plan in place, stating that if Reno did not find any Indians, he was going "to make a double movement, one part of the force going up the Tongue to near its headwaters then crossing to the head waters of the Rosebud & descending that stream the other portion joining Gibbons troops & proceeding up the last named river." (Willert, *Terry Letters*, 19).

43 Willert, *To the Edge of Darkness*, 143-151. The right wing consisted of Companies B, C, E, F, I, and L.

44 Nichols, *Men With Custer*, 315-316.

45 Ibid.,102.

46 Magnussen, Daniel O., *Peter Thompson's Narrative of the Little Bighorn Campaign 1876*, 31-45; Hammer, *Men With Custer*, 346.

47 Hammer, *Men With Custer*,148. Henry Harrington participated in both the 1873 and 1874 expeditions.

48 Ibid.,28-29.

49 Ibid.,106.

50 Ibid.,107.

51 Willert, *To the Edge of Darkness*, 161-163, 176.

52 Magnussen, *Peter Thompson's Narrative*, 74-75.

53 Willert, *To the Edge of Darkness*, 168.

54 Marquis, *Wooden Leg*, 198. The ever-growing numbers of Lakota and Cheyenne were camped along Great Medicine Creek or Sundance Creek [today known as Reno Creek].

55 Ibid.,199.

56 Bourke, *On the Border with Crook*, 311; War Eagle, Red Thunder & Clown, *The Legacy of Crazy Horse*, 7.

57 Willert, *Little Bighorn Diary*, 158.
58 Mangum, *Battle of the Rosebud*, 59-60. Mangum's book is the most comprehensive account of this battle to date.
59 Edmond S. Meany Papers, Undated Pourier interview [1907], Box 13, Folder 20.
60 Mangum, *Battle of the Rosebud*, 61-62.
61 Vestal, *Warpath-The True Story of the Fighting Sioux Told in a Biography of Chief White Bull*, 186-187.
62 Marquis, *Wooden Leg*, 200-201; Mangum, *Battle of the Rosebud*, 69-88.
63 Mangum, *Battle of the Rosebud*, 85-88; Marquis, *Wooden Leg*, 202-203.
64 Marquis, *Wooden Leg*, 202-203; Stewart, *Custer's Luck*, 323.
65 Graham, *The Custer Myth*, 29.
66 Libby, *The Arikara Narrative of the Campaign Against the Hostile Dakotas June, 1876*, 75-76.
67 Merington, *The Custer Story*, 307.
68 Merington, *The Custer Story*, 306-307; Nichols, *Men With Custer*, 324
69 Graham, *The Custer Myth*, 129.
70 Gray, *Centennial Campaign*, 136.
71 Willert, *Terry Letters*, 47.
72 Magnussen, *Peter Thompson's Narrative*, 78.
73 Graham, *The Custer Myth*, 235-237.
74 George Kush Collection, Cooke letter, June 21,1876.
75 Custer, *Boots and Saddles*, 275.
76 Willert, *Little Bighorn Diary*, 195-202.
77 Bradley, *Indian Fights and Fighters*, 367.
78 Evans, *Custer's Last Fight*, 168.
79 Bradley, *The March of the Montana Column*, 143.
80 Stewart, *Custer's Luck*, 239.
81 *New York Herald*, July 11, 1876.
82 Stewart, *Custer's Luck*, 249-250.
83 Manion, *General Terry's Last Statement to Custer*, 39. Another interesting and controversial aspect of Terry's orders is the testimony of Mary Adams, which has been pondered by scholars since General Nelson A. Miles mentioned it in his 1896 autobiography, *The Personal Observations of General Nelson A. Miles*. He claimed to have an affidavit from an unidentified witness to Terry's last meeting with Custer. While some have questioned the credibility of the statement, it has a ring of truth to it, because it repeats the written order and includes something that was likely on Terry's mind—the 'wounded.' While it seems a puzzling addition, it makes perfect sense when placed into the context of recent events. Terry had recently become aware of the controversy attached to the Reynolds attack, in which a wounded soldier was left to his horrible death. Reynolds would be court-martialed and Crook as department commander, likely felt the heat as well. Fresh on his mind, the ever-cautious Terry was in a similar situation, and it was not something that he wished to deal with as a department commander.
84 Willert, *Little Bighorn Diary*, 201.
85 Ibid.
86 Abrams, *James Calhoun's Forgotten Dispatches*, 15.
87 Brininstool, *Troopers with Custer*, 278-279.
88 Ibid.
89 Hammer, *Custer in '76*, 229; Koury, Mike, *Forget the Gatling Guns!*, 55-61.
90 Graham, *The Custer Myth*, 130.
91 Carroll, *The Two Battles of the Battle of Little Big Horn*, 60.
92 Graham, *The Custer Myth*, 134-135.

93 Graham, *The Custer Myth*, 130; Wagner, *Old Neutriment*, 134.
94 Clark, George M., *Scalp Dance-The Edgerly Papers on the Battle of the Little Big Horn*,12; Nichols, *Men With Custer*, 113-114.
95 Hammer, *Camp in '76*, 74.
96 Williams, *Military Register of Custer's Last Command*, 393-394. Twelve of the strongest mules were loaded with *aparejos*, and each was loaded with two boxes of 1,000 rounds of carbine ammunition each. The 24 boxes equated to 24,000 rounds of ammunition. Mules were assigned to the 12 companies based on its number of men and averaged 13 mules to a company.
97 Gray, *Custer's Last Campaign*, 200-202.
98 Hammer, *Custer in '76*, 247. Richard Thompson was with the 6th Infantry and aboard the *Far West*.
99 Langellier, Cox, Pohanka, *Myles Keogh*, 132; Hunt, *I Fought with Custer*, 67; Godfrey, *Custer's Last Battle*, 14.
100 Evans, *Custer's Last Fight*, 173.
101 Wagner, *Old Neutriment*, 136-137.
102 Merington, *The Custer Story*, 307-308.
103 Hunt, *I Fought with Custer*, 68.
104 Ibid., 70.
105 Hanson, *The Conquest of the Missouri*, 247-248.
106 George Kush Collection, Samuel Alcott diary, June 22, 1876.
107 Hunt, *I Fought with Custer*, 68.
108 Upton, *Custer Catastrophe at the Little Big Horn 1876*, 163.
109 Ibid.
110 Stewart, *Custer's Luck*, 253

Chapter 8, Into the Valley of the Shadow of Death

1 Hunt, *I Fought with Custer*, 71.
2 Hardorff, *Camp, Custer, and the Little Bighorn*, 66; Magnussen, *Peter Thompson's Narrative*, 87; Gray, *Last Campaign*, 206; Willert, *Little Bighorn Diary*, 216.
3 Wagner, *Old Neutriment*, 144.
4 Graham, *The Custer Myth*, 134; Greene, *Indian Wars Veterans: Memories of Army Life and Campaigns in the West, 1864-1898*, 129. A demijohn or 'carboy' could contain between five and 15 gallons of fluids. Soldier Jacob Hetler reveals that some of the whiskey made its way to Reno Hill where he received a sip after being wounded in the back.
5 Graham, *The Custer Myth*, 134.
6 Ibid.
7 Gray, *Last Campaign*, 206; A. B. Welch papers, Young Hawk narrative. Along the Powder River, Young Hawk had seen Bloody Knife and Custer having a serious talk. According to the story told to another Arikara, "Bloody Knife was sick and did not want to die in a lodge. He wanted to die fighting. So, he said, 'We should find the Sioux and fight right away. Maybe I will not live long and I want to fight now.' Custer said, ' No, I am going to attack in the morning time or at night. If I see you or any of the scouts running away I will kill you.' Bloody Knife then said, 'Yes, that is right. If you see my back that way you can kill me. But, if I see you, or any of your soldiers, going that way (to the rear) I will kill you, too,'" He disappeared from the command to be alone. (Young Hawk Narrative)
8 Willert, *Little Bighorn Diary*, 217.
9 Evans, *Custer's Last Fight*, 160.
10 Graham, *The Custer Myth* 190-192.

11 Graham, *The Custer Myth*, 177.
12 Evans, *Custer's Last Fight*, 160. The comments were likely aimed at Godfrey and Moylan who had challenged the packing of the mules.
13 Graham, *The Custer Myth*, 177; Mills, *Harvest of Barren Regrets*, 239.
14 Graham, *The Custer Myth*, 177.
15 Mills, *Harvest of Barren Regrets*, 239.
16 Graham, *The Custer Myth*, 134.
17 George Kush Collection, Cooke letter June 21, 1876.
18 Camp Collection, Box 2, Folder 11, 94, I.U.
19 Ibid.
20 Ibid.
21 Ibid.
22 Mills, *Harvest of Barren Regrets*, 239.
23 Graham, *The Custer Myth*, 135.
24 Ibid.
25 Clark, *Scalp Dance*, 24.
26 Klokner, *The Officer Corps*, 59-60.
27 Donovan, *A Terrible Glory*, 118; Klokner, *The Officer Corps*, 105-106.
28 Donovan, *A Terrible Glory*, 125; Nichols, *Men With Custer*, 84.
29 Graham, *The Custer Myth*, 135.
30 Gray, *Centennial Campaign*, 154.
31 Graham, *The Custer Myth*, 177-178.
32 Libby, *The Arikara Narrative*, 78.
33 Hunt, *I Fought with Custer*, 73.
34 Graham, *The Custer Myth*, 135.
35 Willert, *Little Bighorn Diary*, 226-230; Magnussen, *Peter Thompson's Narrative*, 92-93.
36 Graham, *The Custer Myth*, 178.
37 Ibid.
38 Hammer, *Men With Custer*, 208; Noyes, *Custer's Surgeon, George Lord, Among the Missing at Little Bighorn Battle*, 13-19.
39 Graham, *The Custer Myth*, 178.
40 Scott, Willey, and Connor, *They Died With Custer*, 60-79, 272-297. The recruit of the time had to be at least 21 years of age and in good overall health. The typical soldier was in his mid-20s, around 5-foot 7-inches tall with a weight of less than 165 pounds.
41 Ibid., 90-95; Rickey, *Forty Miles A Day on Beans and Hay*, 18-24.
42 Daniels and Davis, *That Fateful Day: Eight More with Custer*, 17.
43 Waddell, O'Connell, Korff, *The Book of Aran*, 231. This holiday is traditionally celebrated on the Aran Islands and other places in Ireland. The word 'bonfire' came from the fact that in its initial rituals, bones were burned making it a 'bone fire.' Young and old men gather for a night of drinking, song, and dance and traditionally light the fire. This is believed to bring prosperity to the family, their home, and their crops for the coming year.
44 Dixon, *The Vanishing Race*, 160.
45 The author has on many occasions been allowed to study these petroglyphs, which were privately owned by the late Jack Bailey. According to Mr. Bailey, the Sundance arbor was still standing in this area in 1883 when his grandmother moved into the area. (Information about Sundance location provided by Thomas M. Heski on the CBHMA Field Trip, Custer Trail document, June 25, 2005, author's collection.)
46 Dixon, *The Vanishing Race*, 160-161.

47 Graham, *The Custer Myth*, 178.
48 Hammer, *Custer in '76*, 78; Camp Collection, Mathey to Camp, April 13, 1910, Folder 12, BYU; Willert, *Little Bighorn Diary*, 236.
49 Hammer, *Custer in '76*, 59.
50 Libby, *The Arikara Narrative*, 78.
51 Graham, *The Custer Myth*, 135.
52 Ibid., 262. Private Augustus Stocker had been killed on May 23, along with another private and a civilian, while hunting away from Gibbon's command.
53 Ibid., 248.
54 Carroll, *Custer's Chief of Scouts*, 60-61.
55 Gray, *Centennial Campaign*, 159.
56 Klokner, *The Officer Corps*, 69-70.
57 Graham, *The Custer Myth*, 30-31.
58 Dixon, *The Vanishing Race*, 162.
59 Willert, *Little Bighorn Diary*, 231.
60 Wagner, *Old Neutriment*, 144.
61 Stewart, *Custer's Luck*, 268. Big Crow and Black Whiteman were the Cheyenne who spotted Custer's camp.
62 Dixon, *The Vanishing Race*, 162.
63 Wagner, *Old Neutriment*, 144-145.
64 Nichols, *Men With Custer*, 390; Taylor, *With Custer on the Little Bighorn*, 30.
65 Graham, *The Custer Myth*, 178-179.
66 Willert, *Little Bighorn Diary*, pg. 244.
67 Graham, *The Custer Myth*, 178.
68 Nichols, *Men With Custer*, 100; Klokner, *The Officer Corps*, 58; Utley, "*Gossip & Scandal*," 18. Charles De Rudio became a revolutionary activist, who along with at least three other men, conspired to kill Napoleon III as he entered the Paris Opera. As the *William Tell Overture* played inside, three bombs went off killing several guards, but only punctured Napoleon's hat. After the failed attempt, De Rudio was to lose his head under a guillotine blade, until a last minute reprieve sent him to life imprisonment in French Guiana. Escaping in a fishing boat, he eventually ended up in England before migrating to the United States.
69 Taylor, *With Custer on the Little Bighorn*, pp. 30-31.
70 Carroll, *Custer's Chief of Scouts*, 86.
71 Gray, Custer's *Last Campaign*, 225; Carroll, *Custer's Chief of Scouts*, 62, 87; Evans, *Custer's Last Fight*, 168; Graham, *The Custer Myth*, 21.
72 Hammer, *Custer in '76*, 78-79.
73 Nichols, *Reno Court of Inquiry*, 85.
74 Daniels and Davis, *That Fateful Day*, 20.
75 Hunt, *I Fought with Custer*, 74; Graham, *The Custer Myth*, 179.
76 Wagoner, *Old Neutriment*, 147.
77 Fougera Collection, *Custer's Last Command*, 2, LBBNM.
78 Magnussen, *Peter Thompson's Narrative*, 95.
79 Barnard, *Ten Years with Custer*, 289.
80 Hardorff, *Camp, Custer, and the Little Bighorn*, 43.
81 Willert, *Little Bighorn Diary*, 245; Evans, *Custer's Last Fight*, 171. The command moved along Davis Creek about six miles; although the exact route and distance can only be approximated.
82 Evans, *Custer's Last Fight*, 171.

83 Barnard, *Custer's First Sergeant John Ryan*, 174.
84 Wagoner, *Old Neutriment*, 148.
85 Liberty, *Cheyenne Memories*, 195.

Chapter 9, Reunion along the Greasy Grass

1 Terry, *Plains Indians Regalia & Customs*, 106.
2 Johnson, *The Sioux-Warriors of the Plains*, 6-25; Hassrick, *The Sioux*, 4-5. Some linguists have recently challenged this translation of Sioux as inaccurate.
3 Marquis, *Wooden Leg*, 155-160.
4 Ibid., 179-185.
5 Vestal, *Sitting Bull*, 148-151.
6 Marquis, *Wooden Leg*, 184-187.
7 Grinnell, *The Cheyenne Indians*, Volume 1, 2. Some other authorities suggest the name Cheyenne may have been derived from the French word *chien* meaning dog. Early French traders noted that Cheyenne villages often had numerous dogs in camp, which sometimes served as food staple. The Cheyenne and Lakota were linguistically different and could not communicate except through sign or an interpreter.
8 Hardorff, *Cheyenne Memories of the Custer Fight-A Source Book*, 95-96.
9 Ibid., 49.
10 Marquis, *Custer on the Little Bighorn*, 35-36.
11 Hardorff, *Indian Views of the Custer Fight*, 97-101.
12 Hassrick, *The Sioux*, 3-6; Marquis, *Wooden Leg*, 211.
13 Marquis, *Wooden Leg*, 211; Hassrick, *The Sioux*, 3.
14 Vestal, *Warpath*, 27.
15 Ibid., 47.
16 Ibid., 52-76.
17 Vestal, *Warpath*, 123-127; Hardorff, *Indian Views*, 150. White Bull camped in the Sans Arc circle with one of his wives, Holy Lodge, and two of his children.
18 Conversation with Oglala Manuel Iron Cloud in June 1993 at Little Bighorn Battlefield. Manuel was a seasonal park ranger and was a full-blooded Oglala, whose great-grandmother, Red Neckerchief Woman, was in the village at the time of the battle. She received her name as she stood and victorious warriors placed on her upraised arms some red bandanas taken from the dead soldiers on the battlefield. Iron Cloud was also a descendent of Knife Chief.
19 Ambrose, *Crazy Horse and Custer*, 37-44; Bray, *Crazy Horse-A Lakota Life*, 5-9.
20 Bray, *Crazy Horse*, 48-49.
21 Ibid.,71-72.
22 Monnett, *Where a Hundred Soldiers Were Killed*, 210-219.
23 Ambrose, *Crazy Horse and Custer*, 338-342.
24 Bray, *Crazy Horse-A Lakota Life*, 170-175.
25 Neihardt, *Black Elk Speaks*, 7-8, 79.
26 Hassrick, *The Sioux*, 3-6. One Indian map notes the Blackfeet as part of the Hunkpapa camp.
27 Utley, *Lance and Shield*, 54-63.
28 Ibid.,107-114.
29 Ibid.,144.
30 Marquis, *Wooden Leg*, 383.

31 Larson, *Gall-Lakota War Chief*, 15-56.

32 Hardorff, *Baliran, Honzinger, and the Custers: The Facts and Fictions of the Rain-In-The-Face Myth*, 6; Donahue, Michael N., *Drawing Battle Lines-The Map Testimony of Custer's Last Fight*, 153.

33 Bray, *Pine Ridge Letters Shed New Light on the Battle of Little Bighorn*, 45.

34 Hassrick, *The Sioux*, 3-6.

35 Michno, *Lakota Noon-The Indian Narrative of Custer's Defeat*. 3-20. Michno does an authoritative job confirming the actual size of the village on June 25. *Wooden Leg* noted the Cheyenne camp as being slightly past the mouth of Medicine Tail Coulee. This is confirmed in the White Bird and Big Beaver drawings, as well as maps by Rain in the Face, Soldier Wolf, Red Hawk, Respects Nothing, Flying Hawk, Standing Bear, Pretty Buffalo Woman and Red Feather. See Donahue's *Drawing Battle Lines*.

36 Stewart, *Custer's Luck*, 243. Warrior numbers have ranged from 1,000 to 10,000. The National Park Service in their 2005 interpretation has taken a more conservative estimate of 6,000-8,000 people in the camp with 1,500-2,000 warriors present. As suggested by the many comments made by villagers, there can only be estimates and there will never be certainty concerning the camp population. Mitch Bouyer, an experienced scout who counted the campfires left behind, convinced his commander there were at least 1,000 warriors and perhaps as many as 1,500. Estimates of warriors who fought Crook a week before Little Bighorn numbered about 1,200. Given the arrival of more reservation warriors, and that some warriors did not participate in the Rosebud fight, the number seems legitimate.

37 Graham, *The Custer Myth*, 109.

38 Ibid.,75.

39 Marquis, *Wooden Leg*, 18-19.

40 Scott, Fox, Connor, Harmon, *Archaeological Perspectives on the Battle of the Little Bighorn*, 117-121.

41 Such items made by non-Indians were discovered and noted in several soldier accounts as they searched the village. Chris Kortlander and Jason Pitsch, who both own property in the vicinity where the Hunkpapa circle was located, have found remnants of the Hunkpapa circle.

42 Hassrick, *The Sioux*, 245-256. Native religion had multiple aspects like the white man's trinity, but whereas the number three was a holy Christian number, the Lakota believed four was sacred—Wakan Tanka was the Chief God, the Great Spirit, the Creator, and the Executive. They believed in four classes of Gods and four elements that embodied Gods: the Sun, Moon, Sky, and Stars. There were four directions: north, south, east, and west. The four aspects of time were: days, nights, months, and years. All plants contained four parts as well: roots, stem, leaves, and fruit. Four classes of animals existed in their belief system: crawling, flying, two legged, and four legged. There were even four groupings of Gods: the Superior, the Associates, the Subordinates and the Gods. And, they would experience everything in their four phases of life: infancy, childhood, maturity, and old age.

43 Rankin, *Legacy-New Perspectives on the Battle of the Little Bighorn*, Flores, *The Great Contraction*, 3-18.

44 Greene, *Indian War Veterans*, 127; Hammer, Custer in '76, 58. Plainsman George Herendeen who saw the pony herd believed it containing about 10,000 animals. Winfield Edgerly believed that there were 20,000 ponies in the herd.

45 Utley, *Lance and Shield*, 136.

46 Marquis, *Wooden Leg*, 203-204.

47 Hardorff, *Indian Views*, 101; Donahue, *Drawing Battle Lines*, 201.

48 Liberty, *Cheyenne Memories*, 193.

49 Graham, *The Custer Myth*, 48-56.

50 George Bird Grinnell Collection, Notebook 348, 270, Gene Autry Western Heritage Museum.

51 Marquis, *Wooden Leg*, 215.

52 Hardorff, *Indian Views*, 137.

53 Utley, *Lance and Shield*, 144.

54 Hardorff, *Indian Views*, 62.

55 Ibid., 82.
56 Marquis, *Wooden Leg*, 215; Neihardt, *Black Elk Speaks*, 108.
57 Liberty, *Cheyenne Memories*, 194. This dance originated with the Cheyenne.
58 Liberty, *A Cheyenne Voice*, 162-163.
59 Liberty, *Cheyenne Memories*, 193-197.
60 Hardorff, *Hokahey! A Good Day to Die!-The Indian Casualties of the Custer Fight*. 17-30.
61 Gray, *Custer's Last Campaign*, 237.
62 McLaughlin, *My Friend the Indian*, 166-167.
63 Neihardt, *Black Elk Speaks*, 108-109.
64 Hardorff, *Indians Views*, 118.
65 Hardorff, *Lakota Recollections*, 81.
66 Hardorff, *Lakota Recollections*, 92.
67 Hardorff, *Indian Views*, 97-101.
68 Marquis, *She Watched Custer's Last Battle*, 36.
69 Liberty, *Cheyenne Memories*, 195.
70 Marquis, *Wooden Leg*, 216.
71 Liberty, *Cheyenne Memories*, 197-198.

Chapter 10, Red Dawn at the Little Bighorn

1 Gray, *Custer's Last Campaign*, 225.
2 Graham, *The Custer Myth*, 15.
3 Carroll, *Custer's Chief of Scouts*, 87.
4 A. B. Welch papers, Red Star's account of the fight as told by his nephew, Miles Horn, 1938.
5 Libby, *The Arikara Narrative*, 195.
6 A. B. Welch papers, Red Star's account.
7 Libby, *The Arikara Narrative*, 88-89.
8 Graham, *The Custer Myth*, 15-21.
9 Hammer, *Custer in '76*, 60.
10 Carroll, *Custer's Chief of Scouts*, 88.
11 Magnussen, *Peter Thompson's Narrative*, 98-99.
12 Wagner, *Old Neutriment*, 148.
13 A. B. Welch papers, Red Star's account.
14 *Leavenworth (Kansas) Daily Standard*, April 18, 1882.
15 Willert, *Little Bighorn Diary*, 254; Graham, *The Custer Myth*, 136.
16 Graham, *The Custer Myth*, 31.
17 Graham, *The Custer Myth*, 34, 136.
18 Libby, *The Arikara Narrative*, 90.
19 Graham, *The Custer Myth*, 289; *The Washington Times*, November 4, 1906. Chief Trumpeter Voss had once again ordered Martin to the duty of orderly trumpeter for Custer. The last time the young Italian immigrant had defied the chief trumpeter, he ended up being punished and was still awaiting due process on his return to Fort Lincoln.
20 Magnussen, *Peter Thompson's Narrative*, 101. Several contemporary researchers believe information by John Burkman, Luther Hare, and scout Red Star suggested that Custer left the camp and rode directly to the Crows Nest, but Varnum and Martin, make it clear that this was not the case. Peter Thompson stated that Custer rode some distance in front

of the troops, which may be the reason that the column did not see Custer and believed he left them behind. Red Star noted that he accompanied John Martin, Tom Custer, Fred Gerard, Bloody Knife, Bobtailed Bull, and Little Brave in this advance party. While trivial and contradictory in nature, this episode has little to do with later outcomes Burkman's memory and chronology failed him, and he sometimes compressed events in his book *Old Neutriment*. Red Star also compressed events; he had Custer's trumpeter blowing a call at the night camp as they left, when in reality, this event occurred when they left the coffee ravine where the command halted for a short time near the Crows Nest. Tom Custer and Martin were left behind at the coffee ravine. Varnum and Martin's accounts coincide with a great deal of consistency. It also seems unlikely that the command would have halted in the coffee ravine without some direction from Custer.

21 Schonenberger, *A Trooper With Custer-August DeVoto's Account of the Little Bighorn*, 68-71.
22 Klokner, *The Officer Corps*, 80.
23 Magnussen, *Peter Thompson's Narrative*, 103; Libby, *The Arikara Narrative*, 90. Someone familiar with the terrain and with authority had to make the decision on the location to hide the troops. Such an important task was obviously made by Custer as he passed the area and left Tom Custer and John Martin behind. There is no direct testimony to this conjecture, but given the circumstances, this is a likely scenario.
24 Magnussen, *Peter Thompson's Narrative*, pg. 103.
25 Hammer, *Custer in '76*, 61.
26 Libby, *The Arikara Narrative*, 172.
27 Willert, *Little Bighorn Diary*, 258.
28 Carroll, *Custer's Chief of Scouts*, 88.
29 Camp Collection, Box 7, Envelope 77, 470, I. U. It is clear that he was referring to the Washita.
30 Graham, *The Custer Myth*, 33.
31 Libby, *The Arikara Narrative*, 171.
32 Hammer, *Custer in '76*, 231.
33 Graham, *The Custer Myth*, 15, 33; Hammer, *Custer in '76*, 176; Libby, *The Arikara Narrative*, 86-87. The Arikara scouts had other names for the Crows—both Half Yellow Face and White Man Runs Him were also known as 'Big Belly. Since Hairy Moccasin recorded that Half Yellow Face was not at the Crow's Nest with them, the Arikara moniker 'Big Belly,' in this case, refers to White Man Runs Him.
34 Hammer, *Custer in '76*, 183.
35 Evans, *Custer's Last Fight*, 193.
36 Clark, *Scalp Dance*, 16-24. It is difficult to know if this humor was intended as a sexual innuendo or a reference to his lack of Indian fighting experience since sabers had been left behind.
37 Graham, *The Reno Court of Inquiry*, 68. Sergeant William Curtis with a detachment had backtracked to retrieve clothing lost during the night march.
38 Clark, *Scalp Dance*, 15-16.
39 Evans, *Custer's Last Fight*, 193.
40 Hammer, *Custer in '76*, 222.
41 Magnussen, *Peter Thompson's Narrative*, 103.
42 Clark, *Scalp Dance*, 16; Daniels and Davis, *That Fateful Day*, 21.
43 Hunt, *I Fought with Custer*, 75-76.
44 Klokner, *The Officer Corps*, 80-81; Nichols, *Men With Custer*, 264.
45 Graham, *The Custer Myth*, 179.
46 Libby, *The Arikara Narrative*, 84.
47 Ibid.

48 Nichols, *Reno Court*, 20. There is little consistency in soldier and warrior accounts as to the time the battle occurred. Some warriors accounts claim it started early in the morning, while most note it was mid-day. While several soldier accounts claim it was around 3:00 p.m. when the battle started, a few soldiers believed it started earlier, which seems to match those of the warriors. This author believes it started around 1:00 p.m. based on the timing of riding times and distances. Reno and Martin believed the command rode over the divide earlier, around 10 a.m. or 11 a.m. rather than noon. Warrior accounts note that attack was in the morning or mid-day, which seems to dispute Wallace's timeline.

49 Graham, *The Custer Myth*, 179.

50 Hammer, *Custer in '76*, 221.

51 Camp Collection, Box 2, Folder 10, 68, I.U.

52 Hammer, *Custer in '76*, 84. Charles De Rudio makes it clear that Custer asked for his field glasses after returning from his trip from the Crow's Nest.

53 Klokner, *The Officer Corps*, 64, Nichols, *Men With Custer*, 140; Utley,"*Gossip & Scandal*,"19.

54 Fougera Collection, *Custer's Last Command*, 3.

55 Nichols, *Reno Court*, 439.

56 Custer, *My Life on the Plains*, (Stewart version), 190.

57 Hardorff, *On the Little Bighorn with Walter Camp*, 184. The soldier who overheard this conversation was Private Freemont Kipp of Company D.

58 Graham, *The Custer Myth*, 179.

59 Nichols, *Reno Court*, 403. While there were three surgeons with the command, none accompanied Benteen.

60 Smalley, *Little Bighorn Mysteries*, 12-3.

61 Evans, *Custer's Last Fight*, 193.

62 Wagner, *Old Neutriment*, 149. While Burkman speculates that Custer joined the five companies by chance, such was not the case.

63 Ibid. 151.

64 Ibid. John Burkman believed in error that Martin was relieved his duty as orderly trumpeter.

65 Ibid. 150-152.

66 Hardorff, *Camp, Custer, and the Little Bighorn*, 81.

67 Barnard, *I Go with Custer*, 138.

68 Wagner, *Old Neutriment*, 155.

69 Carroll, *Custer's Chief of Scouts*, 64.

70 Hammer, *Custer in '76*, 64.

71 Carroll, *Custer's Chief of Scouts*, 64.

72 Magnussen, *Peter Thompson's Narrative*, 108-109; Camp Collection, Box 2, Folder 7, 32 I.U.

73 Graham, *The Custer Myth*, 15; Daniels and Davis, *That Fatal Day*, 21; Hammer, *Custer in '76*, 78.

74 Nichols, *Reno Court*, 512.

75 Wagner, *Old Neutriment*, 156.

76 *New York Evening Post*, February 20, 1897, 12.

77 *New York Evening Post*, Feb. 20, 1897; Graham, *The Custer Myth*, 179-180.

78 *New York Evening Post*, Feb. 20, 1897.

79 Nichols, *Reno Court*, 432.

80 Gray, *Last Campaign*, 258.

81 Burdick, *Army Life*, 27.

82 Hammer, *Custer in '76*, 69.

83 *New York Herald*, September 21, 1876.

84 George Kush Collection, unpublished W. W. Cooke material.
85 Graham, *The Custer Myth*, 289.
86 Hammer, *Custer in '76*, 65.
87 Hammer, *Custer in '76*, 161; Nichols, *Reno Court*, 139.
88 Graham, *The Custer Myth*, 34; Gray, *Last Campaign*, 271.
89 Gray, *Custer's Last Campaign*, 254-255.
90 Graham, *The Custer Myth*, 34; Hammer, *Custer in '76*, 192; Gray, *Custer's Last Campaign*, 255; Sklenar, *To Hell with Honor*, 128. The names 'Sitting Bear,' 'Old She Bear,' 'Plenty Bears,' and 'Little Wolf' have all been used to identify this warrior, who most agree was from the Sans Arc band. George Herendeen also took his knife and cut open the lodge to see the dead warrior wrapped in buffalo hide on a burial scaffold.
91 Libby, *The Arikara Narrative*, pg. 121.
92 Ibid.
93 Magnussen, *Peter Thompson's Narrative*, 110; Hardorff, *On the Little Bighorn with Camp*, 6. In Custer's mind, the village was running, and this blatant action makes clear that he no longer believed stealth was important. It is highly likely that he hoped that Terry was near enough to see the smoke and to let him know their whereabouts.
94 Libby, *The Arikara Narrative*, 121.
95 Nichols, *Reno Court*, 84.
96 Libby, *The Arikara Narrative*, 121-122.
97 Graham, *The Custer Myth*, pg. 34.
98 Nichols, *Reno Court*, 21.
99 Nichols, *Reno Court*, 561; Gray, *Last Campaign*, 275. The 'village' referred to in the orders has been confused with the larger camp later discovered in the valley. It is clear from the distances and circumstances that Custer was referring to the village [or party of warriors] they has just flushed. No one except Varnum knew that it existed for certain at this point, and he had not relayed the important information until after Reno departed. Reno's official report was slightly different, "Adjt. Cooke came to me and said the village was only three miles away and running away; to move forward at as rapid a pace as prudent and to charge afterward, and that the whole outfit would support me." (Gray, *Last Campaign*, 275)
100 Nichols, *Reno Court*, 188.
101 Sklenar, *To Hell with Honor*, 143. Reno's battalion consisted of an estimated 141 officers and enlisted men, supplemented by the detachment of 37 scouts and interpreters.
102 Graham, *The Custer Myth*, 34.
103 Magnussen, *Peter Thompson's Narrative*, 117; Carroll, *Custer's Chief of Scouts*, 64-65.
104 Carroll, *Custer's Chief of Scouts*, 119.
105 Stewart, *Custer's Luck*, 331.
106 Hammer, *Custer in '76*, 148.
107 Ibid., 65.
108 Hardorff, *Hokahey! A Good Day to Die*, 17-30.
109 Nichols, *Reno Court*, 332.
110 Ibid., 366.
111 Ibid., 87.
112 Gray, *Last Campaign*, 277.
113 Nichols, *Reno Court*, 388. The horses watered for about five or ten minutes according to Martin. The author has confirmed both the site of the spring and the travois trail (at present on land that was formerly owned by Bob Eshelman). During blasting in the 1960s, the spring quit running according to the landowner. The remnants of the Indian travois trail

still remain. Both are on the flat between the middle and north fork of Reno Creek and is ¾ mile from the river which is close to the White Man Runs Him's testimony on page 23 of *The Custer Myth.*

114 Graham, *The Custer Myth,* 13.

115 Nichols, *Reno Court,* 397.

116 Hammer, *Custer in '76,* 97.

117 Hardorff, *Indian Views of the Custer Fight,* 174; Hammer, *Custer in '76,* 215.

118 Hammer, *Custer in '76,* 55. Walter Camp noted, "Edgerly thinks Custer really intended to follow Reno into [the] village but thinks when he got near the river he changed his mind. He thinks Knipe's account probably an explanation of Custer changing his mind." (Ibid)

119 Graham, *The Custer Myth,* 25.

Chapter 11, Boys, Hold Your Horses!

1 Hammer, *Custer in '76,* 112.

2 Graham, *The Custer Myth,* 249, 289; Magnussen, *Peter Thompson's Narrative,* 116; Hammer, *Custer in '76,* 92, 156.

3 Nichols, *Reno Court,* 397. This long ridge along the bluffs is known as Sharpshooter Ridge today.

4 Graham, *The Custer Myth,* 13.

5 Carroll, *Custer's Chief of Scouts,* 138.

6 Graham, *The Custer Myth,* 11, 20, 39, 289-290; Nichols, *Reno Court,* 388, 397; Camp Collection, Camp to Thompson, December 21, 1908, LBBNM. John Martin estimated that he and Custer came into view of the camp about 300 yards north of where Reno was later corralled. The southern end of Sharpshooter Ride is about 600 yards from Reno Hill. Goes Ahead noted that the four Crows and an Arikara rode next to the command as they climbed the bluffs next to and slightly ahead of the troops. According to Goes Ahead they all arrived at the point, "where Lieutenant Hodgson's headstone was placed later." [The original marker has been relocated]. The marble stone was originally placed 500 yards from the Reno Hill area and near the southern end of Sharpshooter Ridge. The Arikara testimony recalled, "Custer stood on the bank where Hodgson's stone stands." (Graham, *The Custer Myth,* 20)

7 Hammer, *Custer in 76,* 100; Graham, *The Custer Myth,* 290.

8 Graham, *The Custer Myth,* 13, 39.

9 Graham, *The Custer Myth,* 290. Scores of army officers, writers, scholars, and historians have ignored or simply dismissed Martin's eyewitness testimony.

10 Graham, *The Custer Myth,* 290. Custer was surprised and glad, because he had expected the village to be on the run, as his scouts had predicted a slim-to-none chance of surprise since leaving the Crow's Nest.

11 Hammer, *Custer in '76,* 100.

12 Unger, *Custer's First Messenger!?,* 26.

13 Hammer, *Custer in '76,* 188. It is clear that John Martin either did not see warriors attacking Reno or thought them unimportant, or perhaps he was not looking in that direction. Martin had no personal reason to fabricate some unheralded story from the bluffs. Arikara scout Soldier, Daniel Knipe and Peter Thompson, who also survived the battle and were on the bluff top, did not mention a huge party of warriors rushing to meet Reno either.

14 Graham, *The Custer Myth,* 290.

15 Graham, *The Custer Myth,* 290; Nichols, *Reno Court,* 395; *The Washington Times,* November 4. 1906.

16 Camp Collection, Knipe to Camp, Folder 2, July 14, 1908, B.Y.U.; Graham, *The Custer Myth,* 290.

17 Hammer, *Custer in '76,* 188.

18 Magnussen, *Peter Thompson's Narrative,* 119.

19 Hammer, *Custer in '76,* 100.

20 Graham, *The Custer Myth*, 23, 249; Magnussen, *Peter Thompson's Narrative*, 116, 120; *The Washington Times*, November 4, 1906.

21 Graham, *The Custer Myth*, 249.

22 Magnussen, *Peter Thompson's Narrative*, 120.

23 Hammer, *Custer in '76*, 188.

24 Ibid., 93.

25 Ibid., 92.

26 Nichols, *Reno Court*, 388-389.

27 Nichols, *Reno Court*, 398; Taylor, W. O., *With Custer on the Little Bighorn*, 177. John Martin recalled that only three Crow scouts were in the vicinity.

28 Graham, *The Custer Myth*, 17.

29 Hammer, *Custer in '76*, 84.

30 Ibid., 151.

31 Graham, *The Custer Myth*, 34.

32 Camp Collection, Interview with White Bull and One Bull, Box 5, Envelope 41, 350, I. U.

33 Willert, *Little Bighorn Diary*, 277-278.

34 Barnard, *Custer's First Sergeant John Ryan*, 176.

35 Carroll, *Custer's Chief of Scouts*, 121; Evans, *Custer's Last Fight*, 211.

36 Mangum, *Reno's Battalion in the Battle of the Little Big Horn*, 5, Willert, *Little Bighorn Diary*, 278.

37 Nichols, *Reno Court*, 189.

38 Daniel and Davis, *That Fatal Day*, 26.

39 Hammer, *Custer in '76*, 198.

40 Graham, *The Custer Myth*, 76, 79; Hardorff, *Indian Views*, 35; Vestal, *Warpath*, 192-193; Campbell Collection, Box 111 Folder 2, University of Oklahoma.

41 Willert, *Little Bighorn Diary*, 286; Burdick, *David F. Barry's Indian Notes on "The Custer Battle*, 23; Nichols, *Reno Court*, 157-158. Varnum remembered, "About the time Major Reno's command dismounted in the bottom . . . looking on the bluffs across the river I saw the Gray Horse Company of the regiment moving along those bluffs . . . It was back from the edge of the bluffs." (Nichols, *Reno Court*, 157-158).

42 Gray, *Custer's Last Campaign*, 295- 298.

43 Hardorff, *Camp, Custer, and the Little Bighorn*, 58.

44 Nichols, *Reno Court*, 88.

45 Graham, *The Custer Myth*, 90; Larson, Robert W., *Gall-Lakota Warrior Chief*, 120. It is possible that some of the women, digging turnips near the benchlands, were killed in the area. Former landowner Jason Pitsch showed the author a site on the benchlands where a mass grave of women and children still exists. While probably buried on scaffolds or hidden in the ravine near the bluffs, someone obviously gathered the remains and buried them in one location at a later date. An early newspaper report noted that ten women were killed in a ravine near the village evidently by Ree or Crow scouts. (*Bismarck Tribune*, July 7, 1876) The casualties were later noted as four children and six women.

46 Hardorff, *Hokahey!*, 32-34.

47 Brininstool, *Troopers with Custer*, 48; Barnard, *Custer's First Sergeant John Ryan*, 179.

48 Mangum, *Reno's Battalion*, 5.

49 Neihardt, *Black Elk Speaks*, 120.

50 Donahue, *Drawing Battle Lines*, 126.

51 Nichols, *Reno Court*, 25, 91, 222, 314.

52 Noyes, *The Valley Fight: The Enlisted Man's Perspective*, 40; Nichols, *Reno Court*, 367; Donahue, *Drawing Battle Lines*, 126.

53 Hammer, *Custer in '76*, 106-107.
54 Neihardt, *Black Elk Speaks*, 109-110.
55 Graham, *The Custer Myth*, 75.
56 Hardorff, *Indian Views*, 176.
57 Barnard, *Ten Years*, 293.
58 Hardorff, *Indian Views*, 139, 154.
59 Hardorff, *Lakota Recollections*, 40.
60 Hardorff, *Indian Views*, 148, 176.
61 Hardorff, *On the Little Bighorn with Camp*, 17; Nichols, *Men With Custer*, 178.
62 Hardorff, *Lakota Recollections*, 83; Mangum, *Reno's Battalion*, 5.
63 Nichols, *Reno Court*, 314, 562.
64 Willert, *Little Bighorn Diary*, 287-297.
65 Nichols, *Reno Court*, 562-563.
66 Ibid., 217.
67 Mangum, *Reno's Battalion*, 5.
68 Hammer, *Custer in '76*, 143.
69 Graham, *The Custer Myth*, 290.
70 Gibbon, John, *Gibbon and the Sioux Campaign of 1876*, 33; Carroll, John, *The Benteen-Goldin Letters*, 24. Colonel John Gibbon stated Custer's trail went over the sharp peak, today called Weir Point. He noted, "Custer . . . skirted the hills to our front, passed to the right of the sharp peak, and still on, beyond it and out of sight of where we stand." He also noted Benteen's decision to follow Custer's trail, "One of Colonel Reno's companies is mounted and started for the scene of Custer's fight. It leaves our position, and winding along the rolling hills, ascends the high ground to right of the high peak, and disappears beyond, just as Custer's command would have vanished probably from the sight of an observer standing where we are now." (Gibbon, *Gibbon and the Sioux Campaign of 1876*, 33) Theodore Goldin followed the trail with Martin on June 28 and it matched the Maguire map—the trail went over Weir Point and behind the edge of the first line of bluffs.
71 Nichols, *Reno Court*, 398.
72 Magnussen, *Peter Thompson's Narrative*, 122.
73 George Kush Collection, French letter to Dr. A. H. Cooke, July 4, 1876.
74 Hammer, *Custer in '76*, 103.
75 Hardorff, *Lakota Recollections*, 143; Camp Collection, Interview with Thompson, Box 7, Envelope 129, 610, I. U. This F Company detail was seen by Peter Thompson, Jacob Hetler and recorded by Lakota Two Eagles who saw a "few" soldiers going along the ridge top, while the "main body of troops" went toward the river.
76 Hutchins, *The Papers of Edward S. Curtis Relating to Custer's Last Battle*, 32-33. The three Crow scouts and John Martin were the only surviving participants interviewed about Custer's route to the river, from Sharpshooter Ridge. Photographer E. S. Curtis mapped Custer's route from the Crow testimony. Their map is the same as the Maguire and Benteen maps indicate Custer's advance over Weir Point. See Donahue's *Drawing Battle Lines* for the map evidence of this movement noted by Curtis, 266, Benteen, 66, and Maguire, 26, 31, 38, 41.
77 Graham, *The Custer Myth*, 94.
78 Hunt, *I Fought with Custer*, 84; George Kush Collection, Samuel Alcott papers.
79 Donahue, *Drawing Battle Lines*, 152-155.
80 Graham, *The Custer Myth*, 86.
81 Liberty, *A Northern Cheyenne Album*, 226.
82 Smalley, *More Little Bighorn Mysteries*, 2-4; Evans, *Custer's Last Fight*, 199.

83 The note is now in the collection of the West Point Library. Benteen in the upper right later mistranslated the note and his interpretation points out the last stanza reads, "P.S. Bring pack." A close look at other handwriting by Cooke in the collection of George Kush notes that the "P.S." is actually an ampersand symbol. Cooke also had a tendency to scribble and drop letters off the ends of words when he wrote in a hurry. While the second notation appears to say the word "pack," it actually repeats the words "packs" used earlier in the message. For a clearer and accurate translation, see Vern Smalley's *Little Bighorn Mysteries*, section 8, 10-11.

84 *The Washington Times*, November 4, 1906.

85 Hammer, *Custer in '76*, 100-101.

86 Taylor, *With Custer on the Little Bighorn*, 177.

87 Pickard, *I Fought with Custer*, 5-6, *Portland (Oregon) Journal*, July 31, 1923. As Private Pickard rode away, his 'bunkie' William LeRock told the disappointed horseherder, "All right Ed, if you'll have supper ready for me when we come back, I'll bring you a buffalo robe." Pickard responded, "All right Rockie, see that you get a good one." (Ibid)

88 The Charles Banks information is in the private collection of George Kush, who kindly shared the statement with the author, and wishes to reserve the other details for his long-awaited volume on Custer's adjutant W. W. Cooke.

89 Hammer, *Custer in '76*, 104; Graham, *The Custer Myth*, 290.

90 *The Washington Times*, November 4, 1906.

91 Hammer, *Custer in '76*, 104.

92 *The Washington Times*, November 4, 1906.

93 Hammer, *Custer in '76*, 104.

94 Graham, *The Custer Myth*, 290.

95 Nichols, *Reno Court*, 430.

Chapter 12, Chases Along the Greasy Grass

1 Graham, *The Custer Myth*, 20.

2 Wagner, *Old Neutriment*, 158-159; Graham, *The Custer Myth*, 20. Curly had separated from the other Crow scouts earlier and joined the Arikara horse raiders when they reached the bluff top and rode back in the direction of the pack train.

3 Donahue, *Drawing Battle Lines*, 152; Liberty, Stands in Timber unpublished manuscript, 354; Nichols, *Reno Court*, 299.

4 Hammer, *Camp in '76*, 101.

5 Hardorff, *Camp on Custer*, 122.

6 Winkler, *The Case for a Custer Battalion Survivor-Private Gustave Korn's Story*, 49-50; Hardorff, Richard G., *Markers, Artifacts and Indian Testimony: Preliminary Findings on the Custer Battle*, 27.

7 Winkler, *The Case for a Custer Battalion Survivor*, 49-50.

8 George Kush Collection, Gustave Korn account. While some writers have suggested that Gustave Korn rode through the village of thousands of noncombatants and warriors, such a premise is absurd. Korn himself never claimed that. From his account, it is clear that he turned his horse to the left, and rode up the bluffs in the direction of Weir Point, where he later joined the troops on Reno Hill. Numerous other soldiers claimed Korn told them other stories, but this author has chosen to use only those from Korn himself.

9 Hardorff, *Cheyenne Memories*, 50; Hammer, *Custer in '76*, 86. Pretty Shield's notation of seven groups seems incorrect since Custer had only five companies. The headquarters staff appeared as another unit, and the small F company detachment may have appeared as an additional group. Another possibility is that there were groups of flankers riding with the column that was a tactical formation used by Custer.

10 Hardorff, *Cheyenne Memories*, 52. One reason the Cheyenne casualties were disproportionate to the village population was the fact that they engaged the five companies first and fought them for a longer period of time. Most of the Lakota were fighting Reno as the five companies approached the river.

11 George Bird Grinnell Collection, Bobtail Horse interview, 13, The University of Oklahoma Western History Collection; Hardorff, *Cheyenne Memories*, 142. It seems that Companies I and E were the two companies that charged near the river. The warriors saw the gray horses there, and Korn noted his Company was in the lead and by seniority Keogh would lead such a charge.

12 Stands in Timber unpublished manuscript, 344; Hardorff, *Indian Views*, 40-41. The warriors called Medicine Tail Coulee—Dry Creek and Water Rat Creek.

13 Nichols, *Reno Court*, 138. Soldier and civilian survivors examined the bank on the village side after the fight. Fred Gerard concluded, "The opposite bank was miry and the bank was about 18 inches higher than the water. I found there had been no horses or animals across there. I knew it must be miry or the Indians themselves would use it." (Ibid)

14 Graham, *The Custer Myth*, 96; Nichols, *Reno Court*, 32, 418, 424, 501. Wallace and Godfrey saw no tracks on the banks of the village side of the river, and Benteen admitted it was miry with quicksand but bragged he could cross the river anywhere.

15 Hammer, *Custer in '76*, 198; Hutchins, *The Papers of Edward S. Curtis*, 62. It is probable that Curly's story of the river casualty was from Cheyenne oral history.

16 Liberty, Margot, *A Northern Cheyenne Album*, 226.

17 Hardorff, *Indian Views*, 40-41. Private Edward Pickard who left the command in Medicine Tail Coulee also attested to the difficult ford, noting, "I did not know it at the time, but we discovered later, that where I left them [Medicine Tail Coulee], and where they were going to cross, the river was impassable." (Pickard, *I Rode with Custer*, 5)

18 Donovan, *A Terrible Glory*, pg. 453; Brust, James, "Lt. Oscar Long's early map details terrain, battle positions," *Greasy Grass*, Volume 11, May 1995, pg. 8.

19 Miller, *Custer's Fall*, 128. Much of Miller's book is supposition supposedly based on warrior oral history.

20 Donovan, *A Terrible Glory*, 453-454; Brust, "Lt. Oscar Long's early map details terrain, battle positions," 8. Author James Donovan was the first to suggest James Sturgis as this early casualty. Sturgis's buckskin shirt, underdrawers, flannel under shirt and spurs were found in the village. The top button of his shirt was still buttoned, indicating possible decapitation. One soldier thought he recognized the head of Sturgis in the village. Another private recalled three heads burned beyond recognition, and near the fire found a blue flannel shirt with initials J. G. S. [James Garland Sturgis].

21 E. A. Garlington Papers, Crombey letter to Captain E. A. Garlington, June 10, 1891, John M. Carroll Collection, courtesy of George Kush; Hanson, *The Conquest of the Missouri*, 379; Walter Camp Collection, Mathey to Walter Camp, April 13, 1910, BYU. Sergeant Edward B. Crombey joined Company I of the 7th Cavalry immediately following the battle and was part of the extensive cleanup of the burials in 1877. He noted, "At the spot where General Custer attempted to cross the river but was ambushed we discovered (June 1877) a low patent shoe belonging to Captain Myles Keogh and close to it one of the captain's bloody white canvas leggings." (Crombey letter) The man who actually collected the shoe was Sergeant Caddle—"On the field (June 1877) Sergeant Caddle picked up a shoe that belonged to Captain Keogh." (Hanson, *Conquest of the Missouri*, 379) That some of Keogh's personal items were marked was verified by the discovery of his gauntlets at the later battle of Slim Buttes and it was typical for soldiers to mark their personal gear as noted in numerous accounts. Keogh had his name tags cut from his socks that were found on his body. While many writers have mistakenly designated the horse Comanche as Keogh's mount, the gelding was actually a government owned horse ridden by Lieutenant Mathey right before the battle. 'Paddy' was Keogh's personal horse and Keogh switched to Comanche shortly before the battle.

22 George Kush Collection, Thomas French letter to Dr. A H. Cooke, July 6, 1876.

23 Hardorff, *On the Little Bighorn with Walter Camp*, 8-9. Another soldier stated that Bustard was found near Keogh's body near Horseholder's Ravine.

24 Hardorff, *Camp, Custer, and the Little Bighorn*, 82.

25 Hardorff, *Cheyenne Memories*, 126.

26 Ibid., 52.

27 Ibid., 43.

28 Ibid., 47.

29 Hammer, *Custer in '76*, 213.

30 Hardorff, *Indian Views*, 48. This was likely Company I.

31 Hardorff, *Cheyenne Memories*, 90; J. W. Vaughn letter to Andrew Loveless, July 3, 1965, Sheridan County Library, Sheridan, Wyoming; Hardorff, *Indian Views*, 41. Artifacts left by soldiers were discovered by historian Don Rickey and friend J. W. Vaughn near the mouth of Medicine Tail Coulee. A recent archeological project in 2004 also found soldier cartridges near the mouth of Deep Coulee. Vaughn discovered three expended 45/55 army Springfield cartridges, one 50-70 Springfield cartridge, a soldier's button, an iron tent pin and an iron arrowhead in Medicine Tail Coulee not far from the mouth of Medicine Tail Coulee. Moving toward what he called the North Ford [mouth of Deep Coulee], he discovered two expended 45/55 army cartridges on the west bank [Indian side of river] and an army bullet imbedded in the edge of the riverbank. Vaughn and Don Rickey found on the east side of the river a harness buckle, two .44 Winchester cartridges, a broken horseshoe, 2 45/55 army cartridges that were in slivers with their heads missing, and three Spencer casings.

32 George Kush Collection, Gustave Korn account; Hardorff, *Cheyenne Memories*, 126.

33 Hardorff, *Cheyenne Memories*, 126.

34 This estimate is based on the timing of Reno's fight, retreat to the hilltop, and the arrival of Jacob Hetler who witnessed one battalion riding on Nye/Cartwright Ridge. See Michael Donahue's *The Battle of the Little Bighorn—A New Time Line*, 18th Annual Symposium of the Custer Battlefield Historical and Museum Association, June 2004.

35 Grinnell Collection, Bobtail Horse interview, 13.

36 Nichols, *Reno Court*, 644. Reno saw some unnamed evidence of Companies C and I and part of E in the area near the river. In 2004, an archeological survey was done along the current road and over 200 artifacts were discovered. Army cartridges, matched by firing pin impressions, show that at least two soldiers fired their weapons near the mouth of Deep Coulee and then along a route that included Finckle/Finely Ridge and the Calhoun/Keogh sectors. Such data seems to indicate that Company C soldiers at the river also fought at least on Finckle/Finley Ridge. Since there are no early soldier casualties noted in the retreat, they were likely not captured weapons. The Company C archeology seems to back Reno's report.

37 Hardorff, *Lakota Recollections*, pp. 41-43. Since the gray horse company (E) was seen along Nye/Cartwright Ridge, and that company has been a part of Yate's initial battalion as noted by Knipe, this three-company battalion re-organization with Yates as commander is logical. If Keogh was incapacitated, Yates would be the next in line for command. For more information about seniority and command structure, see Dr. Richard Fox's *Archaeology, History, and Custer's Last Battle* pp. 141.

38 Hardorff, *Camp, Custer and the Little Bighorn*, 85. This description could refer to battle ridge, but since the eyewitness was there when the soldiers attempted to cross the river, the chronology of Nye/Cartwright Ridge seems more probable.

39 Mangum, *Reno's Battalion*, 5; Nichols, *Reno Court*, 368; Personal interview with Jason Pitsch during the summer of 2009. Jason Pitsch metal detected this land owned by his family and mapped 45/55 cartridge finds along the slope helping verify for the first time the actual skirmish line in the timber.

40 Marquis, *Wooden Leg*, 220.

41 Nichols, *Reno Court*, 563.

42 Brininstool, *Troopers with Custer*, 131.
43 Nichols, *Reno Court*, 227; Hardorff, *Lakota Recollections*, 64-65.
44 Barnard, *Ten Years with Custer*, 293.
45 Mangum, *Reno's Battalion*, 5.
46 Daniels and Davis, *That Fatal Day*, 11; Nichols, *Reno Court*, 284.
47 Nichols, *Reno Court*, 284.
48 Ibid., 255.
49 Ibid.
50 Nichols, *Reno Court*, 126; Taylor, *With Custer on the Little Bighorn*, 38.
51 Mangum, *Reno's Battalion*, 5.
52 Hammer, *Custer in '76*, 112.
53 Mangum, *Reno's Battalion*, 5.
54 Carroll, *Custer's Chief of Scouts*, 163; Nichols, *Reno Court*, 317.
55 Marquis, *Wooden Leg*, 220-221; Hardorff, *Lakota Recollections*, 83; Hammer, *Custer in 76*, 119.
56 Barnard, *Ten Years with Custer*, 294.
57 Barnard, *Ten Years with Custer*, 293-294; Daniels and Davis, *That Fatal Day*, 7.
58 Hammer, *Custer in '76*, 223; Hardorff, *Custer Battle Casualties*, 150.
59 Hammer, *Custer in '76*, 119.
60 Dixon, *The Vanishing Race*, 173.
61 Hammer, *Custer in '76*, 223-224; Daniels and Davis, *That Fatal Day*, 9; Hardorff, *Battle Casualties*, 148-150.
62 Hammer, *Custer in '76*, 106-110.
63 Hardorff, *Cheyenne Memories*, 28.
64 Marquis, *Wooden Leg*, 221-222.
65 Ibid., 222.
66 Hardorff, *Indian Views*, 54.
67 Carroll, *Custer's Chief of Scouts*, 66-67.
68 Stewart, *Custer's Luck*, 375; Hardorff, *Battle Casualties*, 131; Camp Collection, Morris letter to Camp, Folder 11, Dec. 24, 1909 & Coolidge letter to Camp, Folder 16, March 14, 1911, BYU. Donald McIntosh's body was later moved some distance from the original death site.
69 Hardorff, *Battle Casualties II*, 99; Daniels and Davis, *That Fatal Day*, 11.
70 Barnard, *Ten Years with Custer*, 296.
71 Hardorff, *Indian Views*, 84.
72 Neihardt, *Black Elk Speaks*, 111-112.
73 Mills, *Harvest of Barren Regrets*, 260.
74 Daniels and Davis, *That Fatal Day*, 7-8.
75 Marquis, *Wooden Leg*, 224.
76 Hardorff, *Battle Casualties*, 155-156; Libby, *The Arikara Narrative*, 126-130.
77 Cowdrey, *Crows Who Scouted for Custer*, 39-44. Cheyenne Brave Bear lanced White Swan while another warrior shot him.
78 Neihardt, *Black Elk Speaks*, 111-112.
79 Nichols, *Reno Court*, 351; Stevenson, *Deliverance from the Little Big Horn*, 52.
80 Walker, *Dr. Henry Porter-The Surgeon who Survived Little Bighorn*, 54-60.
81 Hardorff, *Battle Casualties II*, 106-108.
82 Daniels and Davis, *That Fatal Day*, 27.
83 Smalley, *Little Bighorn Mysteries*, section 17-5; Evans, *Custer's Last Fight*, 448. Another source claims that 36 soldiers

were killed in the valley and the location of some dead will never be proven.

84 Evans, *Custer's Last Fight*, 220.

85 Charles F. Bates Collection, Godfrey letter to J. A. Shoemaker, March 2, 1926, LBBNM; Hammer, *Custer in '76*, 75.

86 Godfrey, *General George A. Custer and the Battle of the Little Big Horn*, 22.

87 Hammer, *Custer in '76*, 93.

88 Ibid., 75.

89 Graham, *The Custer Myth*, 290-291.

90 Nichols, *Reno Court*, 404. For an Italian immigrant, who spoke broken English, to use such terminology is intriguing and unlikely—it sounds more like a 'Benteenism,' as he often used such slang in his personal letters. Such a response suggests Benteen was rationalizing a reason for staying away from the pack mules.

91 Nichols, *Reno Court*, 441.

92 Nichols, *Reno Court*, 405; Marquis, *Wooden Leg*, 222-223.

93 Hardorff, *On the Little Bighorn with Walter Camp*, pg. 21. Private James Rooney stated that Benteen had a fishing pole thrown over his shoulder when he arrived on the hilltop.

94 Graham, *The Custer Myth* 291; Camp Collection; Interview with John Martin, Folder 3, 130, I. U.

95 Hardorff, *Camp, Custer, and the Little Bighorn*, 102.

96 Marquis, *Wooden Leg*, 226.

97 Dixon, *The Vanishing Race*, 174.

Chapter 13, Death on a Pale Horse

1 Taunton, *No Pride in the Little Big Horn*, 19.

2 Greene, *Indian War Veterans*, 129. Hetler's account matches the eyewitness testimony of Wolf Tooth and Peter Thompson. The company F detail were likely the vedettes that Hetler saw and were also noted by Thompson along the northeastern ridges. More importantly, it helps to time the event of the arrival of Company D and the beginning of the Nye/Cartwright fight. Neither Wooden Leg, Wolf Tooth nor Hetler noted that the soldiers along Nye/Cartwright were fighting—both state they were riding without being engaged in combat. This means that there was no early pressure on the soldiers on Nye/Cartwright Ridge. There would soon be volley firing that many on Reno Hill would hear—it was coming from Nye/Cartwright Ridge.

3 Hammer, *Custer in '76*, 66; Camp Collection; Camp letter to Knipe, March 27, 1910, LBBNM. Luther Hare was absent for about twenty minutes. He discovered the lead pack mules on a small hill near the area where Custer had watered his horses on Reno Creek.

4 Nichols, *Reno Court*, 373.

5 Ibid., 351, 354.

6 Ibid., 98, 316, 323, 327, 342. Charles De Rudio, who was in the valley, clarified that he heard firing about 10 minutes after Reno reached the hilltop, which was verified by those on top of the hill. De Rudio said, "[I] heard immense volleys of firing on the other side of the village . . . I don't think it lasted over an hour and a half, the heavy firing . . . The first volley was very plain, then it got further on and then it died out . . . I could hear the volleys and tell they were going away . . . the firing was steady for a long time and in volleys, and after that it was scattering and lasted but a short time . . . Major Reno was about at the top of the hill when that fire started." (Ibid., 316-342) Fred Gerard, who was hiding with De Rudio, articulated that about 10 or 15 minutes after Reno left the timber, "I heard . . . three or four volleys, as if there were 50 or 100 guns at a volley . . . there was a continuous firing all the time the [Custer's] troops were marching down there, not regular volleys, but scattering shots, sometime three or four and sometimes only one, and then it was kept up irregularly, and when it got down below there, where Custer's battlefield was, it became heavy." (Ibid., 98)

7 Ibid., 373, 377.

8 Nichols, *Reno Court*, 566; Carroll, *Custer's Chief of Scouts*, 116.

9 Ibid., 566.

10 Ibid., 581.

11 Ibid., 569.

12 Nichols, *Reno Court*, 407, 430; D. F. Barry Collection; Benteen letter to Barry, March 1, 1895, LBBNM. Frederick Benteen criticized Reno and in a self-protective commentary, noting, "I don't think that Reno pushed the Indians on the bottom on the 25th June '76 quite as hard as he might have done,—not that such 'pushing' would have been of the slightest degree of benefit, but it would have showed more true and soldierly instinct. Nor do I think that he should have retreated at the head of the column in the bottom without even stopping to withdraw the dismounted skirmishers—not even having a trumpet sounded toward such end! These facts exhibit too truly that he was about completely demoralized. However, once on the hill, and after junction of my battalion, all that could be done towards getting to Custer was certainly done by me, but you know there were too many Indians between Custer's point and ours to help him, and the dispatching of Custer's forces, only consumed about 35 minutes, so all of them were killed before Reno's forces had recovered from terror caused by [the] flight from [the] bottom. Some of the violent critics should have been there!" (D. F. Barry Collection; Benteen letter to Barry, March 1, 1895)

13 Nichols, *Reno Court*, 37.

14 George Kush Collection, Gustave Korn account.

15 Nichols, *Reno Court*, 236.

16 Hammer, *Custer in '76*, 70.

17 Nichols, *Reno Court*, 529.

18 Ibid., 574, 625.

19 Ibid., 590.

20 Ibid., 408, 425. Wallace also downplayed any volley firing from Custer's men at the Reno Court of Inquiry. Godfrey, who was almost deaf in one ear, heard the volleys.

21 Ibid., 444-445.

22 Hammer, *Custer in '76*, 55.

23 Liddic & Harbaugh, *Camp on Custer*, 94-95.

24 Ibid..

25 Taunton, *No Pride in the Little Big Horn*, 21-23.

26 Clark, *Scalp Dance*, 19-20. Weir later told Edgerly he had not asked Reno for permission to advance which seems nonsensical given the situation and the other eyewitnesses.

27 Nichols, *Reno Court*, 408.

28 Ibid., 289, 567.

29 Nichols, *Reno Court*, 644; Grinnell Collection; Bobtail Horse interview, 13, University of Oklahoma. The 2004 road dig discovered a soldier 45/55 copper casing [field sample 9697] near Deep Coulee and the road near the river. It's firing pin markings matched cartridges also found at Finckle/Finley Ridge as well as the Keogh sector. Field sample 9696, another 45/55 casing was found near the road in Deep Coulee not far from the river. It matched cartridges found on Finckle/Finley Ridge and Calhoun Hill. It seems that at least two soldiers in the companies angling toward the high ground, fired their weapons near the river and on Finckle/Finley Ridge. Another scenario could be that the weapons were captured at the Finckle/Finley site where soldier fell first and were used against the soldiers on Calhoun Hill and the Keogh sector. While it is impossible to rely totally on the cartridge information alone, such evidence matches warrior oral testimony and also challenges some theories concerning which companies made up the battalion retreating from the river. Which companies were with the two withdrawing battalions is something of a mystery. Reno believed that Companies C and I and part of E attempted to cross the river, but does not give specifics

on why be believed this. The cartridges found on the retreat line, Finckle/Finley Ridge and Calhoun Hill seems to indicate Company C was one of the companies that fell back through Deep Coulee, while the gray horse company was part of the battalion on Nye/Cartwright Ridge as seen by the warriors. The composition of the rest of the battalion's makeup remains conjecture. (Copy of 2004 archeological dig and emails from Doug Scott in author's collection.)

30 Edward Luce letter to J. W. Vaughn, November 12, 1961, Sheridan Library, Sheridan, Wyoming. The timing of the troops along Nye/Cartwright can be deduced by Reno's retreat, Martin's time estimations, and Jacob Hetler's observations. Both Hetler and Wolf Tooth describe the same scene with soldiers nonchalantly moving along the ridge with little Indian pressure. Neither speaks of the soldiers fighting along the ridge in the early stages of the withdrawal. Major Edward Luce did sound tests and was correct in his deductions, noting, "I have never gone along with the idea that the three volleys came from the Custer Battlefield. I've tried rifle firing from Custer Battle field while others were at the Reno-Benteen Battlefield and couldn't hear the rifle firing. The Nye/Cartwright ridge is in the way, but I heard very clearly the firing from the Nye/Cartwright ridge." (Ibid.)

31 Donahue, *The Maps of John Stands in Timber*, 20-29; Stands in Timber unpublished manuscript, author's collection, 355-356.

32 Don Rickey interview with Stands in Timber, Aug. 8, 1956, 2, LBBNM.

33 Greene, *Evidence and the Custer Enigma-A Reconstruction of Indian-Military History, The Trail Guide*, 26-35; Hardorff, *Markers, Artifacts and Indian Testimony*, 33-38; Edward Luce letter to J. W. Vaughn, October 26,1961, Sheridan Public Library, Sheridan, Wyoming; Joe Blummer letter to J. W. Vaughn, August 4, 1964, Sheridan Public Library, Sheridan, Wyoming; Hutton, Marquis, Thomas, *Custer on the Little Bighorn, She Watched Custer's Last Battle*, 35-43; Marquis, *Wooden Leg*, 229. Local rancher Joe Blummer recovered 110 cartridges along Nye/Cartwright Ridge giving 65 of them to the battlefield. A map is noted in the letter. He was also shown the death sites of three horses and four human skeletons in the vicinity of the ridge by another rancher Frank Bethune. Superintendent Luce knew that at least 350 army casings were found along Nye/Cartwright and Luce Ridge and acknowledged he had discovered 50 himself. In the 1940s, Major Edward Luce, superintendent of the battlefield, found a large number of spent army casings on a lower southern extension of Nye/Cartwright today named appropriately Luce Ridge. The artifacts match the Wolf Tooth testimony. Kate Bighead witnessed the fight along the ridge stating, "A few Indians were out there, and shots were being exchanged at long distance." (Marquis, *She Watched Custer's Last Battle*, 37.) *Wooden Leg* remembered, "While they were yet on the far-out ridge a few Sioux and Cheyennes had exchanged shots with them at long distance, without anybody being hurt." (Marquis, *Wooden Leg*, 229)

34 Hardorff, *Lakota Recollections*, 41-43. While Red Hawk suggests some warriors were killed, Wolf Tooth and Wooden Leg do not reinforce any warrior casualties in the fight along Nye/Cartwright Ridge. Red Hawk, with help from his interpreter, drew a detailed map noting the village, Medicine Tail Coulee, the noncombatants, fights at Calhoun Hill, Custer Hill, and Deep Ravine. Many researchers have mistranslated parts of Red Hawk's oral narrative as taking place on the battlefield ridge, but this is clearly not the case. Three dots note 'divisions' or companies on the northern side of Medicine Tail Coulee near Nye/Cartwright Ridge. While the sequence is confusing due to Red Hawk's wording or misinterpretation, it is clear that he witnessed the seesaw fight described by Wolf Tooth. There were soldiers on the ridge who lost their horses and there was no place to conceal the horses from danger—it was broken and open ground. Troopers who lost their horses had to double up on the back of another soldier's already tired mount. This would be a turning point in the fight as it affected mobility of the companies. While the makeup of the three companies is not certain, the gray horse company [E] was one of them.

35 Fox, *Archaeology, History, and Custer's Last Battle*, 144. As many as 800 soldier copper casings may have been recovered along the ridge beginning as early as the 1920s.

36 Hardorff, *Cheyenne Memories*, 171.

37 Ibid., 126, 155, 171. With Keogh down, Yates would have assumed command of three companies due to his seniority in rank.

38 Hardorff, *Cheyenne Memories*, 65; Hardorff, *Lakota Recollections*, 31, 59.

39 Liberty, *A Cheyenne Voice*, 366; Hardorff, *Lakota Recollections*, 31, 75, 165, 180.

40 Hardorff, *Cheyenne Memories*, 102.

41 Hardorff, *Lakota Recollections*, 52; Hardorff, *Indian Views*, 95.

42 Nichols, *Reno Court*, 418; Donahue, *Drawing Battle Lines*, 72, 99; *Hardin Tribune-Herald*, October 14, 1927, 5. Maps created by Lieutenant William Philo Clark and Major Henry Freeman verified this trail noted by Benteen descending from Custer Hill. A civilian visitor, W. A. Allen, saw evidences of the trail of the soldiers from Custer Hill one year later.

43 Hardorff, *Lakota Recollections*, 52, Hardorff, *Custer Battle Casualties*, 115-116; Donahue, *Drawing Battle Lines*, 367. An unpublished letter from the George Kush Collection notes that Sharrow was found one mile from the battlefield. Mike Reynolds was the son of the agent at Crow Agency and had the opportunity to ride over the battlefield with warriors, Crow scouts, and soldiers. An unnamed Sioux warrior told Reynolds that he chased an officer and killed him along Custer Creek north of the battlefield. The location on the Reynolds map is one mile from Custer Hill.

44 Donahue, *Drawing Battle Lines*, 96-104, 382-385, Liberty, Margot, *A Cheyenne Voice*, 368-369. The James Bell map gives some specific information that reinforces the Philo Clark map.

45 Hardorff, *Indian Views*, 95.

46 Ibid., 62.

47 Hardorff, *Cheyenne Memories*, 43.

48 Dixon, *The Vanishing Race*, 174.

49 Hardorff, *Lakota Recollections*, 111-112; Hardorff, *Indian Views*, 154-155.

50 Graham, *The Custer Myth*, 61.

51 Hardorff, *Lakota Recollections*, 54.

52 Hardorff, *Cheyenne Memories*, 101-102.

53 Graham, *The Custer Myth*, 63.

54 Donahue, *Drawing Battle Lines*, 96, 245, Hardorff, *Cheyenne Memories*, 130-131; Stands in Timber unpublished manuscript, 366. This designation "Ford D" was created in a warrior map that was obtained by William Philo Clark.

55 Hardorff, *Lakota Recollections*, 26. Agriculture and homesteading have all but erased this important landmark but remnants of this drainage still exist. Aerial views of the area procured by this author from the Bureau of Indian Affairs at Crow Agency still show the ravine. There may have been as many as 3,000-4,000 women and children hiding in the ravine. It was such an important landmark that it was depicted in at least 10 Indian maps. One contingency of soldiers rode into the valley near the river in an effort to keep the noncombatants from continuing down the valley—from the warrior's perspective, they had been captured or were about to be.

56 Hardorff, *Cheyenne Memories*, 166-173; Liberty, *Cheyenne Memories*, 200.

57 Liberty, *Cheyenne Memories*, 199; Rickey/Stands in Timber interview, 2, LBBNM.

58 Donahue, *Drawing Battle Lines*, 201, 382-384. It is clear from other warrior maps and testimony that some type of conflict took place in the area. The map created by James Bell, based on Indian testimony, exaggerates the casualties but indicates a fight in the area in which 'Custer was killed.' There is little evidence to support this conclusion, and seems improbable since cartridges from his weapon were later found under his body. Custer bloody head wound would have been fatal and the physical evidence indicates someone else was possibly wounded in the area. Algernon Smith is a more likely candidate, because he was also dressed in buckskin and did not die with his company. Two Moons' map seems more accurate in his appraisal of the casualty—"A man in buckskin was shot at 4 [near Ford D] & staggered back to 5 [Custer Hill] & fell." Since several officers wore buckskin shirts it is impossible to know who this soldier was.

59 Jensen, *The Settler and Soldier Interviews of Eli S. Ricker 1903-1919*, Volume 2, 143; Donahue, *Drawing Battle Lines*, 72-75. Interpreter Fred Server, who interviewed both Gall and Two Moons on the 10th anniversary of the battle, indicated

that the soldiers rode from Custer Hill toward the Deep Ravine and could not cross there because of the banks. The ravine could not be Deep Ravine for several reasons. The Kill Eagle, Philo Clark, Two Moons, Stands in Timber, and Bell maps note that the movement was north of the Deep Ravine. The soldiers used the mouth of Deep Ravine as a ford during the burials three days later and Benteen notes it was a ford in his map. It is possible that the ravine referred to was the one that leads to Ford D.

60 Hardorff, *Lakota Recollections*, 46-47; Greene, *Evidence and the Custer Enigma*, 50-53. The area certainly does not refer to Deep Ravine, Deep Coulee or Medicine Tail Coulee, as soldiers and warriors used all these after the battle. In the late 1950s the National Park Service built a new road on the battlefield and not realizing this important part of the battlefield used the ford area as a gravel pit. Thousands of tons of soil were removed as can be seen in before and after aerial photos. Whatever evidence of the battle that occurred there has been lost or displaced. A soldier spur and human leg bone were found in this area.

61 Hardorff, *Indian Views*, 186; Personal interview with Cheyenne Sylvester Knows his Guns, June 23, 2012. Moving Robe Woman's account of the battle is a chronological mishmash with recorded events going back and forth with little regard to timing. She saw a beaver dam somewhere in the river and many researchers have thought it was at Medicine Tail Coulee, but her account does not say that. Early interviewers believed Custer only reached the river once. According to one oral history, some Cheyenne children were placed in holes in the bank created by beavers and were covered with blankets to conceal them near Squaw Creek.

62 Hardorff, *Lakota Recollections*, 94-95. The James Bell maps notes the same testimony of women stampeding horses in the area.

63 Donahue, *Drawing Battle Lines*, 199, 382; Liberty, *A Cheyenne Voice*, 381-382; Hardorff, *Lakota Recollections*, 52. There are four other references to the loss of horses near Ford D. The Bell map reinforces Moving Robe Woman's narrative and shows a wooded area near the river where "Horses left in woods. Stampeded by squaws with blankets." Two Moons noted "horses captured" bear the same vicinity in his sketch. Stands in Timber also knew of an early horse stampede *down below*—a term he used to note the area near the lower fords near the noncombatants. He recorded, "Those [gray] horses were killed there [later near Deep Ravine] but they [warriors] already killed some horses down below there; those broke loose and Sioux warriors came down the [river] bank and were the first to stampede some." (Liberty, *A Cheyenne Voice*, 381-382). Flying Hawk recorded, "Some of the Indians crossed from the place the women fled to and went across at the lower crossing west of Custer Hill, and there they caught a lot of soldiers' loose horses, and some of the Indians went up to the fight." (Hardorff, *Lakota Recollections*, 52) Reinforcing the stampeding of horses is the lack of native testimony noting the bay horses of Company F in the area below Custer Hill. Since the gray horses ended up near Custer Hill, it can be safely deducted that the bays of Company F or some of the mounts of the headquarters staff were lost.

64 Moore, Michael, and Donahue, Michael, *Gibbon's Route to Custer Hill*, 22-31; Donahue, Michael, *Revisiting Colonel Gibbon's Route*, 12-20. The author has discovered convincing evidence including a soldier marble marker that once stood in the ravine where the horses were stampeded. Using Colonel Gibbon's account, the location of Kellogg's body matches this terrain and his testimony.

65 Hardorff, *Indian Views*, 114.

66 Donahue, *Drawing Battle Lines*, 138. A map by Kill Eagle, the captive Blackfoot, shows huge numbers of warriors gathering there and spreading across the battlefield.

67 Marquis, *Wooden Leg*, 229.

68 Marquis, *She Watched Custer's Last Battle*, 37-38.

69 Hardorff, *Cheyenne Memories*, 102.

70 Barry, *Indian Notes*, 27.

71 Graham, *The Custer Myth*, 87, 97; Barry, *Indian Notes*, 27. Mrs. Spotted Horn Bull was with most of the women likely running toward Squaw Ravine. She saw Crazy Horse lead some Cheyenne into the water and up the ravine. Flying

Hawk rode with Crazy Horse and noted, "Came to a ravine; then we followed up the gulch to a place in the rear of the soldiers." He stated, "Crazy Horse went to the extreme north end of the camp and turned to his right and went to another very deep ravine and by following it, which he did, he came very close to the soldiers on their north side." (Barry, *Indian Notes*, 27) Since the end of the village was slightly past Medicine Tail and Deep Coulee, it is clear from the descriptions that he went up Deep Ravine. A ravine below Calhoun Hill, today named Calhoun Coulee, not only turns south, but also would place him north of the soldiers on Finckle/Finley Ridge. Crazy Horse later charged through the Keogh sector. One account suggests that he was haranguing warriors to attack the soldiers on Nye/Cartwright Ridge, but does not note his actual location. Such encouragement could have been made in the valley. One writer theorized he rode up Deep Coulee to attack the soldiers. According to oral history, "He followed a long ravine that went east from the crossing until it passed the ridge [Greasy Grass Ridge]; it then took a southerly direction parallel with and immediately behind the said ridge." (Graham, *The Custer Myth*, 97)

72 Marquis, *Wooden Leg*, 231; Hardorff, *Battle Casualties II*, 33.

73 Marquis, *She Watched Custer's Last Battle*, 38.

74 Hardorff, *Indian Views*, 125; Barry, *Indian Notes*, 27.

75 Vestal, *Warpath*, 195; Hardorff, *Indian Views*, 155.

76 Marquis, *She Watched Custer's Last Battle*, 38.

77 Ibid., 39.

78 Marquis, *Wooden Leg*, 231; Hardorff, *Cheyenne Memories*, 155. Young Two Moons drew a map of Lame White Man's charge and noted that Yellow Nose was close by. He noted they "made the first charge on the side of [the] hill. Both drive soldiers [onto] top." (Hardorff, *Cheyenne Memories*, 155)

79 Marquis, *She Watched Custer's Last Battle*, 39.

80 Marquis, *Wooden Leg*, 232.

81 Hardorff, *Lakota Recollections*, 66.

82 Daniel and Davis, *That Fatal Day*, 8.

83 Dixon, *The Vanishing Race*, 176; Brust, *Lt. Oscar Long's Early Map*, 9. A guidon staff lance head was discovered by Lieutenant Crittenden's body on Calhoun Hill by Captain F. D. Garretty of the 17th Infantry on June 27, National Military Museum, Governor's Island, Collection of Relics and Trophies Catalog 1884, Press of Putnam & Sons, New York. It cannot be proven with certainty that the lance head was from Foley's staff. Another possibility is that if Yellow Nose counted coup on soldiers with his captured guidon, it could have been knocked off in such an act.

84 Hardorff, *Cheyenne Memories*, 90; Sprague, *Hump & Crazy Horse: Defending the Lakota and Cheyenne 1876-1877*, 66-67; Neihardt, *Black Elk Speaks*, 116-117; Dixon, *The Vanishing Race*, 177.

85 Edmond S. Meany Papers, Red Hawk interview, July 19, 1907, Box 13, Folder 20, University of Washington Special Collections, Seattle, Washington.

86 Hardorff, *Lakota Recollections*, 51.

87 Calhoun letter to Custer, April 23,1871 as noted in Thom Hatch's *The Custer Companion*, 120.

88 Graham, *The Custer Myth*, 89; Hardorff, *Cheyenne Memories*, 109.

89 Liberty, *A Cheyenne Voice*, 214, 375.

90 Edmond S. Meany Papers, Red Hawk interview, Box 13, Folder 20, July 19, 1907.

91 Hammer, *Custer in '76*, 201.

92 Barry, *Indian Notes*, 27-29; Hardorff, *Battle Casualties*, 110; Hardorff, *Battle Casualties II*, 50.

93 Fox, *Archaeology, History, and Custer's Last Battle*, 170.

94 Hardorff, *Lakota Recollections*, 112-113.

95 Hardorff, *Cheyenne Memories*, 66.

96 Marquis, *Wooden Leg*, 232-234.

97 Liberty, *A Cheyenne Voice*, 400.

98 Graham, *The Custer Myth*, 110; Hardorff, *Markers, Artifacts and Indian Testimony*, 48; Vestal, *Warpath*, 195-197.

99 Dixon, *The Vanishing Race*, 176.

100 Hardorff, *Cheyenne Memories*, 132-133; Stands in Timber unpublished manuscript, 366; Liberty, *Cheyenne Memories*, 200; Powell, Peter J., *Sweet Medicine*, Volume 1, 116; Graham, *The Custer Myth*, 62 Hardorff, *Cheyenne Memories*, 131.

101 Fox, *Archaeology, History, and Custer's Last Battle*, 195.

102 Hardorff, *Lakota Recollections*, 114.

103 Edmond S. Meany Papers; Red Hawk interview, Box 13, Folder 20, July 19, 1907.

Chapter 14, Last Stands

1 Hardorff, *Cheyenne Memories*, 131.

2 Barry, *Indian Notes*, 27.

3 Scott and Fox, *Archaeological Insights into the Custer Battle—An Assessment of the 1984 Season*, attached map section 6. Army bullets and Indian warrior cartridges were found in proximity on Cemetery Ridge indicating that soldiers were shooting at targets along the ridge.

4 Hardorff, *Cheyenne Memories*, 103. While some have suggested a breakout from Custer Hill, such could not be the case. Tactically it was a repeat of the Washita and Yellowstone fights as well as the offensive thrust by Company C at the other end of the battle ridge. That an entire company would remain to die while another decided to attempt to escape is nonsensical.

5 Hardorff, *Indian Views*, 46.

6 Hardorff, *Cheyenne Memories*, 67.

7 Ibid., 63.

8 Donahue, *Beyond Custer Hill*, 43. Gall indicated this firing line on the 10th anniversary of the battle to photographer Barry who photographed a re-enactment of the line firing at warriors along the Custer Ridge Extension.

9 Dixon, *The Vanishing Race*, 175; Donahue, *Drawing Battle Lines*, 168-171. Red Hawk also drew a map showing three stands—one at Calhoun Hill, one at Custer Hill, and a 'third stand' below Custer Hill in the vicinity of Cemetery Ridge or nearby. Some of the men from Custer Hill retreated to the third stand. "These were surrounded and the Indians rushed the soldiers. Some of the soldiers broke through the Indians and ran for the ravine, but were killed without getting into it." (Donahue, *Drawing Battle Lines*, 168-171) The point that juts out below the monument describes what is today called Cemetery Ridge.

10 Liberty, *Cheyenne Memories*, 201; Neihardt, *Black Elk Speaks*, 122.

11 Powell, *People of the Sacred Mountain*, 117. One Cheyenne historian recorded, "Some of the soldiers jumped on their horses and attempted to ride farther up, toward the site of the monument. However, the suicide boys cut them off, turning them back toward the north about a quarter of a mile from where the soldiers remaining on the monument site were finally killed; after the suicide warriors charged in, the battle did not last long—not even half an hour." Ibid.

12 Hammer, *Men With Custer*, 159-160.

13 Mangum, Neil, *The George Brown Story*, 9-13.

14 Hammer, *Men With Custer*, 98.

15 Hardorff, *Lakota Recollections*, pg. 149.

16 Hammer, *Men With Custer*, 66; George Kush Collection; W. W. Cooke letters.

17 Arnold, *Coming to the Aid of Custer?* 19-28.

18 Carroll, *They Rode with Custer*, names in alphabetical order with no page numbers.

19 Hammer, *Men With Custer*, 169.

20 Liddic & Harbaugh, *Camp on Custer*, 55.

21 Fox, *Archaeology, History, and Custer's Last Battle*, 196.
22 Hutchins, *Great Western Indian Fights*, 165-174.
23 Hammer, *Custer in '76*, 87; Fox, *Archaeology, History, and Custer's Last Battle*, 199.
24 Hammer, *Men With Custer*, 290; Hutchins, *Boots and Saddles at the Little Bighorn*, 28.
25 Hammer, *Men With Custer*, 208-209. It is uncertain whether Dr. George Lord accompanied the headquarters staff or fled the destruction from the Calhoun Hill area.
26 Donahue, *Drawing Battle Lines*, 91-92.
27 Hammer, *Men With Custer*, 359; Wilke, *Broadcasts from the Little Big Horn-1876*.
28 Marquis, *She Watched Custer's Last Battle*, 39.
29 Ibid., 40.
30 Mills, *Harvest of Barren Regrets*, 181-182.
31 Hammer, *Men With Custer*, 110.
32 Hardorff, *Cheyenne Memories*, 63.
33 Hardorff, *Cheyenne Memories*, 56; Hardorff, *Lakota Recollections*, 86.
34 Hardorff, *Cheyenne Memories*, 56; Hardorff, *Indian Views*, 126, 132.
35 Hardorff, *Battle Casualties II*, 61.
36 McGillycuddy, *McGillycuddy Agent*, 77-78.
37 Camp Collection, Unclassified envelope 135, 672, I.U.
38 Hardorff, *Battle Casualties*, 113.
39 Hammer, *Custer in '76*, 56.
40 Ibid.
41 Hammer, *Custer in '76*, 129.
42 Taunton, *No Pride in Little Big Horn*, 21.
43 Clark, *Scalp Dance*, 20.
44 Hammer, *Custer in '76*, 56.
45 Ibid., 76.
46 Ibid., 67.
47 Ibid., 143.
48 Godfrey, *Custer's Last Battle*, 25.
49 Barnard, *Ten Years with Custer*, 297.
50 Liberty, *Cheyenne Memories*, 201-202.
51 Marquis, *Wooden Leg*, 237.
52 Neihardt, *Black Elk Speaks*, 116.
53 Meany Papers, Red Hawk interview, Box 13, Folder 20, July 19, 1907.
54 Ibid.
55 Vestal, *Warpath*, 200.
56 Viola, *Little Bighorn Remembered*, 74-75; Interview with Oglala Manual Iron Cloud, July 1996. Iron Cloud served as a park ranger at the battlefield with the author and heard the story from Red Necklace, sister of his great-grandfather, Eagle Bear, who was present at the battle. Bouyer must not have been killed outright because he had time to plead his case to the Lakota who surrounded him.
57 Scott, Fox, Connor & Harmon, *Archaeological Perspectives*, 79-80. Some of Bouyer's remains, including his lower facial bones, were found during archeological digs in 1984.
58 Marquis, *Wooden Leg*, 238.
59 Hardorff, *Cheyenne Memories*, 150.

60 Graham, *The Custer Myth*, 72-73, 110. A couple of the Arapaho warriors believed a man killed near the end of the fight was Custer, because the man was dressed in buckskin. The problem with such identification is that they did not know Custer and several officers were wearing buckskin on the hill.

61 R. G. Cartwright Collection, Big Beaver interview, item # 45-46b, Hearst Library, Lead, South Dakota; Hardorff, *Cheyenne Memories*, 150. Young Big Beaver witnessed the killing on Custer Hill and looked toward the river where he saw about 15 scared soldiers who "got up over to the west, or the E Troop position." He also stated, "To me, these men made a rush to the river to try to save themselves by hiding in the brush on the riverbank. My idea is that they used all their ammunition and were making this last effort to get away." Ibid.

62 Hardorff, *Cheyenne Memories*, 76.

63 Hardorff, *Indian Views*, 109.

64 Hardorff, *Battle Casualties II*, 56-57, 148.

65 Hardorff, *Battle Casualties II*, 148. Donahue, *Drawing Battle Lines*, 162. Oglala warrior Shot-in-the-Eye saw the same incident and mapped it. Soldier equipment labeled 'J. D.' was found near skeletal remains with an arrow bent through a human vertebra. Darris rode a white horse and is the only man with those initials who fought in the area.

66 Neihardt, *Black Elk Speaks*, 122-123.

67 Hardorff, *Lakota Recollections*, 32.

68 Hammer, *Men With Custer*, 288; Camp Collection, Box 2, Folder 5, 28, I.U.

69 Scott, Fox, *Archaeological Insights into the Custer Battle*, artifact map. The maps shows a concentration of Indian .44 cartridges fired from two different repeating weapons above the curve in Deep Ravine. The author believes the soldiers are still deeply buried there; a map created by Walter Camp and Knipe notes the bodies were discovered there. See Donahue's *On the Battlefield*, 2-15. In the mid-1990s ground-penetrating technology discovered a deeply buried anomaly at the curve in Deep Ravine.

70 Hardorff, *Battle Casualties*, 104.

71 Camp Collection, Box 2, Folder 10, 78, I. U.

72 Hardorff, *Camp, Custer and the Little Bighorn*, 88.

73 Hardorff, *Battle Casualties*, 109.

74 Hardorff, *Hokahey*, 121-130.

75 The author has not included the two civilians or Private Ball, who were not in the fight with Custer.

76 Neihardt, *Black Elk Speaks*, 117, 124.

77 Marquis, *She Watched Custer's Last Battle*, 40.

78 Hardorff, *Indian Views*, 91.

79 Marquis, *She Watched Custer's Last Battle*, 41.

80 Ibid., 41.

81 Neihardt, *Black Elk Speaks*, 126.

82 Hardorff, *Indian Views*, 189.

83 Ibid., 206.

84 Neihardt, *Black Elk Speaks*, 127.

85 Hardorff, *Lakota Recollections*, 190.

86 Grinnell, *The Fighting Cheyenne*, 354.

87 Hardorff, *Indian Views*, 80.

88 Brill, Custer, *Black Kettle, and the Fight on the Washita*, 252.

89 Marquis, *Wooden Leg*, 243.

90 Hammer, *Men With Custer*, 58-59.

91 Hammer, *Custer in '76*, 143-144.

92 Graham, *The Custer Myth*, 142.

Chapter 15, The Rescue of the 7th Cavalry

1 Nichols, *Reno Court*, 409; Carroll, John, *The Benteen-Goldin Letters*, 215.

2 Hammer, *Custer in '76*, 81; Hardorff, *Camp, Custer, and the Little Bighorn*, 60; Carroll, John, *The Benteen-Goldin Letters*, 215-216.

3 Nichols, *Reno Court*, 423.

4 Hammer, *Custer in '76*, 76.

5 Fougera, *With Custer's Cavalry*, 272.

6 Ibid., 269.

7 Hammer, *Custer in '76*, 56-57.

8 Ibid., 130. Farriers were responsible for the maintenance of the horses.

9 Ibid., 56-57.

10 Hardorff, *Camp, Custer and the Little Bighorn*, 62; Hardorff, *Battle Casualties*, 160-161; Liddic and Harbaugh, *Camp on Custer*, 97-98; Neihardt, *Black Elk Speaks*, 117.

11 Graham, *The Custer Myth*, 143. Repeating the same maneuver he had used when chased by the warriors coming from the downstream villages at the Washita, Edward Godfrey ordered his men to withdraw slowly.

12 Ibid.

13 Taylor, *With Custer on the Little Bighorn*, 50; Nichols, *Reno Court*, 192.

14 Clark, *Scalp Dance*, 20.

15 National Park Service publication, *Reno/Benteen Entrenchment Trail Guide*; Graham, *The Custer Myth*, 143; Hardorff, *Camp, Custer and Little Bighorn*, 64.

16 Marquis, *Wooden Leg*, 262.

17 Graham, *The Custer Myth*, 190; Carroll, *The Benteen-Goldin Letters*, 210.

18 Camp Collection, McGuire letter to Camp, Oct. 12, 1909, Folder 9, BYU.

19 Wagner, *Old Neutriment*, 168; Magnussen, *Peter Thompson's Narrative*, 192.

20 Barnard, *Ten Years with Custer*, 298.

21 Hammer, *Custer in '76*, 101.

22 Dixon, *The Vanishing Race*, 178.

23 Wagner, *Old Neutriment*, 164.

24 Carroll, *The Benteen-Goldin Letters*, 243; Liddic and Harbaugh, *Camp on Custer*, 161.

25 Hardorff, *Camp, Custer, and the Little Bighorn*, 77; Donovan, *A Terrible Glory*, 289.

26 Liddic and Harbaugh, *Camp on Custer*, 71-72; Thomas, *White Swan's Great Fight-25 June 1876*, 20-37. Dennis Lynch claimed to have seen the event but was left behind at the Powder River depot. It is possible that he passed the story on from accounts shared by soldiers who were there.

27 Graham, *The Custer Myth*, 25. According to Hairy Moccasin, the Crow scouts helped dig entrenchments and waited until sunset to escape. Their timing and statements seem doubtful for three reasons. Officers such as Benteen would not just allow them to leave unless they were messengers. Secondly, warriors were looking for soldiers to escape toward Reno Creek and had surrounded the soldier position by sunset. Thirdly, they believed their fellow Crows [Half Yellow Face, Curly and White Swan] were dead. In the chaos and uncertainty of the bluecoat move toward Custer, the other three Crow scouts decided to ride down Reno Creek and escape. They rode away before their friends White Swan and Half Yellow Face arrived on the hill and the soldiers were trapped, verifying they departed early.

28 Hardorff, *Hokahey!*, 86.

29 Glenn Sweem Collection, courtesy of Scott Burgan, copy in the author's collection. Whistling Elk's father witnessed the creation as did Twin Woman, the wife of Lame White Man, who was killed during the battle. She had accompanied the group in the hope of gathering more booty from the soldiers if another fight occurred. It is possible that she had the body of her husband, since he was buried far from the battlefield, along the Rosebud.
30 Dixon, *The Vanishing Race*, 177.
31 Hardorff, *Cheyenne Memories*, 81.
32 Evans, *Custer's Last Fight*, 425-445.
33 Nichols, *Reno Court*, 39; Hunt, *I Fought With Custer*, 101.
34 Hunt, *I Fought With Custer*, 102.
35 Hammer, *Custer in '76*, 114.
36 Hunt, *I Fought With Custer*, 102.
37 Graham, *The Custer Myth*, 144.
38 Neihardt, *Black Elk Speaks*, 118.
39 Marquis, *Wooden Leg*, 255.
40 Hardorff, *Indian Views*, 54.
41 Graham, *The Custer Myth*, 123.
42 Wagner, *Old Neutriment*, 170.
43 Nichols, *Reno Court*, 470, 505.
44 Hardorff, *Battle Casualties*, 101.
45 Camp Collection, John Martin to Camp, Jan. 21, 1914, Folder 1, BYU.
46 Hardorff, *On the Little Bighorn with Walter Camp*, 77-78.
47 Camp Collection, John Ryan to Camp, March 27, 1909, Folder 6, BYU.
48 Donovan, *A Terrible Glory*, 294.
49 Hardorff, *On the Little Bighorn with Walter Camp*, 78.
50 Hardorff, *Hokahey!*, 88-91. Long Road was also known as Canku Hanska and Eagle Hat. The Cheyenne knew him as Thunder Shield. Feeling despondent, he was willing to take extreme risks much like the suicide boys. Some accounts note that he was attempting to count coup on a wounded soldier on the line. He was almost naked with the exception of an ammunition belt full of bullets across his chest. Falling from his mouth was a sheath knife, and when killed, he dropped the two revolvers that he carried.
51 Hunt, *I Fought With Custer*, 105-106.
52 Taylor, *With Custer on the Little Bighorn*, 59; Camp Collection, John Ryan to Camp Aug. 11, 1909, Folder 7, BYU. Thomas French later recommended Ryan for a Medal of Honor, but Benteen refused to support it, claiming that Ryan had disobeyed orders from him. JamesTanner's real name was Jacob Gebhart.
53 Graham, *The Custer Myth*, 182.
54 Nichols, *Reno Court*, 471.
55 Carroll, *Custer's Chief of Scouts*, 152.
56 Hardorff, *Battle Casualties*, 163-164.
57 Ibid.,164.
58 National Park Service publication, *Reno/Benteen Entrenchment Trail Guide.*
59 Hammer, *Men With Custer*, 213; Magnussen, *Peter Thompson's Narrative*, 214-215.
60 Hardorff, *Hokahey!*, 91-93.
61 Taylor, *With Custer on the Little Bighorn*, 60.
62 Vestal, *Warpath*, 204-205.
63 Willert, *Little Bighorn Diary*, 232-248. The column consisted of 7th Infantry—four companies of 2nd Cavalry, and a battery of Gatling guns under General Terry.

64 Willert, *Little Bighorn Diary*, 256-277.
65 Gray, *Captain Clifford's Story of the Sioux War of 1876.*
66 Ibid.
67 Bradley, *The March of the Montana Column*, 152-155. The three Crow scouts mistakenly believed the other Crows—Curly, Half Yellow Face, and White Swan were dead.
68 Greene, *Indian War Veterans*, 125.
69 Bradley, *The March of the Montana Column*, 155.
70 Roe, *Custer's Last Battle*, 5-6.
71 Bradley, *The March of the Montana Column*, 156-157. Crow interpreter Barney Bravo, "Exerted himself to induce the Crows to abandon the expedition, representing to them that some of our officers had said we no longer wanted their services. Several of the best Crows were opposed to such a measure, but Bravo, aided by some of the malcontents among them, carried the point against such, and the whole body were seen by the officers at the rear of the column to mount and gallop away together. They crossed the river and proceeded straight to the Crow agency." (Ibid)
72 Roe, *Custer's Last Battle*, 7.
73 Bradley, *March of the Column*, 159; Roe, *Custer's Last Battle*, 7; McBlain, *With Gibbon on the Sioux Campaign of 1876*, 145.
74 Bradley, *March of the Column*, 159-160.
75 Hardorff, *Indian Views*, 126.
76 Barnard, *Ten Years with Custer*, 300-301.
77 Hunt, *I Fought With Custer*, 106.
78 Ibid., 107.
79 Willert, *Little Bighorn Diary*, 417-419; Carroll, *Custer's Chief of Scouts*, 95.
80 Willert, *Little Bighorn Diary*, 416-417.
81 Willert, *To the Edge of Darkness*, 74.
82 Gibbon, *Gibbon on the Sioux Campaign of 1876*, 27.
83 Gray, *Captain Clifford's Story*; Buecker, *A Surgeon at the Little Big Horn--The Letters of Dr. Holmes O. Paulding*, 43-44, Charlie Schofield letters to Lilla Bogert, accession 2406, Montana State University, Bozeman, Montana.
84 Gray, *Captain Clifford's Story.*
85 Paulding, *Diary with Gibbon Expedition to Meet Custer*, LBBNM.
86 *Diary of Matthew Carroll*, 234.
87 Charlie Schofield to Lilla Bogert, accession 2406, University of Montana, Bozeman, Montana. These letters have been mistakenly attributed to Charles Roe in some volumes. Schofield in the letters to his girlfriend, noted that he had celebrated his birthday on the battlefield verifying him rather than Roe as the author.
88 Willert, *Little Bighorn Diary*, 419.
89 Buecker, *Surgeon at the Little Big Horn*, 43.
90 Charles Austin Booth letter to his wife, June 30, 1876, LBBNM.
91 Paulding diary, June 26, 1876.
92 Hardorff, *Battle Casualties II*, 162.
93 Hardorff, *Battle Casualties*, 102-103.
94 Buecker, *Surgeon at the Little Big Horn*, 43; Willert, *Little Bighorn Diary*, 419.
95 Hardorff, *Battle Casualties II*, 162.
96 McBlain, *With Gibbon on the Sioux Campaign of 1876*, LBBNM, 146.
97 Gibbon, *Gibbon on the Sioux Campaign of 1876*, 29.
98 Buecker, *Surgeon at the Little Big Horn*, 44.
99 Carroll, *Custer's Chief of Scouts*, 95.
100 Nichols, *Reno Court*, 64-65.

101 Camp Collection, C. A. Coolidge to Camp, March 14, 1911 & March 9, 1911, BYU; Hynds, & Shay, *Reminiscences of Sgt. H. A. Hynds*, 68.
102 Green, *Indian War Veterans*, 128.
103 Godfrey, *Field Diary*, 19.
104 Wagner, *Old Neutriment*, 180.
105 Windolph, *I Fought With Custer*,109.
106 Hammer, *Custer in '76*, 249-250.
107 Gibbon, *Gibbon on the Sioux Campaign of 1876*, 33.
108 Graham, *The Custer Myth*, 250; Donovan, *A Terrible Glory*, 307; Willert, *March of the Columns*, 12; Hardorff, *Custer Battle Casualties*, 17-18; Hardorff, *Battle Casualties II*, 23. Charles Varnum brought along three Arikara scouts—Forked Horn, Red Foolish Bear and Young Hawk. Several officers that had accompanied Terry's command were allowed to join Benteen as well.
109 Hammer, *Custer in '76*, 105.
110 Nichols, *Reno Court*, 419.
111 Hammer, *Custer in '76*, 86.
112 Hardorff, *Battle Casualties*, 112-113; George Kush Collection, Alcott papers. Both officers' horses were found nearby. Finley horse's name was actually Don Carlos. Finckle's unnamed horse was 15 hands tall with black tips and was a sorrel gelding. Finley had 12 arrows protruding from his body.
113 Hammer, *Custer in '76*, 95.
114 Ellis, *A Survivor's Story of the Custer Massacre on American Frontier*, 227-232.
115 Hardorff, *Some Recollections of Custer and His Last Battle*, 16.
116 Mills, *Harvest of Barren Regrets*, 279.
117 Hunt, *I Fought With Custer*, 110-112.
118 Hardorff, *Battle Casualties II*, 22-23.
119 Carroll, John, *The Benteen-Goldin Letters*, 26.
120 Hardorff, *Battle Casualties II*, 21.
121 D. F. Barry Collection, Benteen to D. F. Barry, Aug. 29, 1895, LBBNM.
122 Liddic, *I Buried Custer*, 125.
123 Carroll, John, *The Benteen-Goldin Letters*, 271-272.
124 Hammer, *Custer in '76*, 95.
125 Mills, *Harvest of Barren Regrets*, 279.
126 Donahue, *Drawing Battle Lines*, 27-30.
127 Godfrey, *Field Diary*, 19.
128 Greene, *Indian War Veterans*, 128.
129 Gray, *Captain Clifford's Story*, part II.

Chapter 16, Reno's Skirmish Line of Horror

1 Willert, *March of the Columns*, 29-30.
2 Powell, *People of the Sacred Mountain*, Volume II, 980-981.
3 Liberty, *A Cheyenne Voice*, 377.
4 Ibid.
5 Willert, *March of the Columns*, 29-30.
6 Nichols, *Reno Court*, 577.

7 Taylor, *With Custer on the Little Bighorn*, 73; Donahue, *Drawing Battle Lines*, 91.

8 Wagner, *Old Neutriment*, 181.

9 Carroll, *Custer's Chief of Scouts*, 138-139.

10 Willert, *March of the Columns*, 27.

11 Nichols, *Reno Court*, 67.

12 Donahue, *Drawing Battle Lines*, 88.

13 Nichols, *Reno Court*, 67; Donahue, *Drawing Battle Lines*, 90.

14 Hardorff, *Battle Casualties*, 113; Donahue, *Drawing Battle Lines*, 91. Myles Moylan's Company A had to act as rear guard and arrived late—in time to begin the burials. His company moved to the right of McDougall's men. On the left nearest the river was Company B under Thomas McDougall, then Moylan with Edward Godfrey's Company K adjacent to its right. Thomas Weir's Company D formed the center of the skirmish line as it advanced. Benteen's Company H rode to the right of Weir with George Wallace commanding Company G to the right of the 'Hsters.' Riding to the far-right flank of the line was Company M under Thomas French. A single map made by Moylan is the only record of the arrangement of the burial detail.

15 Hardorff, *On the Little Bighorn with Walter Camp*, 40-41; Kush, George, *There Lies Foley of Company C!*, 19-22.

16 Kush, George, *There Lies Foley of Company C!*, 19-22.

17 Ibid.

18 Hardorff, *Battle Casualties*, 110, 113-114. There is a stone cairn where Butler's body was discovered near Deep Coulee today.

19 Ibid., 110.

20 Graham, *The Custer Myth*, 376.

21 Hardorff, *Battle Casualties II*, 56-57. Fifty-two years after the battle, a rancher would find the arrow in the spine, cartridges, saddle gear, and most importantly, the initials 'J. D.' etched on the soldier's leather gear and boot. John Duggan was the only other man with those initials 'J. D.,' who was killed with Custer, and his body was discovered in Deep Ravine. The account of a man on a white horse killed near Finckle/Finley Ridge seems to match that of John Darris, who would have ridden a white horse as part of Company E. The warrior testimony, along with the initialed equipment, make a strong case for the death site of Darris. An arrow had struck his spine bending the arrowhead and making it unable to remove.

22 Donahue, *Drawing Battle Lines*, 291; Hardorff, *Battle Casualties*, 108. Daniel Knipe believed he spotted Briody's body between Custer Hill and Deep Ravine. Since Knipe was from a different company, Rooney's description of his company corporal's death site is more probable.

23 Greene, *Indian War Veterans*, 136.

24 Hardorff, *Battle Casualties*, 112-113.

25 Hardorff, *Battle Casualties*, 102.

26 Hardorff, *Battle Casualties*, 113.

27 Liddic & Harbaugh, *Camp on Custer*, 80.

28 Graham, *The Custer Myth*, 376.

29 Donahue, *Drawing Battle Lines*, 91.

30 Hardorff, *Battle Casualties*, 104, Taunton, *Custer's Field: "A Scene of Sickening Ghastly Horror,"* 12; Nichols, *Reno Court*, 67-69.

31 Graham, *The Custer Myth*, 220.

32 Hardorff, *Battle Casualties*, 104; Donahue, *Drawing Battle Lines*, 90; Hardorff, *Battle Casualties II*, 15.

33 Donahue, *Drawing Battle Lines*, 90.

34 Hardorff, *Battle Casualties II*, 15.

35 Hardorff, *Battle Casualties*, 102.

36 Graham, *The Custer Myth*, 345.
37 Hardorff, *Battle Casualties*, 115.
38 Ibid.,116.
39 Nichols, *Reno Court*, 454.
40 Hardorff, *Battle Casualties II*, 49. Camp made a map with Knipe's help and showed Bobo's death site as well as Finckle and Finley. See Donahue, *Drawing Battle Lines*, 276-277. Sergeant James Bustard of Keogh's company may have been found in the area, although Knipe thought he spotted him across the river.
41 Hardorff, *Battle Casualties*, 97-98.
42 Nichols, *Reno Court*, 417.
43 Taunton, *Ghastly Horror*, 26-27.
44 Ibid., 26.
45 Ibid., 1.
46 Greene, *Indian War Veterans*, 130.
47 Liddic and Harbaugh, *Camp on Custer*, 159.
48 Hardorff, *Battle Casualties II*, 169-174. Later, a score of soldiers wanted to claim they 'saved' Comanche from death. Gustave Korn later became one of the horse's caretakers.
49 Nichols, *Reno Court*, 535.
50 Hammer, *Custer in '76*, 136; Graham, *The Custer Myth*, 377. Walter Camp discovered that Dennis Lynch was not with the burial party, but left behind along the Yellowstone River depot. Anything that Lynch said about the burials or battle was hearsay or fiction. Lynch's account of the burials of Company E does not agree with the other soldier testimony. They were not carried out as reported by Lynch, but buried in the ravine.
51 Graham, *The Custer Myth*, 377.
52 Hardorff, *Battle Casualties*, 115.
53 Hardorff, *Battle Casualties II*, 57; Hammer, *Men With Custer*, 168, 242.
54 Mangum, *The George C. Brown Story*, 13.
55 Hardorff, *Battle Casualties II*, 68.
56 Liddic and Harbaugh, *Camp on Custer*, 73; Hammer, *Men With Custer*, 169.
57 Hammer, *Custer in '76*, 72; Hardorff, *Battle Casualties II*, 66-67.
58 Donahue, *Beyond Custer Hill*, 42-43; Hammer, *Custer in '76*, 87; Donahue, *Drawing Battle Lines*, pg. 91. Theodore Goldin, Stanislas Roy and Charles De Rudio all saw gray horses killed near Deep Ravine.
59 Hardorff, *Battle Casualties II*, 33. While many later researchers claimed that Henry Harrington was the soldier who attempted to escape on his sorrel, his identity has been confused with that of John Foley.
60 Taylor, *With Custer on the Little Bighorn*, 76.
61 Hammer, *Custer in '76*, 72, 248.
62 Hammer, *Men with Custer*, 149; Hardorff, *Battle Casualties*, 114.
63 Scott and Fox, *Archaeological Insights into the Custer Battle*, 39; Scott, Fox, Connor & Harmon, *Archaeological Perspectives on the Battle of the Little Bighorn*, 79-82; Hardorff, *Battle Casualties*, 119-120. Of all the bodies, Mitch Bouyer's body location seems to be one of the most disputed.
64 Donahue, *Drawing Battle Lines*, 90-92.
65 Hardorff, *Battle Casualties II*, 26; Hardorff, *Battle Casualties*, 117. Edward Godfrey acknowledged about the same number of dead horses were found there, although photographs taken a year later seem to dispute those numbers.
66 Hammer, *Custer in '76*, 122.
67 Hammer, *Men With Custer*, 191-192.
68 Hammer, *Men With Custer*, 345; Hardorff, *Battle Casualties*, 115.
69 Camp Collection , Fremont Kipp to Camp, Box 2, Folder 11, 92, I. U.; Hardorff, *Battle Casualties*, 116.

70 Hardorff, *Battle Casualties*, 116; Hardorff, *Battle Casualties II*, 80.

71 Gray, *Captain Clifford's Story*, part II.

72 Hardorff, *Battle Casualties*, 101.

73 Hardorff, *Battle Casualties*, 101;Hardorff, *Battle Casualties II*, 44, *Leavenworth (Kansas) Daily Standard*, April 18, 1882.

74 Camp Collection, Knipe letter to Camp, 1908, Folder 2, BYU; Hardorff, *Battle Casualties*, 100; information from the George Kush Collection.

75 Hardorff, *Battle Casualties II*, 31-32.

76 Hardorff, *Battle Casualties*, 29, 98-100, 27; Donovan, *A Terrible Glory*, 309.

77 Hardorff, *Battle Casualties*, 105; Hardorff, *Battle Casualties II*, 38-40.

78 Hammer, *Men With Custer*, 222; Liddic and Harbaugh, *Camp on Custer*, 92.

79 Hardorff, *Battle Casualties II*, 45.

80 Hammer, *Men With Custer*, 359.

81 Hardorff, *On the Little Bighorn with Walter Camp*, 179; Hardorff, *Battle Casualties*, 116-117.

82 Hammer, *Custer in '76*, 87.

83 Hardorff, *Battle Casualties*, 103.

84 Hardorff, *Battle Casualties II*, 26.

85 Bates Collection, Jacob Adams to Godfrey, Jan. 20, 1927, LBBNM.

86 Hardorff, *Battle Casualties*, 29-31.

87 Hardorff, *Battle Casualties*, 28; Hardorff, *Battle Casualties II*, 80-81.

88 Frost, *General Custer's Libbie*, 246.

89 Ibid. One officer saw at least 17 fired cases under Custer's body, while John Ryan saw about four or five cases, and it is possible that souvenir hunters picked up some before the second officer's arrival. Ryan picked up the remaining casings and gave them to Thomas French to forward to Mrs. Custer.

90 Hardorff, *Battle Casualties II*, 27.

91 Donahue, *Drawing Battle Lines*, 92.

92 Hardorff, *Battle Casualties*, 25; Frost, *General Custer's Libbie*, 246.

93 Liddic and Harbaugh, *Camp on Custer*, 93; Hardorff, *Battle Casualties*, 48-49; Donahue, *Drawing Battle Lines*, 27-55. Some soldiers recall that cartridges or pieces of wood with the officer's names were buried with the bodies. It is possible that both techniques of identification were used. One soldier felt sure that Custer's body would be taken back with them, but with the advanced state of decomposition, such thoughts were quickly abandoned. It is clear that the officers thought that the bodies would be recovered at a later date and therefore carefully marked them. Henry Nowlan was not the only person mapping the battlefield. Engineer Edward Maguire, with help from Sergeant Becker, had been attempting to map the topography and body location since the previous day. The sketch was crude, but could be developed when they returned from the expedition. It is highly likely that Nowlan shared his officer locations with Maguire, since the later maps included officers identified by roman numerals.

94 George Kush Collection, Alcott papers; Frost, *Gen. Custer's Libbie*, 246.

95 Taylor, *With Custer on the Little Bighorn*, 93.

96 Graham, *The Custer Myth*, 260; Hammer, *Men With Custer*, 6-7, 229; Hardorff, *Camp, Custer, and the Little Bighorn*, 74; Hardorff, Indian Views, 199. At least two of the heads were identified in the valley. One head belonged to Private John McGinnis and the other to saddler John Armstrong.

97 Gray, *Captain Clifford's Story*, part II.

98 Coon, Homer *Recollections of the Sioux Campaign of 1876*, 5; Paulding diary, June 28, 1876.

99 Donahue, *Drawing Battle Lines*, 40-41; Carroll, John, *The Benteen-Goldin Letters*, 45.

100 Taylor, *With Custer on the Little Bighorn*, 115.

101 Willert, *March of the Columns*, 30-31; Hardorff, *Camp, Custer and the Little Bighorn*, 105-106.

102 Willert, *March of the Columns*, 36-37; Hardorff, *Camp, Custer and the Little Bighorn*, 66.
103 Moore and Donahue, *Gibbon's Route to Custer Hill*, 22-31; Donahue, *Revisiting Colonel Gibbon's Route*, 12-23.
104 Hammer, *Custer in '76*, 79.
105 Gibbon, *Gibbon on the Sioux Campaign*, 38-42.
106 Hardorff, *Camp, Custer, and the Little Bighorn*, 102.
107 Gibbon, *Gibbon on the Sioux Campaign*, 42-45.
108 Donovan, *A Terrible Glory*, 317-320.
109 Collection of Lawrence Frost as recorded by George Kush. The book was published by A. L. Burt Company Publishing, New York.
110 Mueller, *Shooting Arrows & Slinging Mud*, 121-123.
111 Gray, *Centennial Campaign*, 243-249.
112 Ibid., 266-267.
113 Utley, Robert, *Custer Battlefield Handbook*, 79-83; Greene, *American Carnage*, 215-246.

Chapter 17, Nail the Box Up

1 Liberty, *Cheyenne Memories*, 210-211.
2 Marquis, *Wooden Leg*, 283-285.
3 Graham, *The Custer Myth*, 373.
4 Graham, *The Custer Myth*, 373; Hardorff, *Battle Casualties*, 37.
5 Graham, *The Custer Myth*, 374.
6 Ibid.
7 *Chicago Times*, July 25, 1877.
8 Ibid.
9 Letter from August 23, 1877. Letters in the Office of the Adjutant General, 1871-1800, National Archives series M666, roll 273.
10 Hardorff, *Battle Casualties*, 43-44.
11 Brust, *Fouch Photo May be the First*, 2-9; Brust, *Adventurous Detour to the Battlefield*, 8-12.
12 *Bismarck Tribune*, July 13, 1877.
13 Hardorff, *Battle Casualties*, 45.
14 Carroll, *The Benteen-Goldin Letters*, 29.
15 Marquis, *Memoirs of a White Crow Indian*, 283-284.
16 Ibid.
17 *Chicago Tribune*, July 30, 1877.
18 Hanson, *Conquest of the Missouri*, 378-379.
19 Hardorff, *Battle Casualties*, 46.
20 Hutton, *Phil Sheridan and his Army*, 328-329. Joining Phil Sheridan were George Crook, James Forsyth, George "Sandy" Forsyth, and Lieutenant John Bourke.
21 *Chicago Tribune*, August 13, 1877.
22 Graham, *The Custer Myth*, 370.
23 Hardorff, *Battle Casualties*, 54.
24 Ibid., 54-55.
25 *New York Sun*, July 25, 1877.
26 Frost, *General Custer's Libbie*, 242.

27 S. E. Staples to W. W. Rice, November 9, 1877. Letters to the Office of the Adjutant General, 1871-1800, National Archives series M666, roll 273.

28 Hutton, *Phil Sheridan*, 328-329; Letter from P. H. Sheridan, April 8, 1878, Letters in the Office of the Adjutant General, 1871-1800, National Archives series M666, roll 273.

29 Hardorff, *Battle Casualties*, 66-67.

30 Ibid., 39-41.

31 Ibid., 41-42.

32 Elizabeth B. Custer Collection, LBBNM, microfilm roll 1.

33 Hardorff, *Battle Casualties*, 42.

34 Nichols, Reno Court, 2; Carroll, *The Benteen-Goldin Letters*, 212.

35 Carroll, *The Benteen-Goldin Letters*, 209.

36 Hardorff, *Battle Casualties*, 67-68.

37 Rickey, *History of Custer Battlefield*, 60-63; Hardorff, *Battle Casualties*, 70.

38 As a park ranger at Little Bighorn Battlefield National Monument for over two decades, I have personally heard this on many occasions especially from the local Crow tribal members.

39 Brust, Pohanka, and Barnard, *Where Custer Fell*, 138-143.

40 Scott, Fox, Connor, Harmon, *Archaeological Perspectives*, 50; Rickey, *History of Custer Battlefield*, 67-71.

41 Scott, Fox, Connor, Harmon, *Archaeological Perspectives*, 49-50; Rickey, *History of Custer Battlefield*, 69-70.

42 Brust, Pohanka, Barnard, *Custer Fell Here*, 159.

43 Administrative burial records, LBBNM. The Fetterman dead were buried on top of Custer Hill, which caused great confusion, so they were eventually removed and placed in the current National Cemetery.

44 Mueller, *Shooting Arrows & Slinging Mud*, 69.

45 Indian Memorial Dedication "Peace Through Unity" Memorial Booklet, printed by the National Park Service, 10.

46 Ibid., 14.

47 Ibid., front-piece.

Epilogue

1 Utley, *Life in Custer's Cavalry*, 235-247.

2 Ibid., 249-250.

3 Utley, *Life in Custer's Cavalry*, 250-251; Nichols, *Men With Custer*, 24-25; Klokner, *The Officer Corps*, 44-45.

4 Neihardt, *Black Elk Speaks*, 201-274; Steltenkamp, Michael F., *Black Elk-Holy Man of the Oglala*, 132.

5 Klokner, *The Officer Corps*, 46-47.

6 Bradley, *The March of the Montana Column*, 3-6.

7 Utley, *Life in Custer's Cavalry*, 251-252.

8 Nichols, *Men With Custer*, 52.

9 Hardorff, *Washita Memories*, 202-203.

10 Hammer, *Men With Custer*, 70-71.

11 Bray, *Crazy Horse*, 284-398.

12 Bourke, John G., *On the Border with Crook*, 283-491.

13 Hardorff, *Indian Views*, 67.

14 Nichols, *Men With Custer*, 88-89.

15 Reynolds, Arlene, *The Civil War Memories of Elizabeth Bacon Custer*, xi-xv; Leckie, 245-306.

16 Nichols, *Men With Custer*, 100.

17 Nichols, *Men With Custer*, 113-114; Klokner, *The Officer Corps*, 59-60.
18 Johnson, *A Captain of Chivalric Courage*, 47, Nichols, *Men With Custer*, 131; Klokner, *The Officer Corps*, 61-62.
19 Larson, Robert W., *Gall-Lakota War Chief*, 230-238.
20 Woods, *Custer's Arikara Indian Interpreter*, 225-252; Nichols, *Men With Custer*, 141-142.
21 Carroll, *The Custer Autograph Album*, 75.
22 Hardorff, *Lakota Recollections*, 49.
23 Utley, *Life in Custer's Cavalry*, 260.
24 Nichols, *Men With Custer*, 143-144; Klokner, *The Officer Corps*, 65-66.
25 Hammer, *Men With Custer*, 130.
26 Hammer, *Men With Custer*, 141-142. On August 7, 2014, Putt Thompson and I located the site of the cemetery with local landowner, John McCleary. His home stands where St. Ann's Church once stood.
27 Hammer, *Men with Custer*, 143; personal interview with Putt Thompson, summer 2012.
28 Utley, *Life in Custer's Cavalry*, 262-263.
29 Nichols, *Men With Custer*, 164-165; Klokner, The Officer Corps, 69-70.
30 Hardorff, *Lakota Recollections*, 73-74.
31 Nichols, *Men With Custer*, 174.
32 Unger, *Custer's First Messenger? Debunking the Story of Sergeant Daniel Kanipe*, 16-17; Bradshaw, *Palmetto Historical Works*, 4-30.
33 May, *Little Raven*, Oklahoma Historical Society's Encyclopedia of Oklahoma History & Culture (accessed July, 2, 2012 on Wikipedia website)
34 Utley, *Life in Custer's Cavalry*, 267.
35 Grinnell, *The Fighting Cheyennes*, 111, 380, 414, 427.
36 Robinson III, *Sananta*, 193.
37 Hardorff, *Indian Views*, 63.
38 Hardorff, *Washita Memories*, 301.
39 Solimine, *Custer's Bugler*, 81-93.
40 Utley, *Life in Custer's Cavalry*, 267-268; Nichols, *Men With Custer*, 249-250.
41 Nichols, *Men With Custer*, 258; Klokner, *The Officer Corps*, 80.
42 Harrison, *Monahsetah-The Life of a Custer Captive*, 177-185.
43 Duncan and Smith, *The Captives*, 20.
44 Utley, *Life in Custer's Cavalry*, 268-270.
45 Klokner, *The Officer Corps*, 86-87.
46 Michno, *Lakota Noon*, 309-310; LaPointe, Ernie, *Sitting Bull*, 97-98.
47 Utley, *Life in Custer's Cavalry*, 271-272.
48 Nichols, *Men With Custer*, 315-316.
49 Brady, *Indian Fights and Fighters*, 257, 279-292.
50 Hardorff, *Lakota Recollections*, 82.
51 Hammer, *Men With Custer*, 336.
52 Nichols, *Men With Custer*, 329-330; Klokner, *The Officer Corps*, 90-91.
53 Carroll, *The Custer Autograph Album*, 93.
54 Hardorff, *Cheyenne Memories*, 95.
55 Frost, *Custer's 7th Cav & the Campaign of 73*, 16; Carroll, The *Custer Autograph Album*, 31; Lubetkin, *Custer and the 1873 Yellowstone Survey*, 43, death and burial details from Wikipedia—Thomas L. Rosser.
56 Barnard, *Ten Years with Custer*, 6-8; *Nichols, Men With Custer*, 343-344.

57 Robinson III, *Satanta*, 113-194.
58 Hammer, *Men With Custer*, 299-300.
59 Hyde, *Spotted Tail's Folk*, 332-335.
60 Lubetkin, *Custer and the 1873 Yellowstone Survey*, 32-33; Eicher, and Eicher, *Civil War High Commands*, 728; "David S. Stanley-Claim to Fame: Medal of Honor recipients"- Find a grave website.
61 Nichols, *Men With Custer*, 383-384.
62 Utley, *Life in Custer's Cavalry*, 273-274.
63 Ibid., 274.
64 Ibid., 274-275.
65 Utley, *The Lance and the Shield*, 232-313.
66 Hammer, *Men With Custer*, 344.
67 Hammer, *Men With Custer*, 346-347.
68 Nichols, *Men With Custer*, 402-403; Klokner, *The Officer Corps*, 104-105.
69 Nichols, *Men With Custer*, 410-411; Klokner, *The Officer Corps*, 105-106.
70 Nichols, *Men With Custer*, 417; Klokner, *The Officer Corps*, 106-107.
71 Utley, *Life in Custer's Cavalry*, 280-281.
72 Klokner, *The Officer Corps*, 108-109.
73 Duncan and Smith, *The Captives*, 21.
74 Vestal, Stanley, *Warpath*, 217-256.
75 Hammer, *Men With Custer*, 374.
76 Hardorff, *Cheyenne Memories*, 49.
77 Hammer, *Men With Custer*, 375.
78 Hammer, *Men With Custer*, 379-380.
79 Marquis, *Wooden Leg*, 295-384.
80 Hardorff, *Indian Views*, 97.

Appendix 1, Field and Map Research for the August 4, 1873 Yellowstone Fight

1 Copy of the 1877 Maguire map of Ft. Keogh area in the author's possession.
2 A. C. LaBarge papers, 1877, Historical Society of Montana, copy in the author's possession.
3 Thanks to Jerry Greene who has shared this information with the author and plans to publish the Campbell diary at a future date.

Appendix 2, Field Research and Archeology for the August 11, 1873 Fight

1 Thanks to Vance Haynes for corresponding with the author and verifying my maps used here.

Appendix 3, Field Study and the Little Bighorn Battle Timeline

1 Donahue, Michael, *The Battle of the Little Bighorn: A New Time Line*, 18th Annual Symposium, Custer Battlefield Historical & Museum Association, June 25, 2004.
2 Liberty, Cheyenne Memories, 199, 202, Liberty, *A Cheyenne Voice*, 443.

Appendix 4, Field Research on the Trail of George Custer and John Martin

1 Donahue, *Drawing Battle Lines*, 266-269.
2 Ibid., 26-45.
3 Ibid., 66-71.
4 Ibid., 292-296.

Appendix 5, Field Research and Sight Tests for Custer Waving his Hat

1 Graham, *The Custer Myth*, 11, 23, 290,
2 Nichols, *Reno Court*, 337-343.
3 Camp, *Custer in '76*, 133.

Appendix 6, Field and Map Research on the Men Killed in Deep Ravine

1 Scott, Fox, Connor and Harmon, *Archaeological Perspectives on the Battle of the Little Bighorn*, 39-48.
2 Donahue, *Drawing Battle Lines*, 272-278.
3 Kuhlman, Charles, *Did Custer Disobey Orders at the Battle of the Little Big Horn?* 35.
4 Scott & Fox, *Archaeological Insights into the Custer Battle*, 68-71 and map included notes Indian casings 1700,1707, 1704, 1705, 1706, 1708 were found above the bend in Deep Ravine.

Appendix 7, Field Research and the Death Site of Mark Kellogg

1 Barnard, *Photographing Custer's Battlefield*, 208-209.

Bibliography

Books

Abrams, Marc H. *Newspaper Chronicle of the Indian Wars*, Volume 16, *The Final Chapter: Late Additions & New Finds*. Brooklyn NY: Abrams Publications, 2011.

Abrams, Marc H. *Sioux War Dispatches*. Yardley PA: Westholme, 2012.

Ambrose, Stephen E. *Crazy Horse and Custer*. Garden City NY: Doubleday and Company, 1975.

Athearn, Robert G. *Sherman and the Settlement of the West*. Norman OK: University of Oklahoma Press, 1956.

Barnard, Sandy. *Custer's First Sergeant John Ryan*. Terre Haute IN: AST Press, 1996.

Barnard, Sandy. *I Go With Custer-The Life and Death of Reporter Mark Kellogg*. Bismarck ND: *Bismarck Tribune*, 1996.

Barnard, Sandy. *Photographing Custer's Battlefield-The Images of Kenneth F. Roahen*. Norman OK: University of Oklahoma Press, 2016.

Barnard, Sandy. *Ten Years with Custer-A 7th Cavalryman's Memoirs*. Ft. Collins CO: Citizen Printing Company, 2001.

Barnett, Louise. *Touched by Fire: The Life, Death and Mythical Afterlife of George Armstrong Custer*. New York NY: Henry Holt, 1996.

Bates, Charles Francis. *Custer's Indian Battles*, New York NY: Bates, 1936.

Beck, Paul N. *Inkpaduta-Dakota Leader*. Norman OK: University of Oklahoma Press, 2008.

Blake, Michael. *Indian Yell*. Flagstaff AR: Northland Publishing, 2006.

Bordeaux, William. *Custer's Conqueror*. Boston MA: Smith and Co., 1944.

Bradley, James H. *The March of the Montana Column*. Norman OK: University of Oklahoma Press, 1961.

Brady, Cyrus Townsend. *Indian Fighters and Fighters*. Lincoln NB: University of Nebraska Press, 1971.

Bray, Kingsley M. *Crazy Horse-A Lakota Life*. Norman OK: University of Oklahoma Press, 2006.

Brill, Charles J. *Custer, Black Kettle, and the Fight on the Washita*. Norman OK: University of Oklahoma Press, 2001.

Brininstool, E. A. *Troopers with Custer*. Mechanicsburg PA: Stackpole, 1952.

Broome, Jeff. *Custer into the West with the Journal and Maps of Lieutenant Henry Jackson*. El Segundo CA: Upton and Sons, 2009.

Broome, Jeff. *Dog Soldier Justice*. Lincoln KS: Lincoln County Historical Society, 2003.

Brust, James, and Brian Pohanka, Sandy Barnard, *Where Custer Fell*. Norman OK: University of Oklahoma Press, 2005.

Bourke, John G. *On the Border with Crook*. Norman OK: University of Nebraska Press, 1971.

Burdick, Usher L. *David F. Barry's Indian Notes on The Custer Battle*. Baltimore MD: The Proof Press, 1937.

Burdick, Usher L. *The Army Life of Charles "Chip" Creighton.* Paris MD: National Reform Associates, 1937.
Burkey, Blaine. *Custer, Come at Once! The Fort Hays Years of George and Elizabeth Custer 1867-1870.* Hays KS: Ellis County Historical Society, 1976.
Carroll, John M. *Custer's Chief of Scouts.* Lincoln NB: University of Nebraska Press, 1982.
Carroll, John M. *Camp Talk-The Very Private Letters of Frederick W. Benteen of the 7th U.S. Cavalry to his Wife, 1871-1888.* Mattituck NY: J. M. Carroll & Company, 1983.
Carroll, John M. *General Custer and the Battle of the Little Bighorn, The Federal View.* Mattituck NY: J. M. Carroll & Company, 1986.
Carroll, John M. *General Custer and the Battle of the Washita: The Federal View.* Bryan TX: Guidon Press, 1978.
Carroll, John M. *The Benteen-Goldin Letters on Custer and his Last Battle.* Lincoln NB: University of Nebraska Press, 1982.
Carroll, John M. *The Custer Autograph Album.* College Station TX: The Early Press, 1994.
Carroll, John M. *The Two Battles of the Little Big Horn.* Mattituck NY: J. M. Carroll and Company, 1974.
Carroll, John M. *They Rode with Custer.* Mattituck NY: J. M. Carroll & Company, 1987.
Carroll, John M. *Cavalry Bits: One of Seven Volumes: The Yellowstone Expedition of 1873.* Mattituck NY: J. M. Carroll & Company, 1986.
Chambers, Lee. *Fort Abraham Lincoln: Dakota Territory.* Atglen PA: Schiffer, 2008.
Clark, George M. *Scalp Dance-The Edgerly Papers on the Battle of the Little Big Horn.* Oswego NY: Heritage Press, 1985.
Connell, Evan S. *Son of the Morning Star.* San Francisco CA: North Point Press, 1984.
Convis, Charles L. *The Honor of Arms: A Biography of Myles W. Keogh.* Tucson AZ: Westernlore Press, 1990.
Cozzens, Peter. *Eyewitnesses to the Indian Wars-Conquering the Southern Plains,* Volume 3. Mechanicsburg PA: Stackpole, 2003.
Cozzens, Peter. *Eyewitnesses to the Indian Wars 1865-1890-The Long War for the Northern Plains,* Volume 4. Mechanicsburg PA: Stackpole, 2004.
Custer, Elizabeth B. *Boots and Saddles.* Lincoln NB: University of Oklahoma Press, 1961.
Custer, Elizabeth B. *Following the Guidon.* New York NY: Harper & Brothers, Franklin Square, 1890.
Custer, Elizabeth B. *Tenting on the Plains, or General Custer in Kansas and Texas.* Lincoln NB: University of Oklahoma Press, 1994.
Custer, George A. *My Life on the Plains.* Edited by Milo Milton Quaife. Lincoln NB: University of Nebraska Press, 1966.
Custer, George A. *My Life on the Plains.* Introduction by Edgar I. Stewart. Lincoln NB: University of Oklahoma Press, 1962.
Daniels, Karen L., and Davis, Elden E. *That Fateful Day: Eight More with Custer.* Howell MI: Powder River Press, 1992.
Darling, Roger. *A Sad and Terrible Blunder.* Vienna VA: Potomac-Western Press, 1990.
Dippie, Brian W. *Nomad-George A. Custer in Turf, Field, and Farm.* Austin TX: University of Texas Press, 1980.
Dixon, Joseph K. *The Vanishing Race.* New York NY: Poplar Library, 1972.
Donahue, Michael N. *Drawing Battle Lines-The Map Testimony of Custer's Last Fight.* El Segundo CA: Upton and Sons, 2008.
Donovan, James, A. *Terrible Glory: Custer and the Little Bighorn.* New York NY: Little, Brown and Company, 2008.

Donovan, Terrence J. *Brazen Trumpet-Frederick W. Benteen and the Battle of the Little Big Horn.* Lancaster CA: MojaveWest Publishing, 2007.

Dustin, Fred. *The Custer Tragedy.* El Segundo CA: Upton and Sons, 2011.

Eicher, John H. and David J. Eicher. *Civil War High Commands.* Stanford CA: Stanford University Press, 2001.

Evans, David C. *Custer's Last Fight-The Story of the Battle of the Little Big Horn.* El Segundo CA: Upton and Sons, 1999.

Fougera, Katherine Gibson. *With Custer's Cavalry.* Lincoln NB: University of Nebraska Press, 1986.

Fox, Jr., Richard A. *Archaeology, History, and Custer's Last Battle.* Lincoln NB: University of Oklahoma Press, 1993.

Frost, Lawrence A. *Custer's 7th Cav & the Campaign of '73.* El Segundo CA: Upton and Sons, 1986.

Frost, Lawrence A. *General Custer's Libbie.* Seattle WA: Superior Publishing Company, 1976.

Frost, Lawrence A. *Some observations on the Yellowstone Expedition of 1873.* Glendale CA: Arthur H. Clark, 1981.

Frost, Lawrence A. *The Court-Martial of General George Armstrong Custer.* Norman OK: University of Oklahoma Press, 1968.

Frost, Lawrence A. *The Custer Album.* Seattle WA: Superior, 1964.

Gibbon, John. *Gibbon and the Sioux Campaign of 1876.* Bellevue NE: Old Army Press, 1969.

Godfrey, Edward S. *General George A. Custer and the Battle of the Little Big Horn.* Hardin MT: Big Horn County Historical Society reprint, 1996.

Godfrey, Edward S. *The Field Diary of Lt. Edward Settle Godfrey.* Portland OR: Champoeg Press, 1957.

Grafe, Ernest and Paul Horsted. *Exploring with Custer.* Custer SD: Golden Valley Press, 2002.

Graham, Colonel W. A. *The Custer Myth.* New York NY: Bonanza Books, 1953.

Graham, Colonel W. A. *The Reno Court of Inquiry.* Mechanicsburg PA: Stackpole Books, 1995.

Gray, John S. *Centennial Campaign.* Fort Collins CO: Old Army Press, 1976.

Gray, John S. *Custer's Last Campaign.* Lincoln NB: University of Nebraska Press, 1991.

Greene, Jerome A. *American Carnage.* Norman OK: University of Oklahoma, 2014.

Greene, Jerome A. *Evidence and the Custer Enigma-A Reconstruction of Indian-Military History, "The Trail Guide,"* Vol. XVII, March-June 1973, Number 1 and 2. Lawrence KS: Kansas City Posse, The Westerners.

Greene, Jerome A. *Indian Wars Veterans: Memories of Army Life and Campaigns in the West, 1864-1898.* New York City NY: Savas Beatie, 2007.

Greene, Jerome A. *Lakota and Cheyenne-Indian Views of the Great Sioux War, 1876-1877.* Norman OK: University of Oklahoma Press, 1994.

Greene, Jerome A. *Washita-The U. S. Army and the Southern Cheyennes, 1867-1869.* Norman OK: University of Oklahoma Press, 2004.

Grinnell, George Bird. *The Cheyenne Indians,* Volume 1. Lincoln NB: University of Nebraska Press, 1923.

Gwynne, S. C. *Empire of the Summer Moon.* New York NY: Scribner, 2011.

Hammer, Kenneth. *Custer in '76.* Lincoln NB: University of Oklahoma Press, 1976.

Hammer, Kenneth. *Men With Custer: Biographies of the 7th Cavalry.* Edited by Ronald H. Nichols. Hardin MT: Custer Battlefield Historical & Museum Association, Inc., 1995.

Hanson, Joseph Mills. *The Conquest of the Missouri.* Chicago IL: A. C. McClurg, 1909.

Hardorff, Richard G. *Cheyenne Memories of the Custer Fight-A Source Book.* Spokane WA: Arthur H. Clark Company, 1995.

Hardorff, Richard G. *Camp, Custer, and the Little Bighorn.* El Segundo CA: Upton and Sons, 1997.

Hardorff, Richard G. *Washita Memories.* Spokane WA: Arthur H. Clark Company, 1995.

Hardorff, Richard G. *Hokahey! A Good Day to Die!-The Indian Casualties of the Custer Fight.* Spokane WA: Arthur H. Clark Company, 1993.

Hardorff, Richard G. *Indian Views of the Custer Fight.* Spokane WA: Arthur H. Clark Company, 2004.

Hardorff, Richard G. *On the Little Bighorn with Walter Camp.* El Segundo CA: Upton and Sons, 2002.

Hardorff, Richard G. *Lakota Recollections of the Custer Fight.* Spokane WA: Arthur H. Clark Company, 1991.

Hardorff, Richard G. *Markers, Artifacts and Indian Testimony: Preliminary Findings on the Custer Battle.* Short Hills NJ: Don Horn Publications, 1985.

Hardorff, Richard G. *The Custer Battle Casualties.* El Segundo CA: Upton and Sons, 1989.

Hardorff, Richard G. *The Custer Battle Casualties, II.* El Segundo CA: Upton and Sons, 1999.

Harman, Troy. *"Hunterstown-North Cavalry Field of Gettysburg."* Cordorva TN: Little Bighorn Associates, 2008.

Harper, Gordon. *The Fights on the Little Horn-Unveiling the Mysteries of Custer's Last Stand.* Philadelphia PA: Casemate, 2014.

Harrison, Peter. *Monahsetah-The Life of a Custer Captive.* London: The English Westerners' Society, 2014.

Hart, John, editor. *Custer and His Times Book 5.* LaGrange Park IL: Little Big Horn Associates, 2002.

Hassrick, Royal B. *The Sioux.* Norman OK: University of Oklahoma, 1964.

Hatch, Thom. *Black Kettle-The Cheyenne Chief Who Sought Peace but Found War.* Hoboken NJ: John Wiley & Sons, 2004.

Hofling, Charles K. *Custer and the Little Big Horn-A Psychobiographical Inquiry.* Detroit MI: Wayne State University Press, 1981.

Hoig, Stan. *The Battle of the Washita.* Lincoln NB: University of Nebraska Press, 1976.

Horsted, Paul, and Ernest Grafe, Jon Nelson. *Crossing the Plains with Custer.* Custer SD: Golden Valley Press, 2009.

Hunt, Frazier and Robert: *I Fought With Custer-The Story of Sergeant Windolph, Last Survivor of the Battle of the Little Big Horn.* Lincoln NB: University of Nebraska Press, 1947.

Hutchins, James S. *Boots and Saddles at the Little Bighorn.* Fort Collins CO: The Old Army Press, 1976.

Hutchins, James S. *Great Western Indian Fights.* Lincoln NB: University of Nebraska Press, 1960.

Hutchins, James S. *The Papers of Edward S. Curtis Relating to Custer's Last Battle.* El Segundo CA: Upton and Sons, 2000.

Hutton, Paul Andrew. *Phil Sheridan and His Army.* Lincoln NB: University of Nebraska Press, 1985.

Hutton, Paul Andrew. *The Custer Reader.* Lincoln NB: University of Nebraska Press, 1992.

Hyde, George E. *Life of George Bent.* Norman OK: University of Oklahoma Press, 1968.

Hyde, George E. *Red Cloud's Folk.* Norman OK: University of Oklahoma Press, 1979.

Hyde, George E. *Spotted Tail's Folk.* Norman OK: University of Oklahoma Press, 1976.

Innis, Ben. *Bloody Knife: Custer's Favorite Scout.* Bismarck ND: Smoky Water Press, 1994.

Jamieson, Perry D. *Crossing the Deadly Ground-United States Army Tactics 1865-1899.* Tuscaloosa AL: University of Alabama Press, 1994.

Jensen, Richard E. *The Settler and Soldier Interviews of Eli S. Ricker 1903-1919,* Volume 1. Lincoln NB: University of Nebraska Press, 2005.

Johnson, Barry C., and Francis B. Taunton. *More Sidelights of the Sioux Wars.* London: Westerners Publications Limited, 2004.

Johnson, Michael G. *The Sioux-Warriors of the Plains.* Edison NJ: Chartwell Books, 2008.

Kappler, Charles J. *Indian Treaties-1778-1883.* Mattituck NY: Amereon House, 1972.

Katz, D. Mark. *Custer in Photographs.* Gettysburg PA: Yo-Mark Production Co., 1985.

Kazanjian, Howard and Chris Enss. *None Wounded, None Missing, All Dead-The Story of Elizabeth Bacon Custer.* Helena MT and Guilford CT: Twodot Publishers, 2011.

Keim, De B. Randolph. *Sheridan's Troopers on the Borders: A Winter Campaign on the Plains.* Philadelphia PA: David McKay, 1885.

Kelly, Fergus. *A Guide to Early Irish Law.* Dublin, Ireland: Institute for Advanced Studies, 1988.

Kidd, James Harvey. *At Custer's Side: The Civil War Writings of James Harvey Kidd.* Kent OH: Kent State University Press, 2000.

Kidd, James Harvey. *Riding with Custer; Personal Recollections of a Cavalryman with Custer's Michigan Cavalry Brigade in the Civil War.* Ionia MI: The Sentinel Press, 1908; reprint ed., Grand Rapids, MI: The Black Letter Press, 1969.

Kime, Wayne R. *The Powder River Expedition Journals of Colonel Richard Irving Dodge.* Norman OK: University of Oklahoma Press, 1997.

Kinsley, D.A. *Favor the Bold-Custer: The Civil War Years,* Volume 1. New York NY: Holt, Rinehart and Winston, 1967.

Kirshner, Ralph. *The Class of 1861-Custer, Ames, and Their Classmates After West Point.* Carbondale IL: Southern Illinois University, 1999.

Klokner, James B. *The Officer Corps of Custer's Seventh Cavalry-1866-1876.* Atglen PA: Schiffer Military History, 2007.

Kraft, Louis. *Custer and the Cheyenne: George Armstrong Custer's Winter Campaign on the Southern Plains.* El Segundo CA: Upton and Sons, 1995.

Kuhlman, Charles. *"Did Custer Disobey Orders at the Battle of the Little Big Horn?."* Harrisburg PA: The Stackpole Company, 1957.

Ladenheim, J. H. *Custer's Thorn-The Life of Frederick W. Benteen.* Westminster MD: Heritage Books, 2007.

Langellier, John P. and Kurt Hamilton Cox, Brian C. Pohanka. *Myles Keogh-The Life and Legend of an "Irish Dragoon" in the Seventh Cavalry.* El Segundo CA: Upton and Sons, 1991.

LaPointe, Ernie. *Sitting Bull.* Salt Lake, Charleston, Sante Fe, Santa Barbara: Gibbs Smith, 2009.

Larson, Robert W. *Gall-Lakota Warrior Chief.* Norman OK: University of Oklahoma Press, 2007.

Leckie, Shirley A. *Elizabeth Bacon Custer and the Making of a Myth.* Norman OK: University of Oklahoma Press, 1993.

Libby, O. G. *The Arikara Narrative of the Campaign Against the Hostile Dakotas June 1876.* New York, NY: Sol Lewis, 1973.

Liberty, Margot. *A Cheyenne Voice-The Complete John Stands in Timber Interviews.* Norman OK: University of Oklahoma Press, 2013.

Liberty, Margot. *A Northern Cheyenne Album.* Norman OK: University of Oklahoma Press, 2006.

Liberty, Margot. *Cheyenne Memories.* Lincoln NB: University of Nebraska Press, 1967.

Liddic, Bruce R. and Paul Harbaugh. *Camp on Custer.* Spokane WA: Arthur Clark Company, 1995.

Liddic, Bruce R. *I Buried Custer-The Diary of Pvt. Thomas W. Coleman, 7th U. S. Cavalry.* College Station TX: Creative Publishing Company, 1979.

Lieber, Francis. *Instructions for the government of armies of the United States, in the Field.* New York NY: D. van Nostrand, 1882.

Linderman, Frank Bird. *Pretty Shield: Medicine Woman of the Crows.* Lincoln NB: University of Nebraska Press, 2003.

Linn, Brian McAllister. *The Echo of Battle-The Army's Way of War.* Cambridge MA: Harvard University Press, 2009.

Lubetkin, M. John. *Custer and the 1873 Yellowstone Survey.* Norman OK: Arthur H. Clark, Company 2013.

Lubetkin, M. John. *Jay Cooke's Gamble.* Norman OK: University of Oklahoma Press, 2006.

Lummis, Charles F. *General Crook and the Apache Wars.* Outing MN: Northland Press, 1966.

McClernand, Edward J. *On Time for Disaster-the Rescue of Custer's Command.* Lincoln NB: University of Nebraska Press, 1969.

McGillycuddy, Julia B. *McGillycuddy Agent.* Oxford: Oxford University Press, 1941.

McLaughlin, James. *My Friend The Indian.* Boston, MA: Houghton Miffin, 1910.

MacLean, Col. French L. *Custer's Best-The Story of Company M, 7th Cavalry at the Little Bighorn.* Atglen PA: Schiffer Military History, 2011.

Magnussen, Daniel O. *Peter Thompson's Narrative of the Little Bighorn Campaign 1876.* Glendale CA: Arthur H. Clark Company, 1974.

Mails, Thomas E. *The Mystic Warriors of the Plains.* Tulsa OK: Council Oaks Books, 1991.

Mangum, Neil C. *Battle of the Rosebud: Prelude to the Little Bighorn.* El Segundo CA: Upton and Sons, 1991.

Manion, John S. *General Terry's Last Statement to Custer.* El Segundo CA: Upton and Sons, 2000.

Marquis, Thomas B. *Memoirs of a White Crow Indian.* Lincoln NB: University of Nebraska, 1974.

Marquis, Thomas B. *Wooden Leg-A Warrior Who Fought Custer.* Lincoln NB: University of Nebraska Press, 1931.

Marquis, Thomas. *Custer on the Little Bighorn, "She Watched Custer's Last Battle."* Lodi CA: End-Kian Publishing, 1980.

Merington, Marguerite. *The Custer Story-The Life and Intimate Letters of General Custer and his Wife Elizabeth.* New York NY: Devin-Adair Company, 1950.

Michno, Gregory F. *Battle of Sand Creek-The Military Perspective.* El Segundo CA: Upton and Sons, 2004.

Michno, Gregory F. *Encyclopedia of Indian Wars-Western Battles and Skirmishes 1850-1890.* Missoula MT: Mountain Press Publishing, 2003.

Michno, Gregory F. *Lakota Noon-The Indian Narrative of Custer's Defeat.* Missoula MT: Mountain Press Publishing, 1997.

Miller, David Humphreys. *Custer's Fall.* Harmondsworth, Middlesex, England: Meridian Printing, 1992.

Miller, David Humphreys, *Echoes of the Little Bighorn,* Volume XXII, Number 4. New York NY: American Heritage Publishing Company, June 1971.

Mills, Anson. *My Story* (Anson Mills, Brigadier General, U. S. Army). Mechanicsburg PA: Stackpole, 2003

Mills, Charles K. *Harvest of Barren Regrets.* Glendale CA: Arthur H. Clark Company, 1985.

Monnett, John H. *Where a Hundred Soldiers Were Killed.* Albuquerque NM: University of New Mexico Press, 2008.

Mueller, James E. *Shooting Arrows & Slinging Mud-Custer, The Press, and The Little Bighorn.* Norman OK: University of Oklahoma Press, 2013.

Neihardt, John G. *Black Elk Speaks.* Lincoln NB: University of Nebraska Press, 1995.

Neihardt, John G. *Patterns and Coincidences.* Lincoln NB: University of Nebraska Press, 1978.

Nichols, Ronald H. *In Custer's Shadow: Major Marcus Reno.* Ft. Collins CO: Old Army Press, 1999.

Nichols, Ronald H. *Reno Court of Inquiry.* Hardin MT: Custer Battlefield Historical & Museum Association, 1996.

Nichols, Ronald H. (Editor) with Daniel I Bird. *Men With Custer- Biographies of the 7th Cavalry.* Hardin MT: Custer Battlefield Historical and Museum Association, 2010.

Olson, James C. *Red Cloud and the Sioux Problem.* Lincoln NB: University of Nebraska Press, 1965.

Philbrick, Nathaniel. *The Last Stand-Custer, Sitting Bull, and the Battle of the Little Bighorn.* New York NY: Viking, 2010.

Powell, Father Peter John. *People of the Sacred Mountain-A History of the Northern Cheyenne Chiefs and Warrior Societies 1830-1879,* Volume II. New York NY: Harper & Row, 1981.

Powell, Peter J. *Sweet Medicine,* Volume I. Norman OK: University of Oklahoma, 1969.

Puckett, J. L. *History of Oklahoma and Indian Territory and Homeseeker's Guide.* Vinita OK: Chieftain Publishing, 1906.

Rankin, Charles E., editor. *Legacy-New Perspectives on the Battle of the Little Bighorn.* Dan Flores, *"The Great Contraction."* Helena MT: Historical Society Press, 1996.

Reynolds, Arlene. *The Civil War Memories of Elizabeth Bacon Custer.* Austin TX: University of Texas Press, 1994.

Rickey, Jr., Don. *Forty Miles a Day on Beans and Hay.* Norman CO: University of Oklahoma, 1963.

Rickey, Jr., Don. *History of Custer Battlefield.* Fort Collins CO: Old Army Press, 1998.

Robinson III, Charles M. *Satanta.* Austin TX: State House Press, 1997.

Sandoz, Mari. *Crazy Horse-The Strange Man of the Oglalas.* Lincoln NB: University of Nebraska Press, 1942.

Scott, Douglas D. and Richard A. Fox, Jr. *Archaeological Insights into the Custer Battle-An Assessment of the 1984 Season.* Lincoln NB: University of Oklahoma Press, 1987.

Scott, Douglas D. and Richard A. Fox, Jr., Melissa A. Connor, Dick Harmon. *Archeological Perspectives on the Battle of the Little Bighorn.* Lincoln NB: University of Oklahoma Press, 1989.

Scott, Douglas D. and P. Willey, Mellissa A Connor. *They Died With Custer.* Norman OK: University of Oklahoma Press, 1998.

Simson, Jay W. *Custer and the Front Royal Executions.* Jefferson NC: McFarland and Company, 2009.

Sklenar, Larry. *To Hell with Honor.* Norman OK: University of Oklahoma Press, 2000.

Slotkin, Richard. *The Fatal Environment-The Myth of the Frontier in the Age of Industrialization 1800-1890.* New York NY: Atheneum, 1985.

Smalley, Vern. *Little Bighorn Mysteries.* Bozeman MT: Little Buffalo Press, 2005.

Smalley, Vern. *More Little Bighorn Mysteries.* Bozeman MT: Little Buffalo Press, 2005.

Solimine, Leo. *Custer's Bugler-The Life of John Martin (Giovanni Martino).* Boca Raton LA: Universal Publishers, 2012.

Spotts, David L. *Campaigning with Custer 1868-69.* Lincoln NB: University of Nebraska Press, 1988.

Stanley, David Sloane. *Personal Memoirs of Major-General D. S. Stanley, U. S. A.* Cambridge MA: Harvard University Press (reprinted by the University of Michigan), 1917.

Starr, Stephen Z. *Union Cavalry in the Civil War,* Vol II. Baton Rouge LA: State University Press, 1981.
Steltenkamp, Michael F. *Black Elk-Holy Man of the Oglala.* Lincoln NB: University of Oklahoma Press, 1993.
Stevenson, Joan. *Deliverance from the Little Big Horn.* Norman OK: University of Oklahoma Press, 2012.
Stewart, Edgar I. *Custer's Luck.* Norman OK: University of Oklahoma Press, 1955.
Swanson, Glenwood J. *G. A. Custer-His Life and Times.* Swanson Productions Inc., 2004.
Taunton, Francis B. *No Pride in the Little Big Horn.* London: The English Westerners Society, 1987.
Taylor, W. O. *With Custer on the Little Bighorn.* Harmondsworth, Middlesex, England: Penguin Books, 1996.
Terry, Michael 'Bad Hand.' *Plains Indians Regalia & Customs.* Atglen PA: Schiffer Publishing, 2010.
The Book of Aran. Waddell, John, and J. W. O'Connell, Anne Korff, editors.
Ireland: Newtownlynch Tir Eolas, 1994.
Thomas, Colonel Rodney G. *Rubbing Out Long Hair: The American Indian Story of the Little Big Horn in Art and World.* Spanaway WA: Elk Plain Express, 2009.
Turner, Stanfield. *Terrorism and Democracy.* San Diego CA: Houghton Mifflin Harcourt, 1991.
Unger, Arthur C. *Custer's First Messenger! Debunking the Story of Sergeant Daniel A. Kanipe?* Washington NC: Custer 1876 Publisher, 2011.
Upton, Richard. *Custer Catastrophe at the Little Big Horn 1876.* El Segundo CA: Upton and Sons. 2012.
Upton, Richard. *The Battle of the Little Big Horn and Custer's Last Fight Remembered by Participants at the Tenth Anniversary, June 25, 1886, and the Fiftieth Anniversary, June 25, 1926.* El Segundo CA: Upton and Sons. 2006.
Urwin, Gregory J. W. *Custer Victorious.* Lincoln NB: University of Nebraska Press, 1990.
Utley, Robert M. *Cavalier in Buckskin: George Armstrong Custer and the Western Military Frontier.* Norman OK: University of Oklahoma Press, 1988.
Utley, Robert M. *Life in Custer's Cavalry: Diaries and Letters of Albert and Jennie Barnitz 1867-1868.* Lincoln NB: University of Nebraska Press, 1977.
Utley, Robert M. *Little Bighorn Battlefield-A History and Guide to the Battle of the Little Bighorn.* Washington DC: National Park Service, 1994.
Utley, Robert M. *The Lance and the Shield.* New York NY: Henry Holt and Company, 1993.
Van de Water, Frederic. *Glory Hunter: A Life of General Custer.* Lincoln NB: University of Nebraska Press, 1988.
Vaughn, J. W. *The Reynolds Campaign on the Powder River.* Norman OK: University of Oklahoma Press, 1961.
Vestal, Stanley. *Sitting Bull.* Norman OK: University of Oklahoma Press, 1989.
Vestal, Stanley. *Warpath-The True Story of the Fighting Sioux Told in a Biography of Chief White Bull.* Lincoln NB: University of Nebraska Press, 1984.
Viola, Herman J. *Little Bighorn Remembered.* New York NY: Time Books, 1999.
Wagner III, Frederic C. *The Strategy of Defeat at the Little Big Horn-A Military and Timing Analysis of the Battle.*
Jefferson, NC: McFarland & Company, 2011.
Wagner, Glendolin Damon. *Old Neutriment.* Lincoln NB: University of Nebraska Press, 1989.
Walker, Jr., L. G. *Dr. Henry Porter-The Surgeon who Survived Little Bighorn.* Jefferson NC: McFarland and Company, 2008.
Warde, Mary Jane. *Washita.* Cheyenne OK: Washita Battlefield National Historic Site, 2005.
Westfall, Douglas Paul. *Letters from the Field.* Orange CA: The Paragon Agency, 1997.
Willert, James. *The Terry Letters.* La Mirada CA: published by James Willert, 1980.

Willert, James. *Little Bighorn Diary-Chronicle of the 1876 Indian War.* El Segundo CA: Upton and Sons, 1997.
Willert, James. *March of the Columns-Chronicle of the 1876 Indian War June 27-September 16, 1876.* El Segundo CA: Upton and Sons, 1994.
Willert, James. *To the Edge of Darkness-A Chronicle of the 1876 Indian War.* El Segundo CA: Upton and Sons, 1998.
Williams, Roger L. *Military Register of Custer's Last Command.* Norman OK: Arthur H. Clark Company, 2009.
Whittaker, Frederick. *A Complete Life of Gen. George A. Custer.* New York NY: Sheldon & Company, 1876.
Whittaker, Frederick. *Volunteer Cavalry: The Lessons of a Decade.* New York NY: 1871.
Windolph, Charles, as told to Frazier and Robert Hunt. *I Fought with Custer.* Lincoln NB: University of Nebraska Press, 1987.
Woods, Marvin. *Custer's Arikara Indian Interpreter.* New York NY: Vantage Press, 2009.

Periodicals, Booklets, and Pamphlets

Abrams, Marc H. *Custer Last Stand Magazine, Vol. 2,* August 2009.
Abrams, Marc. "James Calhoun's Forgotten Dispatches in the New York World." *The Crow's Nest—The Journal of the Custer Association of Great Britain,* Autumn/Winter 2012, Vol. 12, No. 2. 2012.
Army and Navy Journal. August 30, 1873.
Arnold, Steve. "Coming to the Aid of Custer?" *Research Review,* Vol. 27, No. 2, 2013.
Banfill, W. H. "The Story of John Burkman's Experience with General George Custer and the 7th Cavalry; He Tells of the Characteristics of the General and the Fatal Battle." Thomas Marquis Collection, National Library of Medicine, Bethesda, Maryland, copy at Little Bighorn Battlefield National Monument.
Blake, Bill. "Was Weir A Hero?" *Little Big Horn Associates Newsletter* Vol. XI, No. 5, June 2012.
Braden, Charles. "The Yellowstone Expedition." *Journal of the United States Cavalry Association,* October 15, 1904.
Braden, Charles. "The Yellowstone Expedition of 1873." *Journal of the United States Cavalry Association,* October 16, 1905.
Bradshaw, T.E. "The Pension Application and Controversy of an Alleged Self-Inflicted Wound—Sgt. Daniel A. Kanipe Company C, 7th U.S. Cavalry." Columbia SC: Palmetto Historical Works, 2011.
Bray, Kingsley. "Pine Ridge Letters Shed New Light on the Battle of Little Bighorn." *Ghostherder-The Journal of the Friends of the Little Bighorn Battlefield,* No. 1.
Bradshaw, T. E. "The Pension Application of Daniel Knipe" Palmetto Historical Works, Columbia SC, May 2011.
Brewster, Charles, "Battle of the Washita." *National Tribune,* May 18, 1899.
Brust, James. "Adventurous Detour to the Battlefield." *Greasy Grass,* Vol. 10, May 1994.
Brust, James. "Fouch Photo May be the First." *Greasy Grass,* Vol. 7, May 1991, pp. 2-9
Brust, James. "Lt. Oscar Long's Early Map Details Terrain, Battle Positions." *Greasy Grass,* Vol. 11, May 1995.
Buecker, Thomas. "A Surgeon at the Little Bighorn: The Letters of Dr. Holmes O. Paulding." Montana Historical Society, Vol. 32, No. 4, Autumn 1982.
Connelly, Kevin. "High Flying Cavalier." *Little Big Horn Associates Newsletter,* Vol. XXXV, 2001-5, July.
Coon, Homer. "Recollections of the Sioux Campaign of 1876." Little Bighorn Battlefield National Monument.
Cordry, Dee. "True Heart-The Story of Edmund Guerrier." *The Crow's Nest-The Journal of the Custer Association of Great Britain,* Spring/Summer, Vol. 12, No. 1.

Cowdrey, Mike. "Crows Who Scouted for Custer." *Greasy Grass*, Volume 14, May 1998.

Cowles, David C. "White Swan Crow Artist at the Little Big Horn." *American Indian Magazine*, Autumn 1982.

Custer, George A. "In Memoriam: Brevet Major Louis McLane Hamilton Captain 7th U. S. Cavalry." *The Chronicles of Oklahoma*, Vol. 46, Winter 1968-69.

Custer, George A. "The Black Hills Expedition: Dispatches and Letters." *Army and Navy Journal* 12, No. 2 (August 22, 1874).

Custer National Cemetery Guidebook. Canada, Western National Parks Association, 2009.

Davis, Theo. R. "A Summer on the Plains." *Harper's Monthly Magazine*, Vol. XXXVI, February 1868.

"Diary of Matthew Carroll." Historical Society of Montana, Vol. II, 1966.

Donahue, Michael. "On the Battlefield." *Greasy Grass*, Vol. 14, May 1998.

Donahue, Michael. "Revisiting Colonel Gibbon's Route." *Greasy Grass*, Vol. 19, May 2003.

Donahue, Michael. "The Battle of the Little Bighorn—A New Time Line." 18th Annual Symposium of the Custer Battlefield Historical and Museum Association, June 2004.

Donahue, Michael. "The Maps of John Stands in Timber." 27th Annual Symposium of the Custer Battlefield Historical and Museum Association, June 2012.

Donahue, Michael. "Beyond Custer Hill." Hardin, Montana, 17th Annual Symposium of the Custer Battlefield Historical & Museum Association, June 2003.

Duncan, Charles and Jay Smith. "The Captives." *Research Review*, Vol. 7, No. 2. June 1993

Dunphy, James J. "Measure for Measure." *Research Review*, June 1996, Vol. 10, No. 2.

Ediger, Theodore A. and Hoffman, Vinnie. "Some Reminiscences of the Battle of the Washita." *Chronicles of Oklahoma*, Summer 1955.

Ellis, Horace. "Survivor's Story of the Custer Massacre on American Frontier." *Journal of American History*. Vol. 3, No. 2, April 1909.

Encyclopaedia Britannica, Eleventh Edition, 1911.

Fagan, Roberta. "The Don Juan Runaway Incident." *Research Review*, December 1988, Vol. 2, No. 2.

Godfrey, Edward. *The Cavalry Journal*, October 1928.

Gray, John. "Captain Clifford's Story of the Sioux War of 1876." *The Westerners Brand Book*, Vol. XXVI, December 1969.

Hadley, James A. "The 19th Kansas Cavalry and the Conquest of the Plains Indians." Collections of the Kansas State Historical Society 10 (1908-09).

Hardorff, Richard G. "Baliran, Honzinger, and the Custers: The Facts and Fictions of the Rain-In-The-Face Myth." *Research Review*, December 1988, Vol. 2, No. 2.

Hardorff, Richard G. "Some Recollections of Custer and His Last Battle." *Research Review*, June 1990, Vol. 4, No. 2.

Harrison, Michael. "The Don Juan Runaway Incident Version # 3." *Research Review*, January 1991, Vol. 5, No. 1.

Hills, Louis E. "With General George A. Custer on the Northern Pacific Surveying Expedition in 1873." *Journal of History* 8, 1913.

Hynds, A.A., and William J. Shay. "Reminiscences of Sgt. H. A. Hynds." *Research Review*, Winter 1972, Vol. VI, Number 4.

James, Gary. "Custer's Last Gun." *Guns & Ammo*, December 2009.

Johnson, Barry C. "A Captain of Chivalric Courage." London: *English Westerners' Society*, 1989.

Johnson, Barry C. "With Gibbon Against the Sioux in 1876—The Field Diary of Lt. William English." Vol. 9, July 1966.
Justus, Judith P. "The Saga of Clara H. Blinn at the Battle of the Washita." *Research Review*, Winter 2000, Vol. 14, No. 1.
Keim, DeBenneville Randolph. "The Indian War." *The New York Herald*, Monday, January 4, 1869.
King, Charles. "Custer's Last Battle." *Harpers New Monthly Magazine*, Vol. LXXXI, 1890 (June to November).
Koury, Mike. "Forget the Gatling Guns!" 26th Annual Symposium Custer Battlefield Historical & Museum Association, June 22, 2012.
Kush, George. "There Lies Foley of Company C!." "Custer's Last Flag." *Sotheby's Catalogue*, December 10, 2010.
Indian Memorial Dedication "Peace Through Unity" Memorial Booklet. Washington DC: National Park Service. (Author's collection.)
Langley, Henry. "With Custer at the Washita, 1868." *Winners of the West*, December 15, 1925.
Liddic, Bruce R. "The Book That Never Was-Lemmon's Developing the West and Mary Crawler's Battle of the Little Big Horn Account." *Research Review*, Annual 2011, Vol. 25.
Lubetkin. "Clash on the Yellowstone." *Research Review*, Summer 2003, Vol. 17, No. 2.
McBlain, John F. "With Gibbon on the Sioux Campaign of 1876." *Cavalry Journal*, Vol. IX, June 1896.
McConnell, Ronald C. "Isaiah Dorman and the Custer Expedition." *The Journal of Negro History*, Vol. XXXIII, No. 8, July 1948.
Mangum, Neil. "Reno's Battalion in the Battle of the Little Big Horn." *Greasy Grass*, Vol. 2, 1986.
Mangum, Neil. "The George C. Brown Story." *Research Review*, Summer 1999, Vol. 13, No. 2.
Mann, Charles C. "America, Found and Lost." *National Geographic*, May 2007.
May, John D. "Little Raven." Oklahoma Historical Society's Encyclopedia of Oklahoma History & Culture (accessed July 2, 2012 on Wikipedia website)
Monnett, John. "Reimagining Transitional Kansas Landscapes: Environment and Violence." *Kansas History: A Journal of the Central Texas Plains*, Winter 2011-12.
Moore, Michael and Michael Donahue. "Gibbon's Route to Custer Hill." *Greasy Grass*, Vol. 7, May 1991.
Mueller, Jim. "Swinging Around the Circle with Custer: George Armstrong Custer's Role in the 1866 Election." Annual Symposium of the Custer Battlefield Museum and Historical Association, June 2014.
National Military Museum, Governor's Island, Collection of Relics and Trophies Catalog 1884, *Press of Putnam & Sons*, New York.
Noyes, C. Lee. "Custer's Surgeon, George Lord, Among the Missing at Little Bighorn Battle." *Greasy Grass*, Vol. 16, May 2000.
Noyes, C. Lee. "The Valley Fight: The Enlisted Man's Perspective." 20th Annual Symposium of the Custer Battlefield Historical & Museum Association, Hardin, Montana, June 2006.
O'Neil, Tom. "Two Men of Ohio-Custer and Bingham." *Research Review*, January 1994, Vol. 8, No. 1.
Paulding, Dr. Holmes O. "Diary with Gibbon Expedition to Meet Custer." Little Bighorn Battlefield National Monument.
Pengra, Lilah Morton. "Isaiah Dorman Was Not A Runaway Slave." *The Battlefield Dispatch*, Summer 2014, Vol. 33, No. 3.
Pohanka, Brian. "George Yates-Captain of the Band Box Troop." *Greasy Grass*, Vol. 8, May 1992.
Pohanka, Brian C. "A Study of Company Discipline: General Courts-Martial of Enlisted Personnel, 7th Cavalry,

1875-1876." 9th Annual Symposium of the Custer Battlefield Historical & Museum Association.
Pompeys Pillar National Monument Visitor Guide.
Reedstrom, Lisle E. "Garryowen-Custer's Seventh Cavalry Regimental Fighting Song." *Research Review*, January 1996, Vol. 10, No. 1.
"Reno/Benteen Entrenchment Trail Guide." Washington DC: National Park Service.
Rickey, Don. Interview with Stands in Timber, Little Bighorn Battlefield National Monument, Aug. 8, 1956.
Roe, Charles Francis. "Custer's Last Battle." Ft. Collins CO, *Old Army Press*, No date.
Ryan, John M. "Ten Years with General Custer among the American Indians." *The Newton (Mass.) Circuit*, May 28, 1909.
Schneider, Jim. "Rations, Forage & Ammunition-Logistical Problems Facing the Dakota Column in the Summer of 1876." *Research Review*, June 1991, Volume 5, No. 2.
Schonenberger, Dale T. "A Trooper With Custer-August DeVoto's Account of the Little Bighorn." *Montana: The Magazine of Western History*, Winter 1990, Vol. 40, No. 1.
Sprague, Donovin. "Hump and Crazy Horse: Defending the Lakota & Cheyenne 1876-1877." 24th Annual Symposium of the Custer Battlefield Museum and Historical Association, June 2010.
Steves, Myron. "Benteen from the Tome." 24th Annual Symposium of the Custer Battlefield Historical & Museum Association, June 2010.
Taunton, Francis B. "Custer's Field: 'A Scene of Sickening Ghastly Horror,'" London: The Johnson-Taunton Military Press, 1986.
Taunton, Francis B. "More Sidelights of the Sioux Wars." London: English Westerners' Society Limited, 2004.
Thomas, Rod. "Grabbing Grass in his Hand: White Swan's Great Fight-25 June 1876." 27th Annual Symposium of the Custer Battlefield Museum & Historical Association, June 2013.
Utley, Robert M. "Gossip and Scandal." Research Review, June 1989, Vol. 3, No. 1.
War Eagle, Doug, Red Thunder, Don & Clown Floyd. "The Legacy of Crazy Horse." 23rd Annual Symposium of the Custer Battlefield Historical & Museum Association, June 2019.
Wilke, Lynn. "Broadcasts from the LBH." 1986 script and bugle call tape in the author's collection.
Winkler, Albert. "The Case for a Custer Battalion Survivor-Private Gustave Korn's Story." *Montana: The Magazine of Western History*, Spring 2013.
"1873 Official Report of the Yellowstone Expedition," p. 13, War Department Library.

Newspapers

Bismarck Tribune, July 7, 1876.
Bismarck Tribune, July 13, 1877.
Chicago Times, July 25, 1877.
Chicago Tribune, July 30, 1877.
Chicago Tribune, August 13, 1877.
Kansas City Star, December 4, 1904.
Leavenworth Bulletin, December 22, 1868.
Leavenworth (Kansas) Daily Standard, April 18, 1882.
New York Evening Post, February 20, 1897.

New York Herald, January 4, 1869.
New York Herald, July 11, 1876.
New York Herald, September 21, 1876.
New York Sun, July 25, 1877.
New York Sun, May 14, 1899.
New York Times, April 7, 1869.
New York Times, September 11, 1873.
New York Times, October 13, 1873.
New York Tribune, December 3, 1868.
New York Tribune, July 25, 1873.
New York Tribune, September 6, 1873.
New York Tribune, September 8, 1873.
New York Tribune, September 9, 1873.
New York Tribune, August 25, 1873.
Portland (Oregon) Journal, July 31, 1923.
Sioux City Journal, July 18, 1873.
Sioux City Journal, August 14, 1874
The Cheyenne Daily Leader (Wyoming), October 9, 1875.
The Colorado Transcript (Golden City), Wednesday, January 13, 1869.
The Daily Rocky Mountain News (Denver), Thursday, March 31, 1870.
The Dallas Morning News, August 25, 2009.
The Rapid City Journal, January 28, 2006.
The Washington Times (magazine section), November 4, 1906.
The Weekly Rocky Mountain News (Denver), September 24, 1873.
Washington Post, August 11, 1902.
Watonga Republican, June 18, 1997.

Collections and Archives

A. B. Welch papers, collection of Everett Cox.
Administrative burial records at Little Bighorn Battlefield National Monument, Crow Agency MT.
R. G. Cartwright Collection, Hearst Library, Lead SD.
Charlie Schofield letters to Lilla Bogert, Montana State University, Bozeman MT.
Charles F. Bates Collection, Beinecke Library, Yale University, New Haven CT.
Charles F. Bates Collection at Little Bighorn Battlefield National Monument, Crow Agency MT.
E. A. Brininstool Collection, Center for American History, University of Texas, Austin TX.
William Cary Brown Collection, University of Colorado at Boulder CO.
Edmond S. Meany Papers, University of Washington Special Collections, Seattle WA.
E. A. Garlington Papers, John M. Carroll Collection, courtesy of George Kush AB, Canada.
Charles A. Booth papers, Little Bighorn Battlefield National Monument, Crow Agency MT.
John Burkman files, Little Bighorn Battlefield National Monument, Crow Agency MT.

Joseph G. Masters Collection, Kansas State Historical Society, Topeka KS.
Company C muster rolls dated April 29, 1876, First Cavalry Museum, Ft. Hood TX.
D. F. Barry Collection, Little Bighorn Battlefield National Monument, Crow Agency MT.
Elizabeth Bacon Custer Collection, Little Bighorn Battlefield National Monument, Crow Agency MT.
Elizabeth Bacon Custer Collection, Custer Battlefield Museum, Garryowen MT.
George Bird Grinnell Collection, Gene Autry Western Heritage Museum, Los Angeles CA.
George Bird Grinnell Collection, University of Oklahoma, Norman OK.
Indian-Pioneer History Collection, Archives Division, Oklahoma Historical Society, Oklahoma City OK.
Walter Campbell Collection, The University of Oklahoma Western History Collection, Norman OK.
Walter Camp Collection, Harold B. Lee Library, Brigham Young University, Provo UT.
Walter Camp Collection, Lilly Library, Indiana University, Bloomington IN.
Walter Camp Collection, Little Bighorn Battlefield National Monument, Crow Agency MT.
Katherine Gibson Fougera Collection, Little Bighorn Battlefield National Monument, Crow Agency MT.
George Kush Collection, AB, Canada.
Charles W. Larned Collection, Special Collections and Archives Division, West Point Military Academy Library, West Point NY.
Charles W. Larned "Letters and diary from 1873," Special Collections and Archives Division, United States Military Academy Library, West Point NY.
Glenn Sweem Collection, Courtesy of Scott Burgan, Sheridan WY.
Letters in the Office of the Adjutant General, 1871-1800, National Archives, Washington DC.
Court-martial records of Privates Robert Hamilton, John Fisher and Charles Mooney, Record Group 153, Box 2363, National Archives Building, Washington DC.
Margot Liberty Interviews with John Stands in Timber, Little Bighorn Battlefield National Monument, Crow Agency MT.
J. W. Vaughn Collection, Little Bighorn Battlefield National Monument, Crow Agency MT.
J. W. Vaughn papers, Sheridan County Library, Sheridan WY.

Index

About the Author

Both author and artist, Michael Donahue is chairman of the art department at Temple College in Temple, Texas. He received his B.F.A., M.A., and M.F.A. degrees in Art and History from Stephen F. Austin State University. In 1993, he was awarded the title of Piper Professor for being one of the top 10 professors from all Texas universities and colleges. He has published many articles on the Battle of the Little Bighorn, and his book, *Drawing Battle Lines-The Map Testimony of Custer's Last Fight,* has won numerous awards. Donahue has worked as a park ranger at Little Bighorn Battlefield National Monument for over two decades and has researched the battle for over four decades. Donahue is also a nationally recognized artist with over 50 exhibitions to his credit.

www.michaeldonahueart.com